Distribution	Variable symbol	Quantile $0 \leq q \leq 1$	Observed or calculated value
Chi-squared **Parameter:** *df*, *df* = *1*, *2*, ... **Mean:** *df* **Variance:** 2*df* *Cf.* Chapters 4, 10; Table T-2	X^2 **Domain:** $0 \leq X^2 \leq +\infty$	$\chi_q^2(df)$	χ^2
Normal **Parameters:** μ, $-\infty < \mu < +\infty$ σ, $0 < \sigma$ **Mean:** μ **Variance:** σ^2 *Cf.* Chapter 7	Y **Domain:** $-\infty \leq Y \leq +\infty$	$y_q(\mu, \sigma^2)$	y
Standardized Normal **Parameters:** [*None explicitly*] **Mean:** 0 **Variance:** 1 *Cf.* Chapter 7; Table T-1	Z **Domain:** $-\infty \leq Z \leq +\infty$	z_q	z
Student's *T* **Parameter:** *df*, *df* = 1, 2, ... **Mean:** 0 **Variance:** $df/(df - 2)$, $df > 2$ *Cf.* Chapters 8, 11, 14; Table T-3	T **Domain:** $-\infty \leq T \leq +\infty$	$t_q(df)$	t

[Continued on back flyleaf]

NUREG-1475

United States
Nuclear Regulatory Commission

Applying Statistics

Date published: February 1994

By:

Dan Lurie
Office of the Controller
U.S. Nuclear Regulatory Commission
Washington, DC 20555-0001

Roger H. Moore
Office of Energy Resources
Bonneville Power Administration
Portland, OR 97208-3621

Prepared for:
Office of the Controller
U.S. Nuclear Regulatory Commission
Washington, DC 20555-0001

For sale by the U.S. Government Printing Office
Superintendent of Documents, Mail Stop: SSOP, Washington, DC 20402-9328
ISBN 0-16-043101-8

Available as NUREG-1475 from
The Superintendent of Documents
U.S. Government Printing Office
Mail Stop SSOP
Washington, DC 20402-9328

National Technical Information Service
5285 Port Royal Road
Springfield, VA 22161

Acknowledgments

Collaborating on a book when the collaborators are separated not only by current professional interest and personal focus but by the full width of the North American continent is truly a classic exercise in teamwork. Yes, we did it ... one in Maryland, the other in Oregon/ Washington, with both firmly footed in the Land of Statistics.

Professionally, we stand on the shoulders of giants. To practice daily a profession whose foundation was laid by such geniuses as Pascal, the Bernoullis, Gauss, Maxwell, Boltzmann, Galton, the Pearsons, Hartley, and Neyman is to be constantly reminded that the vein of statistics runs through all sciences.

Our sincere thanks go to Peter J. Rabideau, Director, Division of Budget and Analysis, NRC, who recognized the need for this text and presented us with the challenge of writing the book. Pete always stood behind us to motivate and propel us to complete the assignment and to provide support and encouragement to persevere.

Our sincere thanks go also to David E. Mills, Chief, End-Use Research Section, Bonneville Power Administration (BPA). In accepting NRC's request for cooperation, Dave saw to it that adequate facilities were available for our communication and information needs and always lent the empathetic ear when we needed to bend it.

Staff members in our respective organizational units showed that they knew what we were up to by making sure we never lacked for either teasing or warm-hearted wishes. We now have the answer to the ubiquitous "How many pages?"

Five people were especially giving of their advice and ideas about the technical content of book itself: Lee Abramson, NRC's Office of Research; Hurshell Hunt, Medical University of South Carolina; Wendy Lin-Kelly, BPA's End-Use Research Section; Joe Rivers, Science Applications International Corporation; and Erik Westman, BPA's Policy Development Staff. We—and the book—are all the stronger for their comments and suggestions.

We—and the book—also benefited beyond measure from our editorial encounters and impromptu English-language training sessions with Robin Harris and Bill Maher, both of the NRC. Other authors should be so fortunate as to have their guidance as part of their writing careers.

We are indebted to Reggie Mitchell, NRC, for his insight, advice, and support in making our computer behave. We are also indebted to Lionel Watkins, NRC, for his assistance in cover design and graphics.

Special thanks go to Walter Oliu, Juanita Beeson, Ray Sanders, Mimi Mejac, and Guy Beltz, NRC, for providing editorial support, for navigating us through the formal requirements, for assisting us in setting the style and the appearance of the text, and for the publication of this book.

We particularly appreciated the various offers of help and good will from many others during this adventure. Knowing that to name some may be to slight others, we must nevertheless express our special thanks to Catherine Holzle, Charles Gorday, Elise Heumann, Jackie Silber, and Lori Stadler—NRC—and Jim Cahill, Curtis Hickman, Preston Michie, Ottie Nabors, Judy Rotella, and Sarah Wilson—BPA.

Putting this book together constantly took us back in memory to our many teachers, mentors, and friends. We are particularly grateful to them for the countless ideas and concepts that we have encountered, enjoyed, and exploited in our careers. And we greatly appreciate the need created by our myriad students—the need that this book is intended to fill.

Our wives, Marti Lurie and Eileen Moore, were always there during this effort. Making meals, proof-reading, forcing short breaks, tutoring language use, solving household issues—all their selfless assists and attentions have been ours throughout the last couple of years. Any day now, they'll get the attention from us that they deserve.

M: More than yesterday, but less than tomorrow; thanks—D.

E: It just keeps getting better; always and forever—R.

Finally, with respect to errors—of any sort whatsoever—that may still be present in these pages, we are of one mind: They are the results of something the other one of us did or did not do. No one else can share them.

Dan Lurie
Roger Moore
February 1994

Introduction

Applying Statistics is intended to be both a reference book and a textbook on statistical methods. The broad scope of statistics applies to almost every scientific and technical field; while we had originally planned for the book to be used primarily by practitioners of quantitative methods in the field of nuclear energy, we expanded our range early on to provide and illustrate basic established statistical tools for a full spectrum of quantitative information analysis.

We are convinced that a better knowledge of statistics can prove invaluable for anyone facing the tasks of organizing and displaying information, formulating and resolving quantitative problems, computing and expressing probability. We believe that this book will provide that better knowledge to any reader with a reasonable grasp of college algebra; a good deal of the material is accessible even to readers with a minimal background in mathematics.

Acquiring this knowledge should not be a daunting prospect to the reader. After all, the essence of the discipline called *statistics* is contained in just a few straightforward concepts: thinking, designing, modeling, thinking, counting, measuring, thinking, displaying, computing, thinking, and reporting. We strongly subscribe to the precept expressed by our good friend, Richard J. Beckman of the Los Alamos National Laboratory, who has reminded us on numerous occasions that, in effect, "Statistics is nothing but common sense applied to data."

Since the mid-1970s, we both have enjoyed the opportunity to introduce statistical processes to—and to reinforce them for—hundreds of professional people, representing scores of disciplines, in the nuclear energy and electricity supply communities. As have countless of others, we found that existing texts were never quite what we needed for our particular audiences. So we opted to do what others have done in dealing with Mother Necessity: we decided to invent another text, custom-made.

The title *Applying Statistics* succinctly expresses our rationale for preparing this material and presenting it in the fashion you find here. When we recommend *applying statistics* to your work and to your play, we don't mean that you should have to remember on command such things as regression formulas and critical values and the rules of hypothesis-testing. But we do mean that you should be able to recall the elegance and power of these ideas and to know where to refresh your knowledge of them when you need them.

The material itself is organized into digestible segments, each exploring a particular method of applying statistics. These segments, in turn, are collected into chapters, each of which examines a set of linked statistical ideas. This encourages a pick-and-choose approach, and it allows a discussion-coordinator to point up specific illustrative information, no matter how random the discourse taking place. Many of the bite-sized segments contain examples of data which were encountered in—or simulated from—the application of statistics in the Nuclear Regulatory Commission and the Bonneville Power Administration. Because statistical ideas connect across all of our lives and interests, we also include a number of non-nuclear-regulation-related datasets.

Chapters 1 through 5 introduce some fundamental statistical methods as an approach to dealing with everyday problems. They include definitions, terminology, graphics, probabilistic-statistical linkage, contingency tables, and descriptive statistics. This is the "stuff" of statistics in today's world. These very basic techniques give the reader the beginnings of the vocabulary of statistics and amply demonstrate how much can be done with just a few statistical formalities.

Chapters 6 through 10 describe and expand upon "errors" as a statistical construct. They take you through the fundamental ideas of hypothesis-testing and confidence-interval construction with the focus on a single group.

Chapters 11 through 14 focus on multiple-group situations, addressing analysis-of-variance and simple linear regression and correlation methods.

Chapters 15 through 18 deal with basic ideas of probability and apply them to hypergeometric, binomial, and Poisson processes, which are defined and explained and which are basic to probabilistic risk analysis.

Chapters 19, 20, and 21 discuss the important field of quality assurance and two related and relatively knotty, easily misapplied, probabilistic applications. Our purpose here is to illuminate how slippery is the slope of incorrect or incomplete probabilistic thinking.

All of the special terms used in the book are fully defined and explained, as we get to them. No discipline is without its jargon—its peculiar, sometimes colorful, phraseology. We set off special statistical terms and concepts with *italics*, and index many of them in the left margin for easy return reference. *Indeed, we use italics liberally in the text, whenever we decided something is worthy of your special attention.* We display tables in rectangles bounded by double lines and set up graphics and charts as "Figures," bounded by "rounded-corner rectangles."

Special subsections, indicated by the "For discussion:" tag, are intended to stimulate further exploration, to emphasize particular points, to provide theoretical support, and to connect the associated statistical topics to aspects of our daily lives. Some of the items in these subsections are facts, some are questions, some are commentary, and some are viewpoints presented with varying amounts of tongue-in-cheekiness. Not all the facts are universally accepted, not all the questions have "correct" answers, not all the commentary is widely supported, and not all issues are subjected to authorial viewpoints. If you are not provoked, not challenged, or not exasperated at least once in each chapter, then we have, to that extent, failed you. We have tried hard not to fail you.

The concluding section, called "Afterword: Know what thou art missing," gives some indication of what is *not* included in the book. A number of individually worthy statistical topics are listed and briefly described, and references are given for your further investigation.

This publication includes an extensive bibliography which you may find useful in itself, independent of the text. Most of the bibliographic entries appear explicitly in the text. Some of those entries are general-purpose sources whose titles are sufficiently descriptive to help you decide whether you want to look at them. You will also find a good selection of statistical tables, each illustrated and explained in sufficient self-contained detail that you may use them without having to consult the text itself.

Our notational choices are explained in Chapter 3. At the endpapers of this book, "Notational conventions" relates these choices to the eight basic density functions give prominence throughout the text.

> *An aside to our fellow computer junkies*: We prepared all of the printed material on IBM-PC compatibles of the Intel '386 and '486 vintages, using Microsoft's DOS 5.0 and 6.0 and Windows 3.1. Our software included WordPerfect Corporation's WordPerfect 5.1 for DOS and 5.1 and 5.2 for Windows, Microsoft's Excel 4.0, and Borland's Quattro Pro 4.0 for DOS and Quattro Pro 1.0 and 5.0 for Windows. We used WordPerfect-supplied fonts and printed the pages with a Hewlett-Packard LaserJet 4. And we learned lots more about these modern wonders than we expected to.

Applying Statistics was designed, written, and produced as a collaborative effort by the Nuclear Regulatory Commission (NRC) and the Bonneville Power Administration, a power-marketing agency of the Department of Energy. We both have been synergized by the assignment and the opportunity to reexamine and reflect on our lifetimes' work. We each thank the other's agency for supporting an effort which will benefit both organizations and others seeking to improve their knowledge and use of basic statistical tools.

DL
RHM
February 1994

Table of contents

Chapter contents

Tables

Figures

Examples

Whetting your statistical intuition

What to look for in Chapter 1

Chapter 1 introduces the discipline and the practice of statistics in commonly understood—and often *mis*understood—terms, thereby setting the stage for the ensuing chapters. Along the way, you may find that some of your long-accepted statistical conceptions are, in fact, statistical misconceptions. Be neither dismayed nor deterred. You will soon be equipped with a straightforward set of questions that you can use to appraise any statistical notion or statement. Among the fundamental statistical concepts developed and illustrated in this chapter—and expanded upon in the rest of the book—are:

- *data*
- *statistics (the discipline)*
- *quantitative information*
- *scales of measurement*
- *population*

- *sample*
- *parameter*
- *statistic (a summarizer of data).*

After these several concepts are examined and discussed, they are brought together to define *data* and *the discipline of statistics* to set a consistent tone for the rest of the book. Finally, you will be introduced to OSDAR, your checklist to improved data-acquisition investigations.

On statistical intuition

When we attempt to solve a problem, we often resort to our intuition to find an answer. By calling upon this quick and ready apprehension of the issues and the facts connected to the problem, we save time and effort—and we can move on to the next matter at hand. But intuition is personal and different in each of us; we have different experiences, different training, different interests. Consequently, each of us provides a different intuitive solution to any given problem.

Sometimes we recognize the failure of our own intuition and call on friends and acquaintances for guidance. But we're then usually left with the problem of summarizing an ever-widening collection of others' intuitive solutions. For example, have you ever asked five people what they think about a particular brand and model of automobile?

statistical intuition

This is not to say that your intuition or that of your friends is faulty. Rather, we suggest that everybody's intuition can always stand a tuneup. This book is designed to provide that tuneup to your *statistical intuition*: those insights and ideas and paradigms you bring to bear on problems involving data. There's nothing like a well-honed statistical intuition to get you headed *toward* the solution of any data-related problem. Equally, there's nothing like a slightly-skewed statistical intuition to get you headed *away* from a solution. The rigor—*and* the vigor—of statistical procedures will get you back on track.

Can definitions help?

You can approach an unfamiliar or little-understood subject in many ways. One technique is to settle on some definitions of the basic elements of the subject of discourse, and then build on those definitions as additional topics and details arise. Eventually, the subject and you will develop an affinity and comfort-level that frees you to go on to other subjects.

Here are some definitions of *statistics* that you may have encountered (along with one that you may *not* have seen before):

- A science of collecting and representing data.
- The art and science of treating data.
- An area of science that deals with the collection of data on a relatively small scale to form logical conclusions about the general case.
- An exact science dealing with inexact data.
- A language, a mechanism for creating and communicating quantitative concepts and ideas.
- A science of decision-making in the face of uncertainty.
- A science that supports statisticians and their families *(a declaration made by Mrs. Dan Lurie on a certain graduation day, July 1971).*

For discussion:[1]

- When does a body of knowledge become a *science*? When does its use become an *art*?
- What do you think *data* are?
- What is meant by *exact* and *inexact*?
- Data and statistics: *singular* or *plural*?
- What is *the collection of data on a relatively small scale*?
- What is *the general case*?
- Do you see any problem with *decision-making* in the face of *certainty*?
- What is *uncertainty*? (What is its *face* like?)
- Where and how do *baseball statistics* fit in with these concepts?
- Now, after all this, how would *you* define *data* and *statistics*?

Some popular misconceptions about statistics

The following books, all of which are real and are listed in the bibliography, reflect some of the many misconceptions and many biases held by many people about many statistics (and about many statisticians):

Use and Abuse of Statistics (Reichmann, 1971)
How to Lie with Statistics (Huff, 1954)
Flaws and Fallacies in Statistical Thinking (Campbell, 1974)
How to Tell the Liars from the Statisticians (Hooke, 1983).

[1] As stated in the Introduction, we use these **For discussion:** sections to provoke your further exploration, to emphasize particular points, to provide theoretical support, and to connect the adjacent topics to other aspects of our daily lives. Some of the items are facts, some are questions, some are commentary, and some are viewpoints bearing varying amounts of tongue-in-cheekiness. Not all the facts are universally accepted, not all the questions have "correct" answers, not all the commentary is widely supported, and not all viewpoints are given to all issues. If you are not provoked, not challenged, or not exasperated at least once in each chapter, then we have failed our intent.

Of course, no discourse on statistics escapes the canard[2] that " ... *there are three kinds of lies*—lies, damned lies, and statistics." You really don't have to look very hard or very far to find such statistical wisdom and/or gobbledygook. All you have to do is to pay some attention to the news media, read a few technical journals, listen to your doctors and dentists, take note of your politicians, and participate in everyday life. The fact is that everyone is well acquainted with situations in which statistics is/are used and misused. The real trick is to develop and improve the insights and the tools to sort through the blizzard of "quantitative facts" that surrounds you. Indeed, your load is lightened considerably if you never lose sight of Mosteller's insightful rejoinder: "It's easy to lie with statistics. But it is easier to lie without them."[3]

For discussion:

- Economists practice what has been called "the dismal science." Teachers are constantly up against "those who can, do; those who can't, teach; those who can't teach, teach teachers." Doctors and dentists are "in it for the money and the condo investments." Lawyers "chase ambulances." Statisticians "find problems for other statisticians to solve." Actors "can't get a real job." Engineers design products to last "for the life of the warranty plus one day." How is *your* profession seen by outsiders?

---

[2] Attributed by Mark Twain (Samuel Clemens, 1835-1910), American observer of the human condition, to Benjamin Disraeli (1804-1881), British politician and prime minister. Source: *The Pocket Book of Quotations*, edited by Henry Davidoff, Pocket Books, Inc., New York, NY, 1952.

[3] Frederick Mosteller, quoted in *Chance*, Vol.6, No. 1, 1993, pp. 6-7.

Evaluating a statistical statement

It's not necessary to "be a statistician" to react sensibly and responsibly to a statistical statement. Just about all problems attributed to "bad data" or to "bad statistics" are avoidable. You simply must be aware of pitfalls along the way and take proper precautions to deal with them.

Huff's criteria

The following five questions, adapted from Huff (1954) and hereinafter called *Huff's criteria*, provide a first-line-of-defense against an incorrect interpretation of any statistical statement offered by any advocate of any position on any issue:

- Who says so? (*Does the advocate have an axe to grind?*)
- How does the advocate know? (*Does the advocate have the resources to know the facts?*)
- What's missing? (*Does the advocate give you a complete picture?*)
- Did someone change the subject? (*Does the advocate offer you the right answer to the wrong problem?*)
- Does it make sense? (*Is the advocate's conclusion logical and consistent with what you already know?*)

Throughout these discussions and, even more important, long afterwards, as you go about your regular business, personal or professional, maintain a healthy skepticism about statistical claims and counter-claims. Try to evaluate statistical statements with Huff's criteria in mind. Indeed, you will find them applicable to *non*statistical statements as well, adding vigor and rigor to your analyses and value and meaning to your conclusions.

For discussion:

- How can you use Huff's criteria in your daily professional and personal activities?

A grammar of information

information

You conduct an investigation whenever you seek *information*. For the purposes of this discussion, information is a subset of knowledge:

information: knowledge acquired in any manner; facts; data; learning; lore.[4]

Whether you are placing a telephone call to your local library's reference desk or working over a microscope in a laboratory, your focus is to gather information about a specific item. Your quest will lead you to examine at least one item of interest with the purpose of determining at least one characteristic associated with that item.

variable

value, measurement

observation

datum

That characteristic, which may be either *quantitative* or *qualitative*, is called a *variable*, primarily because its specific value or its specific nature is not known before the item is examined. You determine a *value* for that characteristic by making a *measurement* of it (e.g., how big is it? how strong is it?) with an instrument or by making an *observation* of it (e.g., what color is it? what kind is it?) by looking at it. This single value for this characteristic for this particular item is called a *datum*.

[4] Adapted from *Webster's New Universal Unabridged Dictionary*, 2nd Ed., Dorset & Baber and Simon & Shuster, New York, NY, 1979.

data

Generally, when gathering information, you will examine more than one item and determine the values of more than one variable associated with each item. You then have *data* (the plural of datum) to work with, and the door to statistical ideas and processes is open. It cannot be readily closed.

Figure 1-1 conveys these inter-linked ideas of knowledge, information, and data as a generalized Venn diagram.[5]

Figure 1-1:
Linking knowledge, information, and data

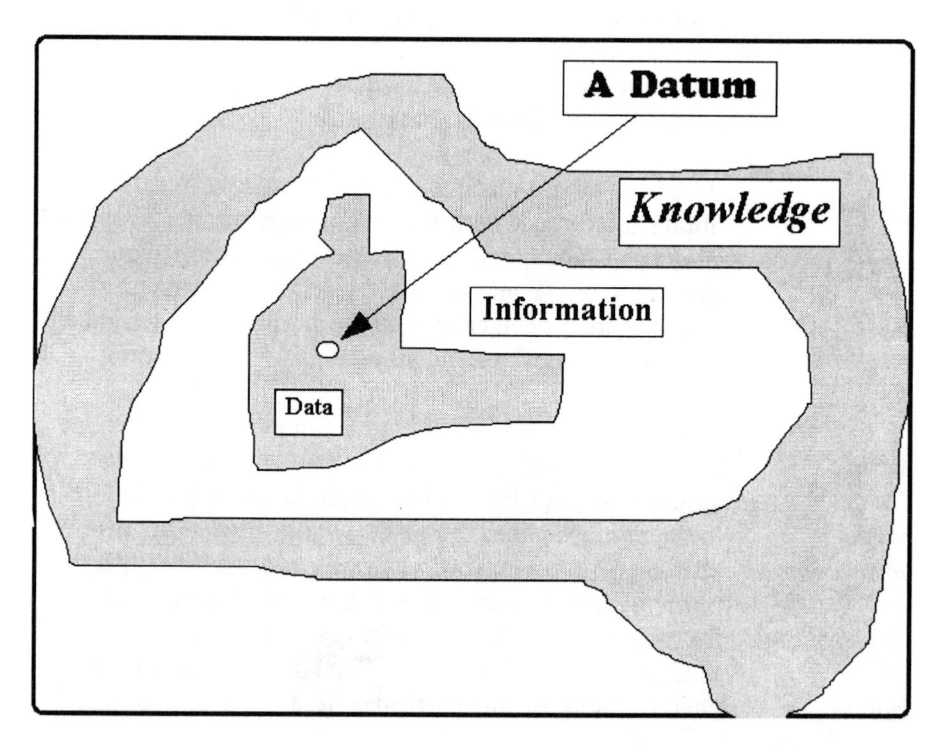

[5] Venn diagrams, named for the English logician John Venn (1834-1923), are graphical depictions used in logic and set theory; they employ two-dimensional figures (e.g., circles, ovals) to represent categorical propositions and to aid in evaluating categorical syllogisms. We have heard it said that "you use Venn diagrams vhen you don't know vhat else to do."

The reason you examine more than a single item is that, as "everyone knows," items identical in name and in definition are not necessarily—or even probably—identical with respect to all of their characteristics. Moreover, the value you obtain for one item seldom provides sufficient information to assign a value for the next item. For example, if the item is a reactor fuel rod and the variable is its weight, it simply is neither good science nor good policy to assume *a priori* that two given rods have predictably related weights.

Working with and discoursing about several items and several variables can be difficult in almost any setting and with almost any audience. A useful device for doing both, with both theoretical and practical benefits, is the *dataset*, literally a set of data. Datasets are commonly displayed in two-way tables with the rows representing the items and the columns representing the variables.

dataset

This idea of a dataset is strongly influenced by modern *database management* theory and *statistical computing* practice. In these disciplines, each row of the table is a *record*, each column is a *field*, each entry at a row-column intersect (often called a *cell*) is a *value*, and the entire table is a *file*.

record, field, value, file

To illustrate these ideas (with an entirely fictitious example), suppose you have determined values for two variables—the weight and the machinist—for each of seven fully identified and distinguishable reactor fuel rods. Table 1-1 is one possible representation of the results of this process.

Table 1-1:
Illustrating the concept of a dataset (fictitious data)

Rod Identification	Weight	Machinist
5X345	600	Billy Smith
7809Q	546.76	Billy
5X345	600	Billy Smith
ABC	59.867	Smith
222-90WT	600.19	
222-91wt	592.83	Smith
ABC	59.857	Smyth

discrete variable

A variable is said to be *discrete* if it can assume only a countable number of values. Examples of discrete variables include the number of defective units in a batch of manufactured items and the number of atomic disintegrations in a sample of radioactive material in a given unit of time. In Table 1-1, the variable labeled "Machinist" is discrete because only a countable number of individuals could be machinists of the fuel elements. (The variable labeled "Rod Identification" also is discrete; it serves the special purpose of uniquely identifying the rod to which the values in the row belong. Two or more rods with identical "Rod Identification" values would leave you in troubling circumstances.)

continuous variable

In contrast, a variable is said to be *continuous* if it can assume any value within its range. Examples of continuous variables include linear dimensions and wave lengths associated with the color spectrum. In Table 1-1, the variable labeled "Weight" is considered continuous because *any* positive number might be reported as the weight of a fuel rod.

For discussion:

- When you pick up a telephone book, are you starting an *investigation*? Are you working with a *dataset*? What are the *items of interest to* you? What are the *variables*? How are the *data* displayed?

- Is the distinction between *quantitative variables* and *qualitative variables* useful in your daily pursuits? Are you comfortable with the distinction?

- Take a careful look at Table 1-1. Have you ever seen a 21-element dataset that raises as many questions as this one does? What are some of those questions?

- The fact that a *discrete variable* can assume only a countable number of values does *not* imply that those values are integers. For example, if a die is thrown four times and variable of interest is the average of the four throws, then the variable can assume any of the 21 values in the set {1, 1.25, 1.5, 1.75, 2, ..., 5.75, 6}; it thus is discrete without being an integer.

- *Qualitative characteristics* often are tags or labels associated with examination of an item; e.g., {good, bad} or {hotel, motel, the Y, campground} or {red, yellow, green}. Tags may be, and often are, recorded as codes, using numbers or characters instead of the tags, such as {1, 0} for {good, bad} or {H, M, Y, C} for {hotel, motel, the Y, campground} or {R, Y, G} for {red, yellow, green}. Without a basis or a legend for the code, your analysis has high potential of leading to interesting adventures. What other codes might you consider for the set {good, bad}? For the set {hotel, motel, the Y, campground}? For the set {red, yellow, green?}

- Suppose, with exceedingly fine and delicate instrumentation, you are able to produce and report the weight of a reactor fuel rod as 10.01093034928 kilograms. Suppose further that you have a pragmatic angel on one shoulder and an idealistic angel on the other. Discuss the consequent dialogue between them with respect to discrete and continuous variables in "the real world."

- How do you think you will react to the potentially synonymous use of *observation* and *value* as you explore the domain of statistics?

Scales of measurement

variables

scales of measurement

Variables (i.e., *measured values* or *observations*) are themselves categorized as arising from one of four *scales of measurement*, beginning with the *nominal* scale. The other three scales, called *ordinal*, *interval*, and *ratio*, are successively built from the nominal scale by adding certain conditions. Without knowing the scale used in making the observations, you risk producing erroneous analyses because each scale's defining conditions determines which set of tools will be used in the analysis.

nominal scale

The *nominal scale* (sometimes called the *categorical scale*) employs words, symbols, or numbers to identify the categories or groups to which the items of interest belong. Thus, no special sequencing or ordering is implied by a nominal scale; that is, the {yes, no} scale carries no more information than the {no, yes} scale. Similarly the {Democrat, Independent, Republican} scale carries no more information than the {Republican, Democrat, Independent} scale. Some other examples of a nominal scale are {chocolate, vanilla, strawberry, bubblegum-butter-brickle, ... }, {helium, uranium, lead, oxygen}, {green, yellow, brown, ... }, {married, divorced, single, widowed}, {physician, engineer, senator, lawyer, statistician, ... }, and {acceptable, not-acceptable}.[6]

ordinal scale

The *ordinal scale* (sometimes called the *relative scale*) places an ordering or ranking of the groups designated by a nominal scale. Thus, something special *is* implied in an

[6] If you feel some discomfort at using the terminology *nominal scale*, you are not alone. The idea here is similar to the use of a zero in real numbers or the empty set in set theory; it completes the concept of scales of measurement by providing a fundamental building block. See Stevens (1946).

ordinal scale; that is, the {truck, auto, motorcycle} scale carries different implications that the {motorcycle, auto, truck} scale. Some other examples of an ordinal scale are {heavy, medium, light}, {helium, oxygen, lead, uranium}, {cute, beautiful, gorgeous, 10}, {nice, handsome, hunk, "beyond category"}, {excellent, good, fair, poor}, and {grade school, high school, college, graduate school}.

interval scale

The *interval scale* (sometimes called the *metric scale*) preserves the characteristics of the *ordinal* scale and, in addition, assigns a *numerical "distance"* between each pair of objects. Thus, the differences between values are important in an interval scale; thus, a scale that measures distances among a group of cities permits the determination of which two cities are closest and which two are farthest apart and which pairs are the same distance apart. Some other examples of an interval scale are {birthdates of a family's members} and {thermostat settings in a group of single-family residences}.

ratio scale

The *ratio scale* is an *interval scale* with a physically meaningful and definable *zero value* (indicated by the symbol 0) which is the value reported when none of the characteristics is detected by the measuring process. The ratio of any two measured values of the same type is independent of the unit of measurement; e.g., the ratio of the width to the length of a rectangle is independent of whether both measurements are made in inches, feet, kilometers, or furlongs. Some other examples of a ratio scale are {weight; e.g., kilograms or pounds}, {length; e.g., meters or inches}, and {volume; e.g., gallons or liters}, all of whose ratios are independent of whether English or metric or other units are used. Note, however, that a ratio of mixed units (e.g., width to length of rectangles) *does* depend upon the units. Thus, a 2-meter by 3-meter rectangle has a width-to-length ratio of 2/3, just as does a 200-centimeter by 300-centimeter rectangle. But a 2-meter by 300-centimeter rectangle gives a width-to-length ratio of 1/150.

It must be noted that, since their introduction by Stevens (1946), these four scales have been the subject of a continuing controversy, summarized in a paper by Velleman and Wilkinson (1993), part of which reads:

> Recent interest in artificially intelligent computer programs that automate statistical analysis has renewed attention to Stevens's work. Computer programs designed to assist in the selection of data analysis methods have been based on his prescriptions. Even some general-purpose programs have used them to structure their interaction with the user.
>
> Unfortunately, the use of Stevens's categories in selecting or recommending statistical analysis methods is inappropriate and can often be wrong. They do not describe the attributes of real data that are essential to good statistical analysis. Nor do they provide a classification scheme appropriate for modern data analysis methods. Some of these points were raised even at the time of Stevens's original work. Others have become clear with the development of new data analysis philosophies and methods.

As with most controversies, there is something to be said for each side. We suggest a middle ground: Use the four scales in your work as long as they are helpful to you and your colleagues—but be especially alert to possible problems arising from blind adherence to them, especially with respect to statistical software applied uncritically to your data.

For discussion:

- Which of the four scales is associated with the following measurements?

  (1) The size (in bytes) of a computer file.
  (2) The size (in kilobytes) of a computer file.

(3) The average number of words per page in this book.
(4) The shortest distance between your home and your office.
(5) The Dow Jones average of 30 industrial stocks as an indicator of New York Stock Exchange activity.

- Why would you be careful to distinguish the zero of the ratio scale from the zero recorded when no observation was made on a particular item?

- Can temperature be measured on a ratio scale? (Hint: Is the ratio of two temperatures measured in degrees Celsius the same as when they are measured in degrees Fahrenheit?)

Four basic concepts in statistics

Four basic concepts underlie and integrate all of statistical thought and processes. Nothing substitutes for addressing these four concepts early and often.

A ***population*** is a collection of items defined by some characteristic of the items.

A ***sample*** is a subset of the population.

A ***parameter*** is a numerical measure of the population; its value is a function of the values of the variables of the members of the population.

A ***statistic*** is a numerical measure of a sample; its value is a function of the values of the variables measured for the members of the sample.

For discussion:

- A population may be a collection of measurements (or responses) rather than individual physical items (or automobiles or houses or employees) which were examined to obtain the measurements (or responses). Does this idea cause you any anxiety or concern?

- In some contexts, you will find such terms as *parent population*, *universe*, and/or *universe of discourse* instead of *population*. These terms serve as a reminder that a sample is an "offspring" of the population of interest.

- Notice especially that the *value* assigned to a parameter is a *function of the values* of a population of interest. This is exactly the same as the use of the term *parameter* in any mathematical discussion. For example, in the mathematics of analytical geometry, $y = mx + b$ is a *parametric* expression of a *general* straight line with a symbolic slope of m and a symbolic intercept of b. In comparison *and* contrast, the expression $y = 3x + 5$ is a *specific* straight line with a slope of 3 and an intercept of 5; thus, given any two points lying on that line, you can determine its slope and intercept unequivocally.

- Contrast this idea of a parameter—and any of its specific manifestations for a straight line (obtained, for example, by simply setting $m = 3$ and $b = 5$)—to that of a statistic which depends upon *measuring* the items in a sample. It follows that the *value* obtained for a statistic *depends upon the elements that appear in the sample*. Thus, while populations and parameters remain creatures of our modeling, our imagination, and our intellectual belief systems, we are stuck in the real world with understanding samples and statistics and how they link those populations and parameters to the treatment of our everyday problems.

- A sample can be the entire population, in which case it is called a *census*.

- The term *random sample* is ubiquitous in statistical discourse. What does *random sample* imply that *sample* does not? Why does it make any difference?

- A population's size (i.e., the number of its elements) can be finite or infinite. Give examples of each.

- A population can, and usually does, have more than one parameter.

- How would you go about obtaining a value for a population's parameters? Is the problem easier if it's limited to *finite populations*? Give an example of a population for which you can determine values for its parameters.

- A sample can, and usually does, have more than one statistic.

- Can a sample be infinite?

A definition of *data* for this book

Data are those elements of information that either
are quantified by their basic nature
or
are capable of quantification.

A definition of the *discipline of statistics* for this book

The discipline of statistics is
the science* and *the art of the treatment of data:
their collection,
their analysis,
their summarization,
and
their presentation.

Introducing *OSDAR*

When you contemplate conducting an investigation, one of the first questions you must decide is whether you are going to collect data; i.e., will you require quantification of any of your findings? If the answer is *no*, you're off the statistical hook, and we won't nag you any more.

However, if the answer is *yes*, then these two pages have special meaning for you and your ways of dealing with numerical information. The basics are contained in an easy-to-remember five-letter acronym, OSDAR.[7] These five letters serve as reminders of five essential elements that must be considered whenever you plan an investigation with the purpose of collecting data:

Objective(s)
Scope
Data Collection
Analysis
Report.

These five elements are important in two ways: as planning devices *and* as checkpoints during your study.

Notice especially that there is nothing in OSDAR that inherently implies a laboratory-with-instruments setting. These five elements apply equally as well when you are involved in a literature search as when you are involved in a high-tech, no-funding-limit investigation.

Each of the five elements is expanded upon on the facing page. Make a copy of the following page. Keep it around your workspace. You never know.

[7] We are especially indebted to James Cahill, Bonneville Power Administration, for a series of discussions in the spring and summer of 1992 that resulted in OSDAR's codification.

O S D A R

Your checklist for investigations involving data acquisition

Objective(s) State the ***objective(s)*** of the investigation carefully, clearly, and completely. What target population(s) is/are of interest, and under what conditions and for which time interval? Which variable(s) do you wish to measure? Which hypothesis(es) do you intend to test? How is your new work linked to that of other investigators? Do you have agreement from all participants in the study?

Scope Specify the ***scope***—the limits and the limitations—of the investigation. What sampling population(s) will you use? Will it/they yield the variables you seek? What can you accomplish with your resources—personnel, equipment, material, budget? Do you have the proper analytical tools to address the **objective(s)**?

Data Collection Ensure that your methods of ***data collection*** are planned and carried out realistically. Are your procedures feasible? Can they be conducted properly in terms of your resources? Are the variables you can obtain the same as those you expected to collect when you set your **objective(s)** and specified the **scope** of the work? Are the data properly organized and labeled? Will they be sufficient to accommodate your analytic tools? What data-quality monitoring will be performed?

Analysis Establish the methods of your ***analysis***—assumptions, models, algorithms, computing processes. Does your proposed analysis address the **objective(s)** *and* the **scope** of the investigation? Will you be giving the correct answer to the wrong question?

Report Plan the preparation and issuance of a ***report*** of your findings, keying those findings to your investigation's **objective(s)** and its **scope**. In which medium (e.g., internal report, journal article, book) will you issue your report? What are the limitations of your study and your analysis? Will your text, tables, and charts be properly labeled and integrated and easily interpreted? Does your report aid in understanding your work, or have you created a contest in which your audience must labor to appreciate your cleverness? What did you find? What did you not find? How are your results linked to those of other investigators? Do your data survive application of Huff's criteria?

What to remember about Chapter 1

You now have a number of statistical concepts and ideas with which to conduct investigations that involve the acquisition of data.

Keep these concepts and ideas in mind (or refer back to this chapter) when you encounter the following terms in subsequent chapters:

- *Huff's criteria for evaluating statistical statements*
- *quantitative information*
- *discrete variables*
- *continuous variables*
- *scales of measurement*
- *populations*
- *samples*
- *parameters*
- *statistics.*

Check your regular task assignments against Chapter 1's definitions of data and the discipline of statistics.

Revisit OSDAR. Always remind yourself of OSDAR's precepts when you are digging into data-acquisition matters.

Above all, as you peruse the following pages, *enjoy the journey*.

2

Some statistical graphics methods

What to look for in Chapter 2

Chapter 2 reminds you of some of the popular—and venerable—statistical graphs used to convey information, including:

- *pie charts*
- *bar charts*
- *histograms*.

A relatively new special purpose graph, called the *box plot*, is demonstrated. Some rules for the effective use of these four types of graphics are offered.

From out of the past ...

Since the dawn of man's record-keeping on the walls of caves, graphical methods and statistical procedures have been intertwined. Charts of all types, using all sorts of colors and designs, are part of the fabric of modern life. In a variety of forms, some certainly more useful than others, *statistical graphics* appear in newspapers, magazines, television, and other media. As simple as some of these chart types may be, they scarcely can be equalled as elegantly powerful devices to convey information contained in a collection of data.

Because statistical graphics are so powerful and so influential, they are subject to a range of ill-use, from deliberate hiding of facts to inadvertent "chart junk," to use a concept emphasized by Tufte (1983). You will encounter four basic types of graphical displays in this chapter. You'll start with the venerable and ubiquitous *pie chart*, remind yourself of the revealing *bar chart* and the always-important *histogram*, and finish with the late-20th-century *box plot*.

But statistical graphics do not stand alone; they can start their value-added function only after some data are collected. You will find two keys to quality statistical graphics:

(1) they must display their source data in a readily apparent fashion, and
(2) they must be free of irrelevancies.

The pie chart

The *pie chart* is a graphical display designed to show and emphasize the relative proportions of several values of a nominal scale (recall the discussion in Chapter 1 on scales of measurement). Consider the data displayed in Table 2-1 (NRC, 1992, p. 19).

Table 2-1:
1990 U.S. electric capability by energy source

Source	Gigawatts	Percent of total
Coal	300	43
Gas	120	17
Nuclear	100	14
Hydro	91	13
Petroleum	77	11
Other	4	1
Totals	692	99*

* A rounding discrepancy, something featured in many tabulations.

Table 2-1 is a typical set of data that leads to the use of a pie chart. For each of several values of a nominal scale—in this case, the energy source: {Coal, Gas, Nuclear, Hydro, Petroleum, Other}—you have a value of an interval scale (in this case, the amounts of energy). Convert each of these interval values to a percentage of their total. Then convert those percentages into proportional segments of a circle to form the "slices of the pie." The data in Table 2-1 are rendered as a pie chart in Figure 2-1.

Figure 2-1:
Pie chart derived from data in Table 2-1

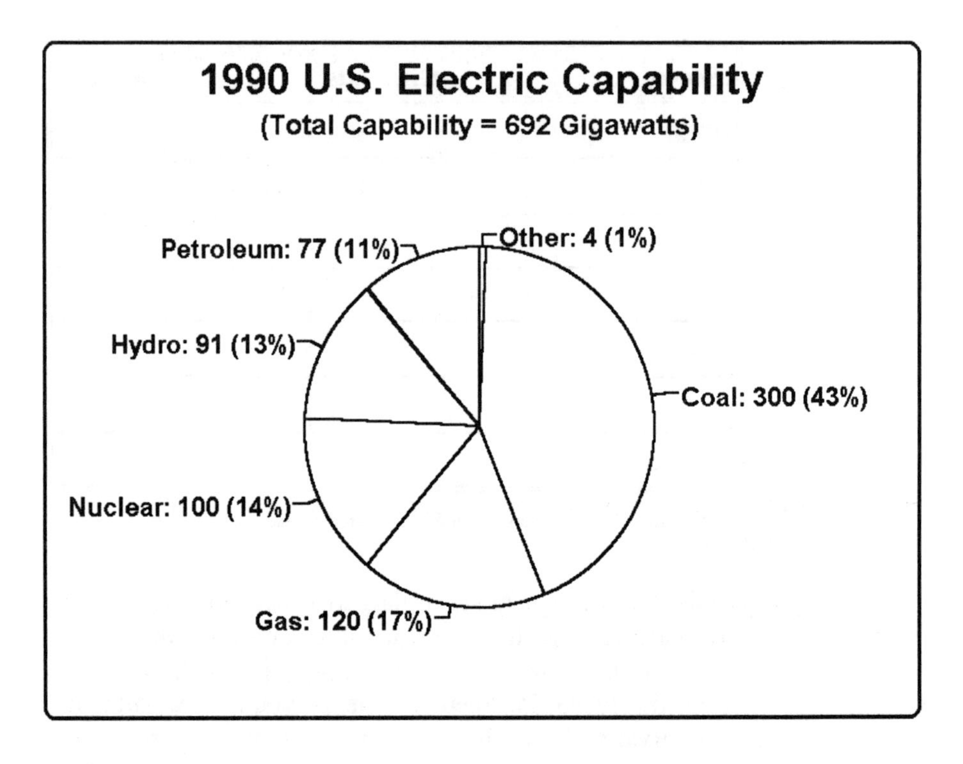

Suggestions for constructing pie charts

- The number of slices of the pie should be between 4 and 10. Having too many slices clutters the chart, while having too few insults your audience.
- Arrange the slices in increasing or decreasing order of magnitude.
- Begin the first slice at the 12 o'clock position, followed clockwise by the other slices.
- Show the "raw" data as well as the percentages.

- Despite the pie chart's popularity, venerability, and ubiquity, it is not universally endorsed as a statistical display of data. For example, Stein, *et al.* (1985), list the pie chart, among others, as *not* recommended. They go on to say specifically:

> The pie chart has the sole advantage of indicating the data must add up to 100 percent. At the same time, it is difficult to see the relative size of two slices, especially if they are within 10 percent of each other.

For discussion:

- Comment on the construction of pie charts, especially the suggestions given in this chapter.

- Look at the pie chart in Figure 2-1a. Compare it with the pie chart in Figure 2-1. Is one format more informative or useful than the other? Note the differences, for example, in the shadings of the slices and the angle of view. What are some other differences between the two charts? Which of them enhance—or detracts from—the presentation of the information?

Figure 2-1a:
A second pie chart derived from Table 2-1

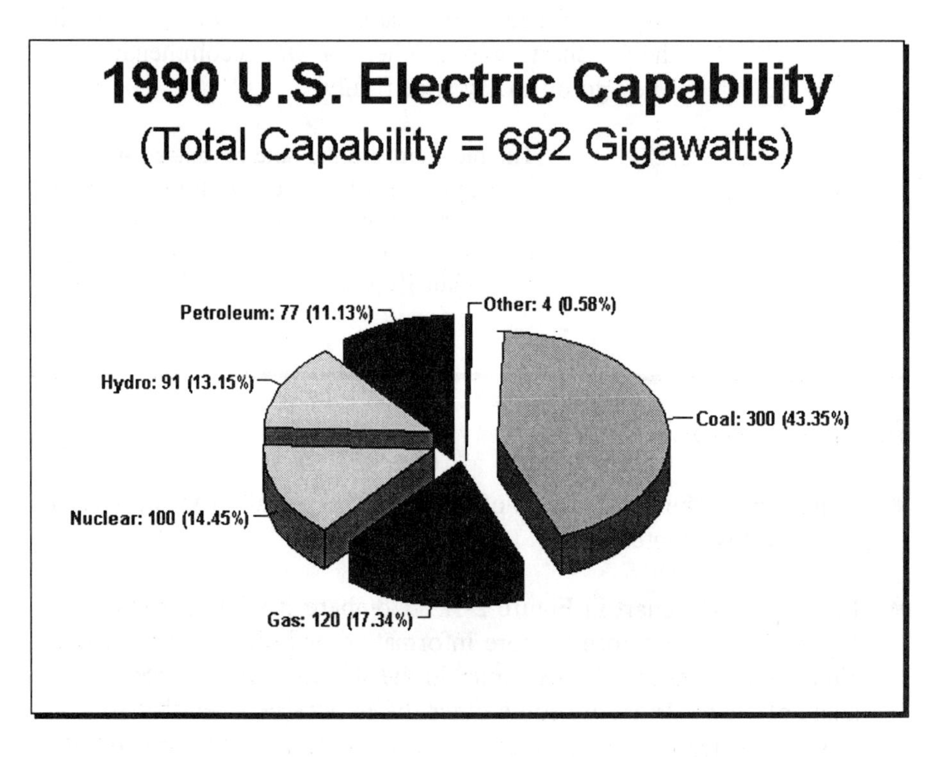

- Is Figure 2-1 or Figure 2-1a the easier to read and interpret?
- Comment on this statement: *It is easier to distort a graph than to distort a table.*

The bar chart

bar chart

The *bar chart* is a graphical representation used to display data that come primarily from nominal or ordinal scales (recall the discussion on measurement scales in Chapter 1). Consider the data displayed in Table 2-2 (Bowen and Bennett, 1988, p. 12).

Table 2-2:
Monthly rejects of fuel rods

Month	Jan	Feb	Mar	Apr	May	Jun
Rejects	6	4	4	2	4	7
Month	**Jul**	**Aug**	**Sep**	**Oct**	**Nov**	**Dec**
Rejects	6	2	2	5	4	4

Table 2-2 contains a set of data that typically leads to a bar chart, such as that shown in Figure 2-2. Each group (or category) has a frequency-of-occurrence. Each bar in the bar chart is intended to represent one group (in this case, Month), with the bar's height representing that group's frequency (in this case, Rejects). The width of the bars is constant, as is the space between each pair of bars; if either of these requirements is violated, the bar chart's "message" is subject to misinterpretation. If the categories are on an ordinal scale (as they are in Table 2-2), the bars are placed in sequence, usually smallest to largest. The bars may be oriented horizontally or vertically.[1] The data from Table 2-2 are rendered as a vertical bar chart in Figure 2-2.

[1] In some contexts (e.g., Microsoft's spreadsheet Excel), vertical bar charts are called "column charts."

Figure 2-2:
Bar chart obtained from data on monthly fuel rod rejects (Table 2-2)

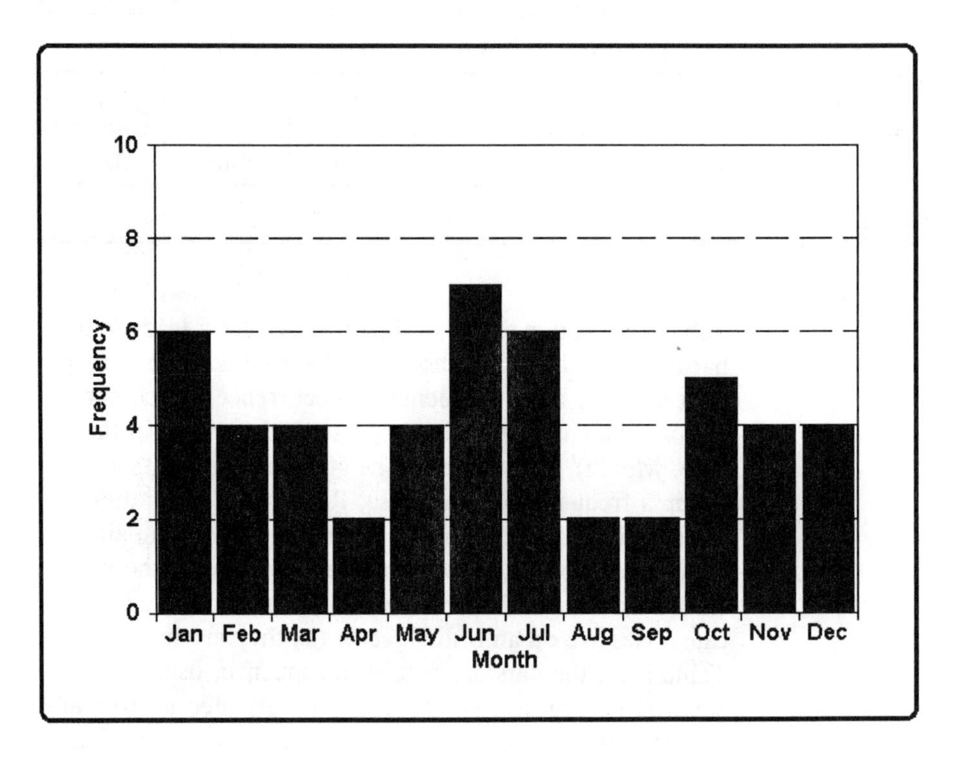

For discussion:

- What interpretation(s) do you give to the bar chart in Figure 2-2a?

Figure 2-2a:
Fuel rod reject rates: the second year

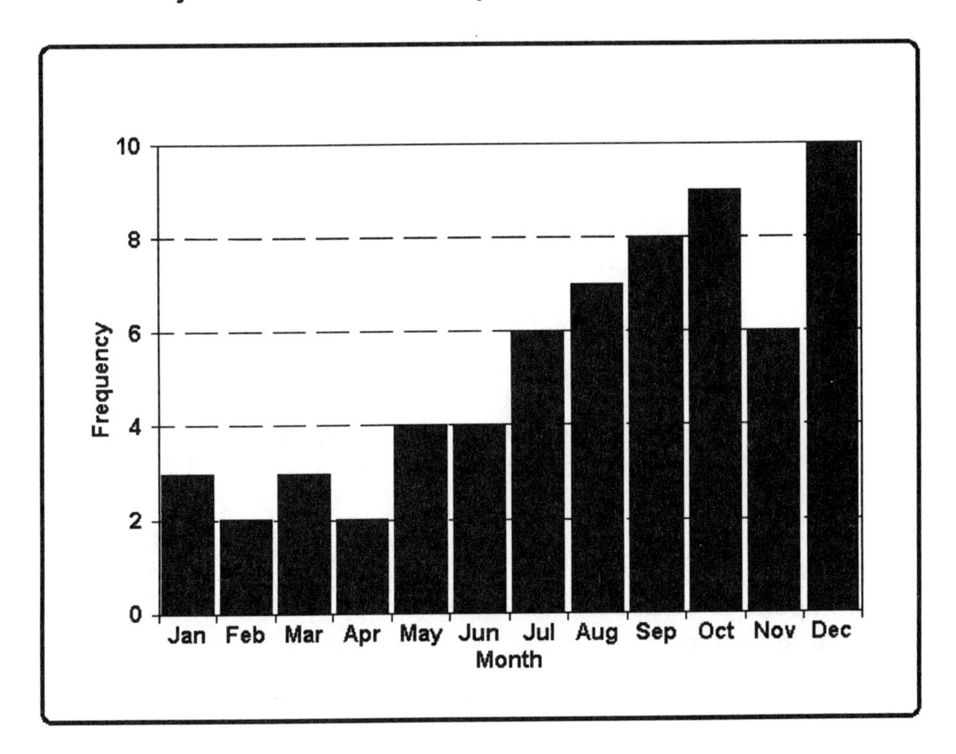

- Compare the bar chart in Figure 2-2a with that in Figure 2-2. How have things changed? Does the chart necessarily describe a deteriorating fuel rod quality as the year progresses?

- Sketch these two charts as *horizontal* bar charts. Is either form, horizontal or vertical, more informative than the other?

The histogram

histogram The *histogram* is a specialized type of bar chart; it is used primarily for displaying frequencies of a variable measured on the interval (metric, continuous) scale or on

class intervals

class marker

the ratio scale. The measurements are grouped into *class intervals*, usually of equal widths. The middle of each interval is called the *class marker* for that interval. For the purpose of display, then, all measurements falling into a particular class interval are considered to have the same value as the class marker. A histogram usually shows the frequencies and/or the relative frequencies (i.e., the fractions of the measurements that "fall" into each class interval) on the vertical axis. If the values are continuous, no spaces are used between the bars of the histogram.

Consider this fictional-yet-realistic example: One month after a nuclear reactor shutdown, 25 buckets of water were drawn out of a nearby lake and inspected for impurities. Water impurities for each of the buckets, measured in parts per million (ppm), are given in Table 2-3.

Table 2-3:
Water impurities measured for 25 buckets in parts per million (ppm)

42.50	42.61	42.49	43.17	42.63	42.45	42.61
42.31	42.83	42.32	42.82	42.91	42.42	42.97
42.48	42.62	42.83	42.93	42.67	42.40	42.85
42.65	42.72	42.45	42.59			

To build a histogram from the data in Table 2-3, you must first determine the smallest and the largest values and decide upon a class interval. Here is the ordered set, from smallest to largest: {42.31, 42.32, 42.40, 42.42, 42.45, 42.45, 42.48, 42.49, 42.50, 42.59, 42.61, 42.61, 42.62, 42.63, 42.65, 42.67, 42.72, 42.82, 42.83, 42.83, 42.85, 42.91, 42.93, 42.97, 43.17}. The smallest value in this dataset is 42.31 ppm and the largest is 43.17 ppm.

Next, suppose you decide upon a class interval width of 0.1 ppm, and you set up a scheme of class intervals that have "nice" (i.e., easily set and managed) class markers. For these particular data, you might decide on this set of class markers {42.3, 42.4, 42.5, ..., 43.2}. Given this choice, the first class interval contains all measurements less than or equal to 42.35 ppm; the second contains all those greater than 42.35 ppm *and* less than or equal to 42.45 ppm, and so on. In this fashion, every one of the 25 values can be placed unambiguously into one of the 10 intervals so defined. This process yields the frequencies which are the basis for the heights of the bars.

Here, for example, you have two values associated with 42.3, four values associated with 42.4, ..., and one value associated with 43.2. With these associations, you can proceed to construct the histogram. It should look something like Figure 2-3.

Figure 2-3:
Histogram derived from 25 values of water impurities recorded in Table 2-3

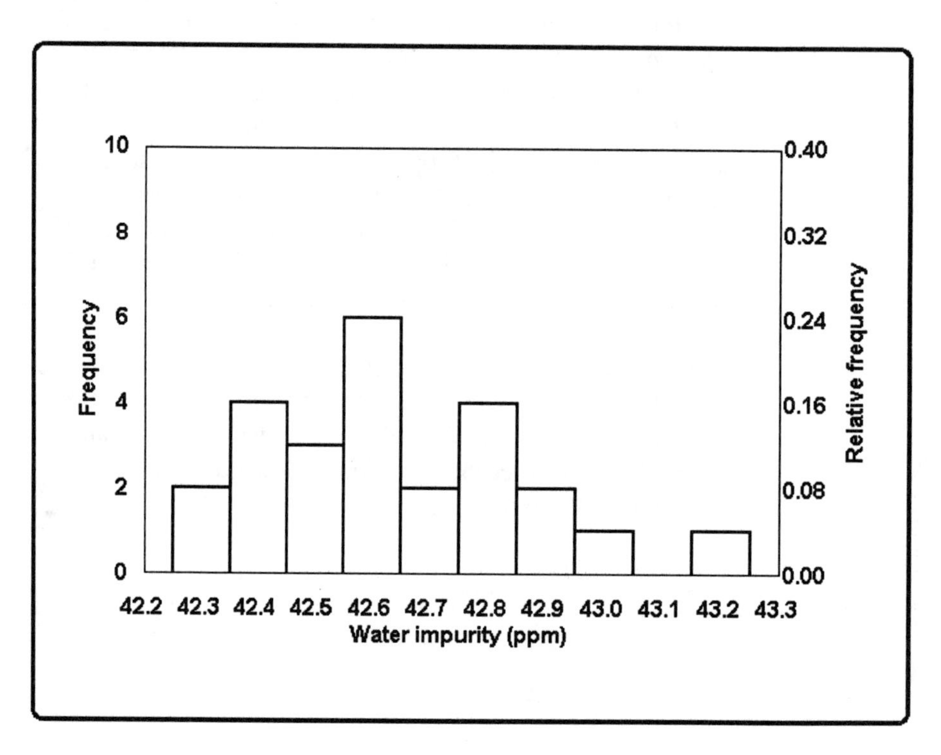

The box plot

box plot

A *box plot* is a graphical display designed to represent a set of data in a relatively small amount of space *and yet* give a picture of the extent of the data as well as its distributional features. Figure 2-3a is another view of the impurity data in Table 2-3, this time displayed as a box plot.

Figure 2-3a:
An example of a box plot (derived from 25 water impurity measurements recorded in Table 2-3)

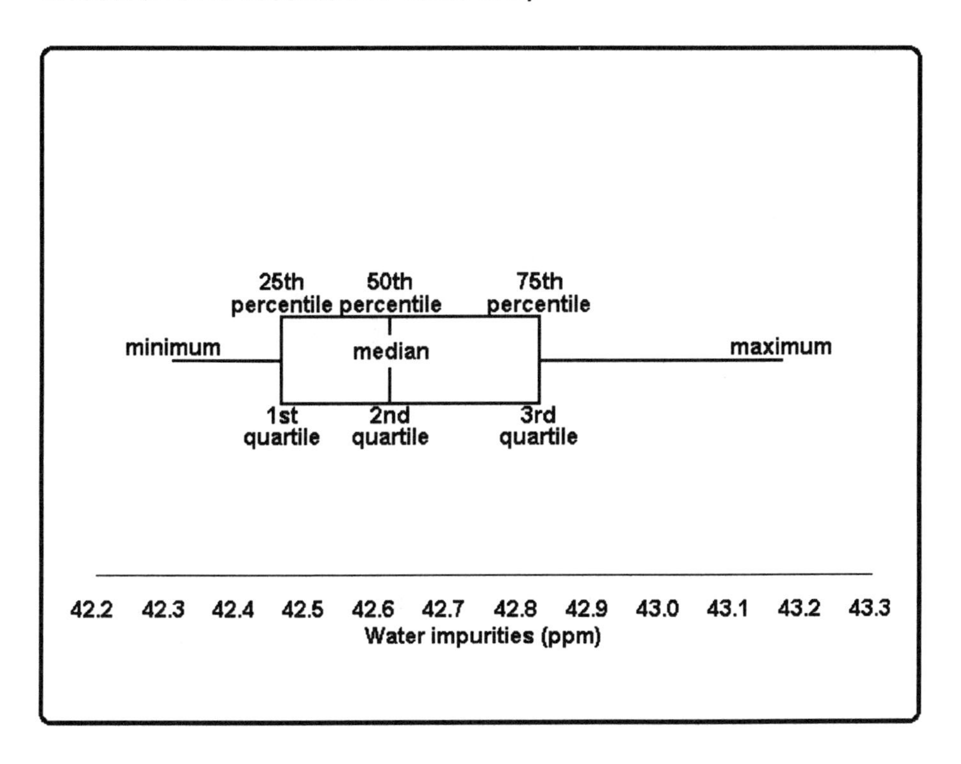

The display itself consists of a rectangular box that represents the bulk of the data and lines extending from each end of the box that indicate the range of the data.

The construction of a basic box plot requires the determination of five particular quantities from the data: the *minimum*, the *maximum*, the *median*, the *25th percentile*, and the *75th percentile*. The 25th percentile and the median and the 75th percentile divide a set of data into four equal parts. Thus, you will sometimes find them called, respectively, the 1st, 2nd, and 3rd quartiles; they are so indicated in Figure 2-3a.

The five determining quantities for a box plot are obtained in the following fashion (and are numerically illustrated with the 25 water impurity values in Table 2-3):

The *minimum*: Find the smallest value in the group.
The minimum of the 25 water impurities values is 42.31.

The *maximum*: Find the largest value in the group.
The maximum of the 25 water impurities values is 43.17.

The *median*: Arrange the observations in increasing order of magnitude. Divide them into two equal-sized groups by taking
(a) the middle observation *if the number of observations is odd* or
(b) the average of the two middle observations *if the number of observations is even.*
The median of the 25 water impurities values is 42.62.

The *25th percentile*: Find the median of the group containing the smaller values.
The 25th percentile of the 25 water impurities values is (42.45 + 42.48)/2 = 42.465.

The *75th percentile*: Find the median of the group containing the larger values.
The 75th percentile of the 25 water impurities values is (42.83 + 42.83)/2 = 42.83.

Owing to its particular power as a data-analysis tool, the box plot has been expanded and refined in a number of its aspects. For example, the box plot can include indicators of "unusual" values and/or other characteristics of the dataset. See Chambers, *et al.* (1983), pages 21-24, for an extended discussion.

Box plots are of special value when you are comparing two or more sets of data with respect to the same variable. For example, consider the data in Table 2-4 in which 10 consecutive months of a shipper's reported net weights of uranium fluoride (UF_6) cylinders are displayed beside those reported by a receiver (Bowen and Bennett, 1988, pages 18-19). Box plots of these data are given in Figure 2-4.

Table 2-4:
Net shipper's and receiver's weights (measured in kilograms) of UF_6 cylinders for 10 consecutive months

Month	Shipper	Receiver
1	1471.22	1468.12
2	1470.98	1469.52
3	1470.82	1469.22
4	1470.46	1469.26
5	1469.42	1465.96
6	1468.98	1470.80
7	1469.10	1467.89
8	1470.22	1472.28
9	1470.86	1469.02
10	1470.38	1470.16
median	1470.42	1469.24
minimum	1468.98	1465.96
25th percentile	1469.42	1468.12
75th percentile	1470.86	1470.16
maximum	1471.22	1472.28

Figure 2-4:
Box plots of net shipper's and receiver's weights (measured in kilograms) of UF_6 cylinders for 10 consecutive months recorded in Table 2-4

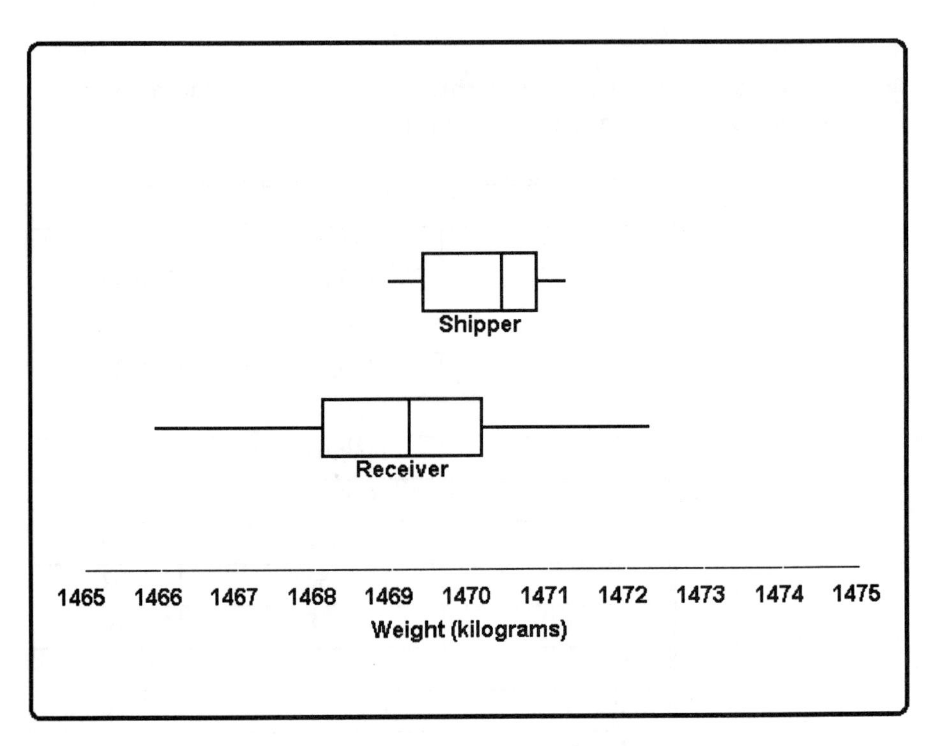

It doesn't require a great deal of data-analytic acumen to observe from Figure 2-4 that the Receiver is reporting fewer kilograms of material than the Shipper is claiming. Also, the Receiver's values are considerably more variable that the Shipper's. The message is there for all to see—and without any particular statistical sophistication to muddy up the discourse.

Recall that the vertical line inside each box is the median for the corresponding values of the variable, while the left and right ends of each box are the 25th percentile and the 75th percentile, respectively, of the variable. In box plot parlance, the 25th and the 75th percentiles are called

hinges, *whiskers*

hinges. The horizontal lines extending from the ends of the box are called *whiskers*. Indeed, you often will find that *box plots* are called *box-and-whiskers* plots. You will sometimes encounter box plots that are erected vertically with the whiskers extending upward and downward.

For discussion:

- Discuss the advantages and disadvantages of a box plot compared with a histogram.
- Box plots have numerous variations in appearance. For example, unusual or extreme values may be shown as asterisks with the whiskers curtailed according to a rule of practice. In other cases, the box's width may be proportional to its sample size. What changes in box plots would enhance their value in your data presentation?
- The 25th, 50th, and 75th percentiles, which are necessary for the construction of a box plot, are not necessarily members of the dataset. How can this be?
- What interpretation do you have for the whiskers of a box plot?

What to remember about Chapter 2

You now have been introduced to—and, of course, thoroughly assimilated—a number of graphical tools for conveying statistical information. Each of these tools has its own strengths and weaknesses, and each has the potential to inform or to *dis*inform.

The five statistical graphical tools mentioned in this chapter are:

- *pie chart*
- *bar chart*
- *histogram*
- *box plot.*

Some specific statistical terms used in connection with these graphs include:

- *class interval*
- *class marker*
- *median*
- *25th percentile*
- *75th percentile*
- *hinges*
- *whiskers.*

Statistics and probability

What to look for in Chapter 3

Chapter 3 aims directly at the hearts of two tightly interlaced subjects: statistics and probability. This material contrasts and links the two topics, paying particular attention to the ideas of:

- *experiment*
- *sample space*
- *random variable.*

In addition, the fundamentals of statistical decision-making are described, along with such specialized terminology as:

- *statistical significance*
- *Type I error*
- *Type II error*
- *level of significance.*

Is there a need to make a *real* distinction between statistics and probability?

Of course there's a need to make a distinction—or we wouldn't have brought up the subject. It's just that all of us are so used to seeing and thinking of statistics and probability in close juxtaposition, like "wine and cheese" and "milk and cookies," that we sometimes neglect to work through the implications. We tend to forget that each concept in the pair is individually important, even though they are clearly interdependent in our use of them; e.g., *wine* really is different from *cheese ... but together they are something special.*

Similarly, *probability* really is different from *statistics ... but together they are very special.* It may seem that, as you delve into these matters, these are twin tigers that you have by their tails. Nevertheless, they *are* distinguishable tigers. More than that, you *must* be able to distinguish between them to progress in the analysis and interpretation of data. Consider how that distinction is made in the following two scenarios, based on material in Gilbert (1976, pp. 59-60).

Scenario 1: Consider a box containing 100 colored marbles, 40 of them are red and the other 60 are blue. Blindfolded, you remove 10 marbles from the box. These 10 marbles are a *sample* of size 10 taken from a *population* of size 100. *Before* you examine your holding (i.e, the marbles in your sample), you may ask such questions as:

What is the probability that all 10 of the marbles in the sample will be red?

What is the probability that the sample will contain exactly 9 red marbles and 1 blue marble?

What is the probability that at least 7 marbles in the sample will be red?

What is the probability that not more than 3 marbles in the sample will be blue?

What is the probability that the sample will contain exactly 4 red marbles and 6 blue marbles?

The processes by which these types of questions are addressed are directly within the domain of the discipline of science called *probability*. Its practitioners are called *probabilists*.

The essence of ***Scenario 1 (the Probability Scenario)*** *is that you* ***know*** *the makeup of the population (the marbles in the box and their pertinent characteristics—in this case, the numbers of each color), but you* ***don't see*** *the sample data (the colors of the marbles in your hand).*

Scenario 2: You *do not* know the nature of the population; at least, you don't know that 40% of the 100 marbles are red and 60% are blue. All you have to work with is the set of data derived from the sample. For example, suppose you know only that you have 3 red and 7 blue marbles in your hand. Once you've examined your holding, your knowledge is reversed from **Scenario 1**; now you may ask such questions as:

What is the fraction (proportion) of red marbles in the box?

How "confident" are you of your estimate of that fraction?

How many marbles are in the box?[1]

The processes by which these types of questions are addressed are directly within the domain of the discipline of science called *statistics*. Its practitioners are called *statisticians*.

*The essence of **Scenario 2 (the Statistics Scenario)** is that you **don't know** the makeup of the population (the marbles in the box and their pertinent characteristics—in this case, the numbers of each color), but you **do see** the sample (the marbles in your hand) and the information contained in it.*

A short—but necessary—discourse on a notation for probabilities

Probabilities are real numbers. They are never smaller than zero or larger than one. But, like a myriad of other scientific concepts, they can be difficult to deal with without some kind of shorthand or notation.

Probabilities are indicated in this book by a special symbol: *Pr*{○}. The ○ between the braces is a placeholder; i.e., it must be replaced with a meaningful description of an event before the symbol *Pr*{○} has meaning. You read the symbol as: "The probability of the event ○." Thus, $Pr\{H\} = 0.5$ might denote the probability of obtaining a head, symbolized by H, in a single throw of a "fair" coin. Similarly, the expression *Pr*{5R, 5B | 100 marbles with 40R and 60B} might denote the probability of drawing 5 red marbles and 5

[1] This problem is real. Such varied professionals as wildlife managers face it when they set out to estimate the size of a population of particular species as do census takers when they attempt to estimate the number of people *not* counted in a census operation. You will find material on these problems and their solutions in Brownlee (1965, pp 162-163), Chapman (1951), and Hogan (1992).

blue marbles from a batch containing 40 red marbles and 60 blue marbles.

You will see more of these concepts as you work through this chapter and on into the rest of the book.

Linking probability and statistics: Experiments, sample spaces, and random variables

As teachers and promoters of statistical techniques, we are ever-sensitive to introducing any concept at any time that risks bringing on an attack of *YEGO*.[2] At the same time, certain ideas must be faced unflinchingly because, once conquered, they are keys to deeper understanding and appreciation. In law, one faces contracts and torts. In accounting, it's debits and credits. In engineering, it's wave guides and two-port parameter conversions. In medicine, whatever the subject, it tends to be unpronounceable. Yet, in each discipline, understanding and progress are in the details.

In statistics, it's *experiments*, *sample spaces*, and *random variables* that serve both as barriers and as bridges.[3] All five of these italicized words have common, everyday meanings and values. In the next few paragraphs, they are given precise probabilistic/statistical meaning. Grappling with them at this point will bring satisfying rewards later.

---

[2] *YEGO* = Your Eyes Glaze Over.

[3] The development of these ideas owes a great deal to careful and more detailed expositions in Brownlee (1965, pp. 1-86) and Meyer (1970, pp. 1-116); either source will reward the interested reader.

experiment, *outcomes*

An *experiment* is a planned inquiry to obtain new facts or to confirm or deny the *outcomes* of previous experiments, where such inquiry will aid in decision-making.[4]

Some examples of experiments are:

- *the tossing of 5 coins and observing the* absolute value *of the difference between the number of heads and the number of tails*

  One of the canons of the scientific method is that an experiment is capable of being repeated an indefinite number of times under essentially unchanged conditions. Fanciful as it seems, the main idea here is that you *could*, if you had the longevity and the stamina, repeatedly toss those 5 coins and observe the outcomes a decillion (i.e., 10^{33}) times—or more.

[4] This definition is adapted from Steel and Torrie (1980, pp. 122-136). Although their focus is on agricultural experiments, their precepts are adaptable across all experimental situations, as shown by these excerpts from their material:

> If we accept the premise that new knowledge is most often obtained by careful analysis and interpretation of data, then it is paramount that considerable thought and effort be given to planning their collection in order that maximum information be obtained for the least expenditure of resources...
>
> ... experiments fall roughly into three categories, namely, preliminary, critical, and demonstrational, one of which may lead to another. In a preliminary experiment, the investigator tries out a large number of treatments in order to obtain leads for future work; most treatments will appear only once. In a critical experiment, the investigator compares responses to different treatments using sufficient observations of the responses to give reasonable assurance of detecting meaningful differences. Demonstrational experiments are performed when extension workers compare a new treatment or treatments with a standard. In ... the critical type of experiment ..., it is essential that we define the population to which inferences are to apply, design the experiment accordingly, and make measurements of the variables under study.
>
> Every experiment is set up to provide answers to one or more questions. With this in mind, investigators decide what treatment comparisons provide relevant information. They then conduct an experiment to measure or to test hypotheses concerning treatment differences under comparable conditions. They take measurements and observations on the experimental material. From the information in a successfully completed experiment, they answer the questions initially posed. Sound experimentation consists of asking questions that are of importance in the field of research and in carrying out experimental procedures which answer these questions...."

- *the selecting of a painted automobile fender from a production line and observing the fender's paint color and the number of flaws in the paint*

  Although you may not be able to state what a particular outcome will be, you are able to describe the set of all possible outcomes of an experiment. Thus, for any single fender, you *could* observe 0 or 1 or 2 or ... flaws.

- *the selecting of 10 commercial establishments in a (sufficiently large) community and observing the amount of fuel used for heating each of them*

  As you perform an experiment repeatedly, the individual outcomes are unpredictable. However, as you repeat the experiment a large number of times, you expect a pattern in the frequencies of the outcomes. You *could* select 10 commercial buildings many different ways, each time observing 10 different values for the amounts of fuel. Thus, each set of 10 commercial buildings yields a different histogram of fuel use; although the histograms are different, you expect them to bear similar shapes, locations, and widths.

sample space

A *sample space* is the set of all possible outcomes of an experiment.

You must have a clearly stated idea of the set of outcomes (i.e., what you are observing) of the experiment. This means that you must be ready to record what happens when you conduct an experiment. How do you decide upon the sample space for the difference between the numbers of heads and tails in the toss of 5 coins?

You must be able to conceptualize and articulate the *number* of outcomes in the sample space. A given sample space may be one of the following:

- *finite* (i.e., have exactly k possible values, where k is a positive integer)
- *countable* (i.e., have an infinite number of possible answers, each of which corresponds to the set of positive integers {0, 1, 2, ... })
- *continuous* (i.e., take on any value in a range of real values)

How, for example, might you conceptualize and articulate the number of outcomes for the difference-between-heads-and-tails experiment? For the fender-examining experiment? For the fuel use experiment?

An outcome of an experiment is not necessarily a number. The color of the paint of the fender is not a number. But this is not to say that a number cannot be assigned to each color. For example, you might assign the number 1 to red and the number 2 to blue. Or you might decide to code each color according to its position in the spectrum.

random variable

A *random variable* is a function that takes a defined value for each point in the sample space.

Although your intuitive understanding of "random variable" is sufficient for some situations, it is important to know that the idea can be given precise definition. Thus, to bite this bullet just a little harder:

> Let S be a sample space associated with an experiment. The *function* X which assigns a real number, $X(s)$, to every element s contained in S, is called a *random variable*.

The degree to which this formality is intimidating is a direct measure of the need to understand it. So let's take another look at the toss-of-5-coins experiment.

The sample space for this experiment consists of six ordered pairs; i.e., $S = \{(5H, 0T), (4H, 1T), (3H, 2T), (2H, 3T), (1H, 4T), (0H, 5T)\}$.

Now define the random variable X as the absolute value of the difference between the number of heads and the number of tails in a toss of five coins. Thus, you have $X(5H, 0T) = 5$, $X(4H, 1T) = 3$, $X(3H, 2T) = 1$, $X(2H, 3T) = 1$, $X(1H, 4T) = 3$, and $X(0H, 5T) = 5$. Note especially that, even though the sample space contains six points, this random variable has only three distinct values: 1, 3, and 5. See Figure 3-1 for a graphical representation of this process.

Figure 3-1:
The 5-coin experiment illustrated

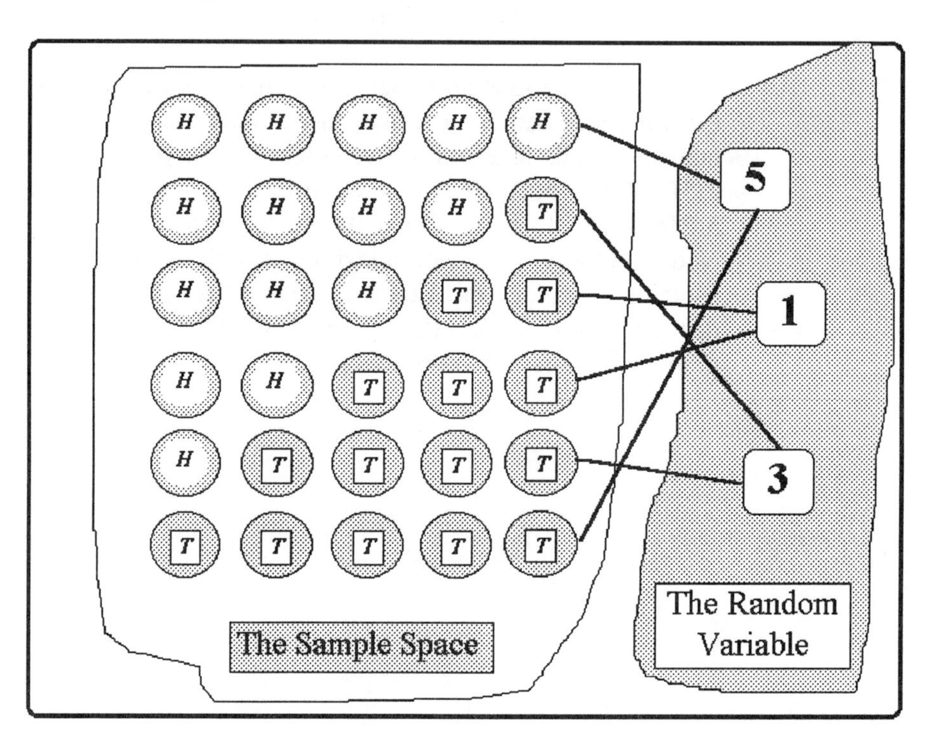

With only minor modifications to keep notation consistent, we concur with Meyer (1970, pp. 56-57) when he notes:

> In referring to random variables we shall, almost without exception, use capital letters such as X, Y, Z, etc. However, when speaking of the *value* of these random variables assume we shall in general use lower case letters such as x, y, z, etc. This is a *very important distinction* to be made and the student might well pause to consider it. For example, when we speak of choosing a person at random from some designated population and measuring his height (in inches, say), we could refer to the *possible* outcomes as a random variable X. We might then ask various questions about X, such as $Pr\{X \geq 60\}$. However, once we actually

choose a person and measure his height we obtain a specific value of X, say x. Thus it would be meaningless to ask for $Pr\{x \geq 60\}$ since his height either is or is not ≥ 60. This distinction between a random variable and its value is important, and we shall make subsequent references to it.

As you explore statistical/probabilistic literature, you will encounter some writers who use the term *variate* as a synonym for *random variable* and/or use $P(\bigcirc)$ as a symbol for the probability of the event $\bigcirc$.

For discussion:

- Consider an experiment consisting of independently tossing two "fair" coins. Define a random variable, say Y, as the total number of heads observed. Let the values of Y be denoted by y. Show that the sample space consists of these four points: (0, 0), (0, 1), (1, 0), and (1, 1). Show further that, at these points, it is reasonable to declare that y takes the values 0, 1, 1, and 2, respectively.

- Define the sample space and corresponding random variable(s) for the fender-examination example.

- Define the sample space and corresponding random variable(s) for the fuel-use example.

- Do you have any experiments, sample spaces, and random variables where you work? In your life? What are they? How do you characterize them for yourself and for others?

Statistical significance

A couple of Webster-like definitions of *significant* are:

1. important, weighty.
2. caused by something other than mere chance.

The first definition has no place in statistical analysis. Importance and weightiness derive primarily from the participants' circumstances and their perception of an event; they do not derive from processes that lead to statistical assessments of that event. Moreover, when you are dealing with probabilistic/statistical matters, you must avoid emotional involvement with the data.

Suppose that a guy named Joe buys an airplane. To Joe, it's a *personally significant* event. No doubt about it. But the event itself has nothing to do with *statistical significance*. The event is not connected to a set of data or to deciding what to do about the implications and consequences of the purchase.

But now suppose that Joe and 50 of his neighbors all buy airplanes in the same month. They live in the same community, population: 2,567. Now you have data—the event is now an *informed event*, and you can begin to consider implications and consequences: a secret society? a clever airplane sales pitch? sudden riches ... and you're not a part of it? is there some way to join them?

This informed event now invokes the second definition: that "mere chance" is rejected as a reason for the occurrence of a specific event (a probable answer); i.e., something other than "the laws of chance" is believed linked to the event. That's the crux of statistical inference.

Consider the possibly apocryphal tale of the lady who bought tickets for several raffles at the Texas State Fair in the fall of 1957 and then won an automobile in three of those raffles. This is certainly *personally significant* to the ticket buyer, but is it a *statistically significant* event? Which explanation suits you: (1) she had "help" in matching her tickets to the winning numbers, or (2) she truly was "lucky" and benefited from materialization of the very small probability of winning three raffles.

And what do you make of the fact that all seven of Edith and Ed's children became undertakers? This fact has all the makings of a significant event, doesn't it? And, just to spice things up, it contains a touch of the macabre. One explanation is that all the children were educated and motivated similarly. Another explanation is that the children were not influenced similarly, but they independently (i.e., "by chance alone") chose the same career. (Could there be such a thing as *mortician's genes*?) Because the second explanation seems so unlikely, you might opt to accept the first.[5]

In everyday life, we all tend to be fairly loose in our use of terms such as *likely* and *unlikely* and *probable* and *improbable*. However, one of the major contributions of *statistics* to the 20th century is a body of knowledge and processes with which you can tackle these matters. It boils down to this question: When do you justifiably classify an event as *unlikely*?

For discussion:

- Consider these words and phrases: *likely*, *unlikely*, *plausible*, *implausible*, *probable*, *improbable*, *possible*, *impossible*, *always*,

[5] *The untold story*: Both Edith and Ed were themselves undertakers. Every evening before the children went to bed, the parents regaled them with tales of how much fun they had that day, going on and on about their exciting professional activities. Now, is this significant, or is this significant?

never, *certain*, *uncertain*, *now-and-then*, *once-in-a-while*. What sequence of these words and phrases would you use to indicate increasing degrees of belief? Which indicates the lowest level of belief and which indicates the highest level? Where would you place the word *significant* in your sequence?

- What are some other words and phrases that convey similar degrees of belief? Where do they rank in your list?

- What, indeed, are *degrees of belief*?

- Comment on the pessimist's credo: *Just about everything is 8-to-5 against.*

- The following appeared as "Fisherman's Luck" in *Sports Illustrated*, January 16, 1978:

> A bass fisherman in the State of Washington worked out a surefire formula that would allow him to fish without buying a license, yet not get caught. Into a computer he fed data on his county's population, the total miles of roads in the area, and the number of wildlife agents in the region. The computer gave back the answer that the odds were 10,000-to-1 against his being caught.
>
> The fisherman later explained all this with some embarrassment to the wildlife agent who caught him and charged him with fishing without a license.

What, if anything, is *significant* in the "Fisherman's Luck" story?

Preliminary concepts of decision and risk: Coin-tossing and statistical decision-making

Consider an experiment which is a game in which you toss a single coin one time:

Heads: you win a dollar	**Tails:** you lose a dollar

You expect to play with a "fair" (or "unbiased") coin, not an "unfair" (or "biased") one. Of course, the ultimate biased coin would have two tails, and you would expect to have a difficult time winning.

Consider a second game: You toss a coin 4 times, and you lose on every toss. You are out $4. Does this sequence of 4 straight losses seem "unlikely?" That is, do you have reason to be suspicious that the coin is unfair (biased)? What's the evidence?

How do you determine the "unlikeliness" of 4 straight losses with a "fair" coin? Suppose, admittedly quite arbitrarily at this point, you decide to declare the coin to be "unfair" only if you observe something unlikely when you play the game. Moreover, before examining the coin, you decide that "unlikely" means a result whose probability is not larger than 5% (i.e., one chance in 20; usually written as a decimal 0.05). With 4 tosses, you have the following 16 possible sequences:

```
HHHH   HHHT   HHTH   HHTT   HTHH   HTHT
HTTH   HTTT   THHH   THHT   THTH   THTT
TTHH   TTHT   TTTH   TTTT
```

Of the $2^4 = 16$ sequences enumerated here, only the very last one (TTTT) corresponds to 4 losses in a row. Based upon all sequences being *mutually exclusive* and *equally likely*[6] to occur, the probability of such a loss (*before* the game started, of course!) is exactly 1 out of 16, or 1/16 = 0.0625. Because this result is greater than your 0.05, you conclude that, with a "fair" coin, a sequence of 4 tails out of 4 tosses is not unlikely.

Consider now a third game: You toss the coin 10 times, and you lose on every toss. You are out $10. *Now*, do you suspect that the coin is biased? "Probably, yes." With a "fair" coin, winning 0 out of 10 seems unlikely. What's the evidence? How do you measure this unlikeliness?

As with the 4-toss case, you could set out to list all 2^{10} = 1,024 possible sequences from HHHHHHHHHH, HHHHHHHHHT, HHHHHHHHTH, ..., through ..., TTTTTTTTHT, TTTTTTTTTH, TTTTTTTTTT.

If you lay out these 1,024 different sequences very carefully, you will find that only the last one (TTTTTTTTTT) corresponds to a no-win result. Hence, you now know that the probability of losing 10 out of 10 is exactly 1 out of 1024, roughly 0.001. Because this result is considerably *less than* the 0.05 used—arbitrarily, it must be admitted—to identify unlikeliness, you conclude that this 10-for-10 loss streak is very unlikely. The basis for action is thus laid, although the details of the action taken depend upon many additional factors.

[6] As with the concept of *random variable*, your intuitive understanding of *mutually exclusive* and *equally likely* will serve you well enough in this discussion. Both terms are more fully developed in Chapter 15.

For discussion:

- What do you suppose is meant by the terms *mutually exclusive* and *equally likely*?
- What's magic about 0.05 as the level of significance?
- The example of 10 straight tosses without the appearance of a head may seem to be an extreme case. Would you reject the coin if 9 tails showed up in 10 tosses? Would you reject the coin if 8 tails showed up in 10 tosses? 7 in 10? 6 in 10? ... 2 in 10? 1 in 10?
- We admittedly flinched when it came to enumerating all 1,024 possible outcomes of 10 tosses. The methodology for doing so, along with the derived probabilities is discussed in Chapter 16. Pending that discussion, speculate on how you might go about finding the probability of getting more than 8 tails in 10 tosses?

Type I and Type II errors

From a statistical point of view, you may regard the games just described as processes designed to decide whether a coin is fair or biased. You'd like to believe that your decision is correct. But you know that 10 tails in 10 tosses of a fair coin is not impossible—it's just unlikely. So, if you call the coin biased when it's not, you make an error. On the other hand, a biased coin *might* give a nearly equal heads/tails split, say 55-45, and you might declare it to be fair when it's not, thus committing an error—although it is a different kind.

Rather than submit to paralysis by analysis, consider the following pair of statements:

If a coin is fair and you call it biased, you make an error. This error (which is not necessarily your fault!) is called a *Type I error*.

Type I error

If a coin is biased and you call it fair, you make an error. But it's a different kind of error. To keep track, this error (which also is not necessarily your fault!) is called a *Type II error*.

Type II error

If you find yourself struggling with the concepts of Type I and Type II errors, be not of faint heart. You'll find them in greatly extended form in Chapter 9. Until we reach that point, consider this algorithm for dealing with such situations:

- Determine the "standard state" of the situation.
  Example: *A particular coin is fair; i.e., the probability of getting a head in a single toss is 0.5.*

- Establish a procedure to examine the "standard state." This procedure provides data, upon the analysis of which you will base a decision about the correctness of the "standard state."
  Example: *Toss the coin 10 times. Observe the number of tails. Reject the correctness of the "standard state" if you observe 10 tails.*

- If the "standard state" pertains *and* you declare that it does not, you commit a Type I error.
  Example: *The coin is fair, but you declare it biased. This is a Type I error.*

- If the "standard state" does *not* pertain and you declare that it does, you commit a Type II error.
  Example: *The coin is biased, but you declare it fair. This is a Type II error.*

Now, let's restate the rules of the coin-tossing game and display general expressions for its Type I and Type II errors: You will toss a coin n times (4, 10, 57, or whatever) and reject the coin if each of the n tosses results in a tail. The probability of getting exactly n tails in n tosses when the coin is unbiased, i.e., when $Pr\{\text{head in one toss}\} = Pr\{\text{tail in one toss}\} = 1/2$, is written as

$$Pr\{n \text{ tails in } n \text{ tosses} \mid \text{coin is unbiased}\} = (1/2)^n.$$

In a similar fashion, when the coin is biased, $Pr\{\text{tail in one toss}\} \neq Pr\{\text{tail in one toss}\}$ which means that $Pr\{T\} \neq 1/2$. It follows that the probability of getting exactly n tails in n tosses can be expressed as:

$$\text{Pr}\{n \text{ tails in } n \text{ tosses} \mid \text{coin is biased}\} = [Pr\{T\}]^n.$$

From a statistical point of view, you make your decision about the bias or the non-bias of the coin *only after* you've observed the results of the n tosses; that is, you wait until the coin is tossed n times *and* observe the results *before* you decide either (a) the coin is unbiased or (b) the coin is biased. (What you do after making this decision is your own nonstatistical business.)

The number of tosses you will make depends on the risk you are willing to take in rejecting a coin which in reality is fair; i.e., the coin's "standard state" is that it is fair. In terms of the definitions developed here, this risk is measured as the probability of making a Type I error. The nearly universal statistical symbol for this probability is the lower-case Greek letter alpha α. In mathematical symbols, for the coin-tossing experiment in which n tails appear in n tosses, this is expressed as

$$\begin{aligned}\alpha &= Pr\{\text{Type I error}\}\\ &= Pr\{n \text{ tails in } n \text{ tosses} \mid \text{coin is unbiased}\}\\ &= (1/2)^n.\end{aligned}$$

Thus, if you are willing to play the game with $\alpha \leq 0.05$, go for a five-toss game because $(1/2)^n = (1/2)^5 = 1/32 = 0.03125$; a four-toss game is inappropriate because $(1/2)^4 = 1/16 = 0.0625 > 0.05$. Similarly, for $\alpha \leq 0.01$, play 7 tosses; for $\alpha \leq 0.001$, play 10 tosses.

level of significance

The probability you are willing to take in committing a Type I error is called the *level of significance* of the game. Thus, you often encounter such phrases as "the α level of significance" or "α-level" when these types of studies are conducted.

As for the Type II error, in which the probability of a tail is *not* 0.5, i.e., $Pr\{T\} \neq 1/2$, the nearly universal statistical symbol for its probability is the lower-case Greek letter beta β. In symbols, β is expressed as

$$\begin{aligned} \beta &= Pr\{\text{Type II error}\} \\ &= Pr\{\text{at least 1 head in } n \text{ tosses} \mid \text{coin is biased}\} \\ &= 1 - [Pr\{T\}]^n. \end{aligned}$$

For discussion:

- The probabilities associated with the coin-tossing game are discrete. This means that you may not be able to exactly match an arbitrarily chosen level of significance. That is, you wind up adopting identical strategies, whether your *level of significance* is 4%, 5%, or 6%. Statisticians cannot tell you what risk you should take (for that matter, neither can probabilists), but they can help you make an informed choice.

- For certain classes of problems, many federal agencies and private industries have adopted the 5% level of significance level. In most of this book, we set the probability of Type I error at the 5% level, unless we have compelling reasons to do otherwise. Do you foresee any problems arising from this "one α fits all" approach?

- Describe the coin-tossing game as an *experiment*. What is its *sample space*? What *random variable* is involved? Can the experiment be repeated? Can you get different decisions (i.e., whether the coin is biased) from different repetitions of the experiment? What might you do about that?

- The following paragraph appeared in *The Sunday Oregonian*, January 31, 1993, p. D2, in a review of *Assembling California*, published by Farrar Straus Giroux:

> "The Geological Survey sees a 67 percent chance of another major San Francisco Bay Area earthquake on either the San Andreas or the Hayward Fault before the year 2020," writes John McPhee in this fourth and concluding book in his excellent series on geologists and geology, *Annals of the Former World*.

Using whatever means you wish, whether based on this chapter or derived from other knowledge, comment on the meaning and interpretation of McPhee's statement as displayed. What questions come to mind? Is "chance" a synonym for "probability"? What happens when you apply Huff's criteria from Chapter 1 to the statement? How might you interpret the statement for your family, your colleagues, your favorite teacher? Is it a statement that leads to action?

What to remember about Chapter 3

Chapter 3 contrasted and linked the two topics of statistics and probability, emphasizing the importance of

- *experiments*
- *sample spaces*
- *random variables*.

Statistical decision-making was described, along with such specialized terminology as

- *statistical significance*
- *Type I error*
- *Type II error*
- *level of significance*.

These concepts are further illustrated and implemented with respect to contingency tables in Chapter 4.

Contingency tables

What to look for in Chapter 4

Chapter 4 introduces contingency tables and their analyses. They serve as a vehicle for introducing the formalities of statistical inference. Special topics and ideas to look for include:

- *2×2 contingency tables*
- *chi-squared statistic*
- *general two-dimensional contingency tables*
- *some sample-size considerations*
- *McNemar's test statistic*
- *Simpson's paradox.*

The chapter concludes with a general protocol for two-dimensional contingency table analysis and makes a connection with OSDAR's guidelines from Chapter 1.

On the value of contingency tables

This chapter breaks a long-standing tradition honored in many books on applied statistics; they tend to relegate their discussions and analyses of *contingency tables* to a position near the end of the text—if they bring up the topic at all. Here you will find material on the construction, analysis, and interpretation of contingency tables, which are of special value in the development of this text for the following reasons:

- You can describe and demonstrate contingency tables with very few, and often uncomplicated, data.
- You can use contingency tables to illustrate many of the concepts required in making statistical inferences without being bogged down by complex calculations.
- You will find that contingency tables arise "naturally" in many data-driven situations, that analysis of them is a straightforward procedure, and that they provide you with a valuable tool that can be used in dealing with every-day problems.
- You will discover how contingency table processes vividly illuminate situations in which statistical naivete leads to incorrect inference.

We choose to delay a formal definition of contingency tables until we have introduced a few underlying ideas by means of examples.

Example 4-1:
On the connection between intelligence and wealth (Or: If you're so smart, why ain't you rich?)

Suppose you wish to investigate whether intelligence and wealth are related. As used here, the word "related" suggests a mutual connection or a linking between two variables; in this case, the variables are *intelligence* and *wealth*. But *do not* mistake this idea of connection or linkage as being cause-and-effect. Thus, in this example, you will not address the question "If you're so smart, why ain't you rich?" any more than you will address the symmetric question "If you're so rich, why ain't you smart?"

In order to legitimize this study of a difficult—even emotional—matter, you must at least:

- Define the population of interest (single, males, aged 18-35 years, ...).
- Define "rich" precisely (say, gross annual income greater than $100,000).
- Define "smart" precisely (say, Intelligence Quotient greater than 120).
- Establish the sample size needed for the investigation (say, 100).
- Determine a procedure for obtaining a sample from the agreed-upon population.
- Have the resources (time, money, ...) to get the needed data.

- Know what analyses are available and appropriate.
- Know how to interpret and present the results of the analysis.

Suppose that you are able to meet all of these conditions. Suppose further that the sample you collect gives the following data:

24 individuals who are smart and rich

6 individuals who are not-smart and rich

56 individuals who are smart and not-rich

14 individuals who are not-smart and not-rich.

You begin your analysis by forming a 2×2 (read as *two-by-two*) contingency table. That is, you place the data (which are counts or frequencies) into 4 cells of a table with 2 rows and 2 columns, as shown in the shaded cells in Table 4-1. The row totals and column totals usually are added at the right and bottom as shown in Table 4-1. The total in the lower right corner is called the

Grand Total

Grand Total, which sometimes is also denoted by n to indicate the number of individuals represented by the data.

Table 4-1:
A 2 × 2 contingency table for the smart-rich study

	Column 1: Smart	Column 2: Not-smart	Row totals
Row 1: Rich	24	6	30
Row 2: Not-rich	56	14	70
Column totals	80	20	100

For discussion:

- Which of ***OSDAR***'s elements have been included and discussed so far?

- Look again at Table 4-1. Is there anything interesting or unusual or curious about these data?

- If you pick a person at random from the population of interest, what is the probability that he will be smart? rich? smart *and* rich? What assumption went into your answers to these questions?

- Some probability-based terminology is useful in discussing contingency table analyses. In terms of the smart-rich study, for example:

    - ☐ The probability that a selected individual is smart, irrespective of his being rich or nor, is called a *marginal probability*.

    - ☐ The probability that a selected individual is both smart and rich is called a *joint probability*.

- ☐ The probability that a selected individual is smart, given that he is rich, is called a *conditional probability*.

- ■ You will find an extended discussion of probability in Chapter 15.

Example 4-1 (continued)

Note that, in your sample of 100 individuals, you find 30 (30%) classified as rich. Then note that 24 out of the 80 (30%) smart participants are rich. Similarly, 30% of the not-smart participants are rich. This indicates that, whether smart or not, 30% of the participants in the study are rich. Hence, without any special statistical weaponry or particular computational difficulty, you conclude that wealth is independent of intelligence.

In a similar manner, you can see that 80% of the participants are smart, regardless of whether they are rich or poor. Hence, you conclude that intelligence is independent of wealth.

Restated in a more generic formulation, the rows (representing two different levels of wealth) and the columns (representing two different levels of intelligence) are independent. The independence is demonstrated by the fact that the proportions of participants in the individual cells are identical to the proportions given in the row and column totals.

Your inevitable conclusion from Table 4-1 is that wealth and intelligence are independent.

Example 4-2: Weld quality

A certain welding operation is performed by two teams, A and B. A regional inspector periodically evaluates their welds for acceptability. The inspection results for 100 welds, selected at random from a single day's work, are summarized in Table 4-2. You note that the row and column totals are the same as those in Table 4-1, but the tallies in the individual cells are different.

Table 4-2:
A 2 × 2 contingency table for weld inspections

	Column 1: Acceptable	Column 2: Not acceptable	Row totals
Row 1: Team A	30	0	30
Row 2: Team B	50	20	70
Column totals	80	20	100

Note that 80% (80 of 100) of all welds are acceptable. Assuming row and column independence and again judging by the row and column totals alone, you expect 24 (80% of 30) acceptable welds by Team A, but you get 30 instead. You also expect: 56 acceptable welds from Team B, but you get 50; 6 unacceptable welds from Team A, but you get 0; and 14 unacceptable welds from Team B, but you get 20.

For discussion:

- Look again at Table 4-2. Is there anything unusual about these data? More particularly, do they indicate a relationship between teams and weld quality?
- Do you think it is "fair" that only 30 welds from Team A are rated, while 70 from Team B are rated? "Fair" (or "unfair") to which team? What could be done to make things "fair"? Is there really a contest going on?
- How important is it that the day's welds made by both teams are examined by the same inspector? Or is it important?

Some contingency table terminology

contingency table

Paraphrasing Kendall and Buckland (1971, p. 32), a *contingency table* results when the items in a sample are set out in a two-way table, the rows and the columns of which are any two qualitative characteristics possessed by the items, provided that no row or column is completely empty. For example, if the first characteristic A is r-fold (i.e., r categories) and the second characteristic B is c-fold (i.e., c categories), then the resulting display is said to be an $r \times c$ contingency table; that is, it is a two-way table with r rows and c columns.

Grand Total

Adding a row at the bottom to contain column totals and a column at the right to contain row totals yields an augmented table with $(r + 1)$ rows and $(c + 1)$ columns. The lower-right cell in the augmented table is called the *Grand Total*.

With reference to Table 4-1, let characteristic A be wealth and characteristic B be intelligence. Both are 2-fold characteristics; i.e, each characteristic has two *categories* which accommodate every member of the population being studied. Table 4-1 is therefore a 2×2 contingency table, whose Grand Total = 100.

Returning now to Table 4-2, note the following:

expected frequency

- In a strict mathematical sense, if the rows and the columns are independent, then the proportions suggested in the totals are reflected in identical proportions within every column and within every row. Thus, the count in any cell would be the fraction of the cell's column total that is equivalent to the ratio of the cell's row total to the Grand Total. The resulting value is the *expected frequency* (sometimes called the *expected count* or *expectation*) of the cell.

- The expected frequency in any cell is not necessarily an integer.

- A quick way to calculate the expected frequency of a cell is to multiply the cell's row total by the cell's column total and then divide the product by the Grand Total. To illustrate with the data in Table 4-2, the expected frequency in the Row 1, Column 1 cell is (30)(80)/100 = 24.

contingency

- Recall that the 2×2 table containing the raw data is called a *contingency table.* The term itself can be related to the use of the table as a tool to investigate possible "mutual dependency" (i.e., a contingency) between the two factors denoted by the rows and the columns. Kendall and Buckland (1971, p. 32) state: "The *contingency* is the difference in the cells of a contingency table between the actual frequency and the expected frequency on the assumption that the two characteristics are independent...." How does *your* dictionary define *contingency*?

- The expected frequency in a cell of a contingency table *need not be* an integer.

- Rarely do you obtain cell frequencies in a contingency table that agree exactly with their expected frequencies. Cell frequencies fail to agree with their expectations for one of two reasons:

  (1) random fluctuations arising from the process under scrutiny give the appearance of a relationship, or

  (2) the rows and the columns are indeed related.

  But how are you to determine which of these reasons prevails for a given contingency table? That question is treated in the next section.

Toward a general analysis of contingency tables

fluctuation index, coefficient of contingency

The general statistical approach to examining a relation between the rows and the columns of a contingency table involves the use of some kind of *fluctuation index* or *coefficient of contingency*. This fluctuation index measures (i.e., is a function of) the discrepancies between the cell frequencies and the cell expectations and takes into account the number of cells in the table.

The fluctuation index is constructed to be exactly zero if the cell frequencies are exactly equal to their expectations and positive otherwise. If the calculated value for this index is "small" (close to zero), then you are tempted to conclude that the rows and the columns are independent; i.e., you have insufficient evidence to disprove the independence of the table's rows and columns. On the other hand, if the calculated value is "large" (considerably larger than zero), then you are tempted to conclude that the rows and the columns are *not* independent; i.e., based on the data at hand, further belief in row-column independence is untenable.

chi-squared statistic

The fluctuation index most commonly associated with the analysis of contingency tables is called the *chi-squared statistic.*[1] The calculated chi-squared statistic is matched against a criterion that is designed to determine whether it (the calculated chi-squared statistic) is sufficiently large to suggest that the rows and columns are *not* independent.

Recall the discussion in Chapter 3 where it was argued that many decisions are based upon the 0.05 level of significance. Hence, the calculated chi-squared statistic is termed "large" if the probability of obtaining such a value (or a higher value), when the rows and columns *are* independent, is less than 0.05.

Turn now to Table T-2: *Quantiles of the chi-squared distribution.* You seek a specific value in Table T-2 with which you compare your calculated chi-squared statistic. Somewhere in the top row of the body of the table (the row labeled with 1 in the column labeled *df*), you find the value 3.84. Now, locate the column heading above this number. The heading is listed as 0.950. Thus, the value 3.84 is a value that is critical in the assessment of any calculated chi-squared statistic with 1 degree of freedom at the 0.05 level of significance. (The concept of *degree(s) of freedom* associated with a calculated chi-squared value is addressed later in this chapter when tables with more than 2 rows and/or columns are described. For the time being, as an aid in dealing with other examples in this chapter, mark, highlight, or simply remember this particular column as the "0.05 level of significance" column.)

[1] In this expression, *chi* is the Greek letter (upper-case X and lower case χ), pronounced like the *chi* in *chi*-ropractor.

level of significance

In the context of contingency table analysis, the *level of significance* is the measure of the risk you are willing to take in concluding the rows and columns are related when in reality they are not. The number 3.84 is a "cutoff point" for 1 *df*, a value beyond which the assumption of no relationship is *not* supported by the data in the table. That means that, if the calculated chi-squared statistic is larger that 3.84 in Example 4-1, you claim that the fluctuations of the observed frequencies from their expectations are excessive, and you conclude that wealth and intelligence are related. If the calculated chi-squared statistic is smaller than 3.84, you conclude that you have no evidence to contradict the independence of wealth and intelligence. The value 3.84 is the *critical value* (sometimes called the critical point), for a calculated chi-squared statistic with 1 *df*.

critical value

Some statistical rationale

The calculated chi-squared statistic derived from a 2×2 contingency table "behaves" like a *chi-squared random variable with 1 degree of freedom* when the rows and the columns are not related. In this book, we use the square of the upper-case Greek letter (X^2) as the symbol for a chi-squared random variable.[2] A chi-squared variable with 1 degree of freedom variable has this particular characteristic: 95% of its values are less than 3.84 and 5% of them are larger than 3.84. Thus, some authors place a label of "0.05" on the "0.05 level of significance" column, while others label it with "0.95."

[2] Many authors indicate the chi-squared random variable with the square of a lower-case Greek letter called *chi*, which looks like this: χ^2.

For discussion:

- Consider a 2×2 table in which one column is empty; that is, its sum is zero. Discuss the implications. What's different about a 2×2 table in which one row is empty?
- Suppose a calculated chi-squared statistic for a 2×2 table in a weld-quality study turns out to be 9.667. What conclusions might you draw about the relation between team affiliation and weld quality? Could you be wrong? In what way(s)?
- Suppose the calculated chi-squared statistic were 2.087 (instead of 9.667). What do you conclude?
- Suppose you calculated a chi-squared statistic exactly equal to 3.84. What do you conclude?
- Suppose you calculated a chi-squared statistic exactly equal to -6.926. What do you conclude?

Example 4-2a:
Continuing the weld-quality study

Now combine the observed frequencies from Table 4-2 with the expected frequencies of the individual cells. Table 4-2a shows both sets of frequencies, with the expected frequencies in parentheses.

Table 4-2a:
A 2 × 2 contingency table for the weld-quality study, displaying observed and (expected) frequencies

	Column 1: Acceptable	Column 2: Not acceptable	Row totals
Row 1: Team A	30 (24)	0 (6)	30
Row 2: Team B	50 (56)	20 (14)	70
Column totals	80	20	100

The next step is the calculation of the chi-squared statistic itself. The details are displayed in Table 4-2b, where several quantities for each cell in the body of Table 4-2a are repeated for clarity.

(1) The first column of Table 4-2b contains the row indices, repeated for each column. Symbol: i.

(2) The second column contains the column indices, repeated within each row index. Symbol: j.

(3) The third column contains the observed frequencies for each combination of row-column indices. Symbol: O_{ij}.

(4) The fourth column contains the expected frequencies for each combination of row-column indices. Symbol: E_{ij}.

(5) The fifth column contains the differences between the observed and expected frequencies. Symbol: $(O_{ij} - E_{ij})$.

(6) The sixth column contains the squares of the values in the fifth column. Symbol: $(O_{ij} - E_{ij})^2$.

(7) The seventh column contains the ratios of the elements of the sixth column to the elements of the fourth column. Symbol: $(O_{ij} - E_{ij})^2/E_{ij}$.

(8) The cell at the bottom contains two important quantities:

(8a) The sum of the four values in the last column is called the calculated chi-squared statistic. Symbol: x^2.

(8b) The degrees of freedom associated with the chi-squared statistic. For this example, there is 1 *df*.

Note that all calculations reported in this chapter's analyses are recorded to four decimal places. If you round or truncate your intermediate values, you may reach a wrong conclusion because the calculated chi-squared statistic x^2 may be incorrect.

Table 4-2b:
Calculations for the 2 × 2 contingency table in Table 4-2a

Row	Col	Observed	Expected	(O - E)	(O - E)²	(O - E)²/E
1	1	30	24.0000	6.0000	36.0000	36.0000/24.0000 = 1.5000
1	2	0	6.0000	-6.0000	36.0000	36.0000/6.0000 = 6.0000
2	1	50	56.0000	-6.0000	36.0000	36.0000/56.0000 = 0.6429
2	2	20	14.0000	6.0000	36.0000	36.0000/14.0000 = 2.5714
						x^2 = 10.7143 with 1 *df*

Note the following facts that apply to all contingency table analyses:

- The sum of the cell expectations (fourth column) equals the sum of the cell frequencies (third column).
- The sum of the column labeled "(O - E)" is equal to zero, apart from rounding errors.

These provide checks on your calculations.

For discussion:

- Draw your conclusions from Table 4-2b. Could you be wrong? What is the probability of that?
- All of the contingency tables studied thus far are of the 2×2 variety. Extensions to larger tables are made later in the chapter. Do you think the calculations will be more complicated?
- Why is it necessary to record more significant figures in your intermediate calculations than appear in the given data in your calculation of the chi-squared statistic? That is, why not just round everything to integers at every step?
- Which **OSDAR** elements have been covered up to this point?
- Reconsider the data in Table 4-1. Why are the expected frequencies identical to the observed frequencies?

Example 4-3: Mastering technical material

Suppose you wonder whether the outside-of-class studying of technical material improves a student's mastery of the material. You might categorize a class of 170 students into the four groups shown in Table 4-3.

Table 4-3:
A 2 × 2 contingency table for the study-mastery study

	Column 1: Master	Column 2: Not-master	Row totals
Row 1: Study	46 (22.9412)	14 (37.0588)	60
Row 2: Not-study	19 (42.0588)	91 (67.9412)	110
Column totals	65	105	170

Table 4-3a:
Calculations for Table 4-3

Row	Col	Observed	Expected	$O - E$	$(O - E)^2$	$(O - E)^2/E$
1	1	46	22.9412	23.0588	531.7093	531.7093/22.9412 = 23.1771
1	2	14	37.0588	-23.0588	531.7093	531.7093/37.0588 = 14.3477
2	1	19	42.0588	-23.0588	531.7093	531.7093/42.0588 = 12.6420
2	2	91	67.9412	23.0588	531.7093	531.7093/67.9412 = 7.8260
						x^2 = 57.9928 with 1 *df*

For discussion:

- Draw your conclusions for Example 4-3. Could you be wrong? What is the chance of that?

Example 4-4:
Computer training and the need for computer support

Suppose you are on an office staff facing this problem: Does formal computer training of staff reduce the number of calls for computer-assistance services? Suppose you are told of interviews of 198 employees during a specific month. You develop the contingency table shown as Table 4-4.

Table 4-4:
On the relationship between computer training and the need for computer support

	Column 1: Given training	**Column 2: No training**	**Row totals**
Row 1: Support required	36 (37.1717)	124 (122.8283)	160
Row 2: Support not required	10 (8.8283)	28 (29.1717)	38
Column totals	46	152	198
			$x^2 = 0.2507$ with 1 *df*

For discussion:

- Draw your conclusions from the data obtained in Example 4-4.
- Example 4-4 produces some curiosities:

    The Grand Total = 198 is suspicious—at least worthy of exploration. Were two more employees interviewed but not reported? Would the data from the two "missings" change the conclusion? Irrespective, what *was* the planned-for total sample size?

    How were the employees selected? What about those who didn't even use their computers during the month studied?

- What would OSDAR say about the sample-selection methodology?

A shortcut calculation for 2 × 2 contingency tables

For any 2×2 contingency table analysis, a shortcut calculation formula is available. Consider a generic 2×2 table with the cell frequencies designated by a, b, c, and d, as shown here in Table 4-5. The marginal sums are denoted by e, f, g, and h. The Grand Total is denoted by n.

Table 4-5:
A generic 2 × 2 contingency table

	Column 1	Column 2	Row totals
Row 1	a	b	$e = a + b$
Row 2	c	d	$f = c + d$
Column totals	$g = a + c$	$h = b + d$	n $= e + f$ $= g + h$ $= a + b + c + d$

The chi-squared statistic for this generic 2×2 contingency table can be calculated directly from the cell entries a, b, c, and d, as shown here:

$$\chi^2 = \frac{n(a \times d - b \times c)^2}{e \times f \times g \times h}.$$

This formula applies only to 2×2 contingency tables. The calculation is illustrated in Example 4-5.

Example 4-5:
Political party affiliation and support for nuclear power

Interviews of 100 randomly selected residents of a rural community were conducted to examine the relation between political party affiliation and support for nuclear power. The results appear in Table 4-5a.

Table 4-5a:
Political party affiliation and support of nuclear power

	Column 1: Supports nuclear power	Column 2: Does not support nuclear power	Row totals
Row 1: Party A	$a = 38$	$b = 2$	$e = a + b = 40$
Row 2: Party B	$c = 37$	$d = 23$	$f = c + d = 60$
Column totals	$g = a + c = 75$	$h = b + d = 25$	$n = e + f = g + h = a + b + c + d = 100$

The calculated chi-squared statistic is:

$$x^2 = \frac{100[(38)(23) - (2)(37)]^2}{(40)(60)(75)(25)} = 14.2222.$$

For discussion:

- Calculate the chi-squared statistic for the Table 4-5 in the formal way. Draw your conclusions. Could you be wrong? What is the chance of that?

- Do these results reflect the sentiments of the state? Of the entire country? What sentiments do they reflect?

- How would you *prove* that the shortcut calculation holds for *all* 2×2 contingency tables? That is, what steps are needed to show that the simplified calculation works for any 2×2 contingency table? (*Hint*: Some algebra is required.)

The formulation of $r \times c$ contingency tables

Up to this point, contingency tables have been of the 2×2 variety. No matter what the situation (assuming, of course, it is appropriately treated in a 2×2 contingency table), the calculated chi-squared statistic is compared to the value 3.84 for decision-making purposes because 3.84 is the critical value at the 0.05 level of significance with 1 degree of freedom.

It is now time to extend the process to any arbitrary number of rows and columns. To this end, let r be the number of rows and let c be the number of columns. This is called an *$r \times c$ contingency table*. Before going further, however, the concept of *degrees of freedom* must be defined for this generic $r \times c$ contingency table.

degrees of freedom

In an $r \times c$ contingency table analysis, the term *degrees of freedom (df)* refers to a parameter whose value is the product of two terms: (number of rows minus 1) and (number of columns minus 1); i.e., $df = (r - 1)(c - 1)$.

For example, for all 2×2 contingency tables, $df = (2 - 1)(2 - 1) = (1)(1) = 1$. Similarly, for a contingency table with 9 rows and 6 columns, $df = (9 - 1)(6 - 1) = (8)(5) = 40$.

But what do you do with this *df*? Recall that, in a 2×2 contingency table analysis, you declare that a row-column relationship exists when the chi-squared statistic exceeds 3.84 (i.e., with a 5% chance of being wrong when the rows and columns truly are unrelated). Recall also that the number 3.84 is obtained from the first row ($df = 1$) of the table of the quantiles of the chi-squared distribution. For contingency tables of different dimensions, different table values are used to determine whether the chi-squared statistic is "large."

Thus, for a general $r \times c$ contingency table, you consult your chi-square table for the row carrying $df = (r - 1)(c - 1)$ and the column that corresponds to the predetermined level of significance.

To fix the idea, consult Table T-2 and verify that, for the 0.05 level of significance, the table values are 12.6 if $df = 6$ and 55.8 if $df = 40$.

Although "degrees of freedom" has the mixed redolence of obscurity and complexity, it is a concept that occurs in many contexts in statistical theory and practice. For contingency table analysis, it's sensible and useful to regard "degrees of freedom" as an index which tells you which row to use in the chi-squared table.

Example 4-6: A 2×3 contingency table

Table 4-6 offers an example of a 2×3 contingency table, and Table 4-6a displays its corresponding calculations.

Table 4-6: A 2×3 contingency table

	Column 1: Day shift	Column 2: Swing shift	Column 3: Graveyard shift	Row totals
Row 1: No defects	155	180	90	425
Row 2: Some defects	16	60	4	80
Column totals	171	240	94	505

Table 4-6a:
Detailed chi-squared calculations for Table 4-6

Row	Col	Observed	Expected	$O - E$	$(O - E)^2$	$(O - E)^2/E$
1	1	155	143.9109	11.0891	122.9683	0.8545
1	2	180	201.9802	-21.9802	483.1291	2.3920
1	3	90	79.1089	10.8911	118.6158	1.4994
2	1	16	27.0891	-11.0891	122.9683	4.5394
2	2	60	38.0198	21.9802	483.1291	12.7073
2	3	4	14.8911	-10.8911	118.6158	7.9656
						$x^2 = 29.9582$ with $(2 - 1)(3 - 1) = 2$ *df*

For discussion:

- What might be a context for the data in Table 4-6? What value from the chi-squared table do you select for comparison with the calculated chi-squared statistic? What conclusions do you draw? Could you be wrong? What is the chance of that?

- How might you display graphically the results of the analysis of Table 4-6?

Two more contingency tables

Contingency tables provide many opportunities to explore and appreciate statistical issues. These two examples are designed to extend your understanding of these matters.

Example 4-7:
Age discrimination

Suppose you are investigating whether your organization practices age discrimination in its annual performance appraisals. The data you collect are displayed in Table 4-7.

Table 4-7:
Age and performance appraisal

	Needs improvement	Good performer	Wonderful performer	Row totals
Age: Under 36	16	12	8	36
Age: 36-50	68	48	20	136
Age: Over 50	61	50	16	127
Column totals	145	110	44	299
				x^2 = 2.3857 with 4 *df*

For discussion:

- Draw your conclusions from the data in Table 4-7. Could you be wrong?
- Is something missing in the description of the study or in Table 4-7? What is the population under study? What is the sample?
- What problems do you see in the way the rows and columns are constructed in Table 4-7?
- What happens to the calculated chi-squared statistic and the degrees of freedom if you reverse the roles of the rows and columns in Table 4-7?
- How does Example 4-7 stand up to Huff's criteria from Chapter 1?

Example 4-8:
Age and accidents

Suppose you are investigating a suspected relationship between drivers' ages and their tendencies to be in accidents. The data you collect are displayed in Table 4-8. This example and its data are from Gilbert (1976, p. 229).

Table 4-8:
Drivers' ages and accidents

	Ages: 18-25	Ages: 26-40	Ages: 41-50	Row totals
No accidents	75	120	105	300
1 accident	50	60	40	150
2 or more accidents	25	20	5	50
Column totals	150	200	150	500
				$x^2 = 19.4439$ with 4 *df*

For discussion:

- Draw your conclusions from the data in Table 4-8. Could you be wrong? What is the chance of that?
- Is anything missing in Table 4-8? What is the population under study?
- Do you see any problem with the partition of the data in Table 4-8? Could the data be set in a matrix of 3 rows by 4 columns? 4 rows by 5 columns? Could you get different conclusions from different partitions?

Some sample-size considerations in contingency table analysis

The use of the chi-squared statistic to examine the independence of rows and columns may be inappropriate for small sample sizes. A number of recommendations for tackling this problem exist in the statistical literature. They often include cautions about interpreting the results. Curiously, these rules are stated in terms of the cells' expected frequencies rather than the number of observations. One such widely respected rule states that:

> In none of the $r \times c$ cells should the expected frequency be smaller than 5.

Using this rule alone, the sample size required, even for the smallest contingency table (2×2), must be at least 20.

Some authors offer alternatives to this "rule-of-5." For example, Dixon and Massey (1983, p. 277) tolerate contingency tables which meet one of two conditions:

(1) No expected frequency smaller than 2 for any cell in the table

or

(2) No more than 20% of all the cells have expected frequencies less than 5 with *none* of the expected values less than 1.

This second condition is endorsed by Ott and Mendenhall (1984, p. 436).

Even with the most careful planning, you cannot always be sure that the *expected* frequency of a given cell will be sufficiently large. This is particularly true for so-called "rare events." For example, in studying the relationship between driver's age and number of accidents (Table 4-8), you may find very few persons with two or more accidents. This may be caused by the study's limitations. It may also be caused by very few people with two or more accidents being licensed to drive. Remember: You can speculate all you want about these sorts of things—but you'll never really know the reason(s) until you seek them out.

In some instances, you may be able to deal with small expected frequencies by "collapsing" rows and/or columns. Consider Table 4-8 again. Combine the second row (1 accident) with the third row (2 or more accidents) to form a single row of "one accident or more." This process, clearly, also reduces the degrees of freedom.

Fisher exact probability test

When your sample size is quite small (say, 30 or less), a special technique for 2×2 contingency tables can be used. Called the *Fisher exact probability test*, this procedure is especially useful when one of the four cells in the table is empty (i.e., zero tally). Otherwise, the computations can be quite extensive and inhibiting. See, for example, Rosner (1989, 336-342) or Siegel (1956, pp. 96-104).

A contingency table look-alike: McNemar's test statistic

McNemar's test statistic

McNemar's test statistic (McNemar, 1947) is designed for situations in which you are interested in the *change* of a set of individuals after they receive some sort of treatment or stimulus. In this case, all individuals in the sample serve as their own before/after *controls*.

In its setup, McNemar's test bears a strong resemblance to the usual contingency table discussed in this chapter. But this resemblance applies only to the use of a 2×2 table as a means of assembling the data. The actual purposes, calculations, and interpretations are quite different.

Example 4-9:
Effect of laundering on protective mask filters

Suppose you want to test the effect of laundering on a mask filter used to protect inspectors who enter a contaminated area. Each of 70 filters is first exposed to the same contaminated environment. A chemical test is then conducted to determine the acceptability (on a pass/fail basis) of each filter by comparing the contaminant's contents on the two surfaces (i.e., the exposed and the protected surfaces) of the filter. The filters are then laundered, and the exposure and the chemical test protocols are repeated. You find that some filters pass both pre- and post-laundry tests, others fail both, and still others pass one test and not the other.

The issue under consideration here is whether the laundering has an effect on the rate of filters passing the chemical examination. Restated, you wish to determine whether the passing rate is independent of when the examination was made, either before or after the laundering.

Table 4-9 gives the schematic representation of these kinds of data, whereas Table 4-9a provides fictional frequencies for Example 4-9. McNemar's test statistic employs the sensible principle that it (the test statistic) is a function only of those cases that show change from one state to the next (fail followed by pass or pass followed by fail).

The data are arranged in a 2×2 table similar to that used in 2×2 contingency analyses. The numbers in the cells are designated a, b, c, and d, just like in the 2×2 generic contingency table we've all come to love. McNemar's test statistic is given by the expression $(b - c)^2/(b + c)$. As you can see, the statistic employs only cases that show differences and, thus, may be attributed to the treatments they receive. Because McNemar's test statistic behaves like a chi-squared variable with 1 degree of freedom, a value larger than 3.84 is significant at the 0.05 level of significance.

Table 4-9:
Schematic representation for McNemar's test

After	Before		Totals
	Pass	**Fail**	
Pass	a	b	a+b
Fail	c	d	c+d
Totals	a+c	b+d	a+b+c+d

McNemar's test statistic is calculated as $(b-c)^2/(b+c)$; it is compared with the appropriate chi-squared value with 1 *df* (i.e., 3.84 for $\alpha = 0.05$).

Table 4-9a:
McNemar's test applied to filter laundering example

After Laundry	Before Laundry		Totals
	Pass	**Fail**	
Pass	24	16	40
Fail	13	17	30
Totals	37	33	70

McNemar's test statistic is calculated as $(16-13)^2/(16+13) = 9/29 = 0.3103$; it is compared with the appropriate chi-squared value with 1 *df* (i.e., 3.84 for $\alpha = 0.05$).

Because McNemar's test statistic in Table 4-9a falls short of the 0.95 quantile for a chi-squared variable with $df = 1$ (i.e., 0.31 is less than 3.84), you cannot claim statistical evidence that the laundering changes the filter's effectiveness, neither for the better nor for the worse.

Simpson's paradox—you better watch out!

Consider the following example in which two departments—indeed the *only* two—of Fictitious University are under investigation for sex discrimination in their admission procedures. Each department processes 600 applications. Their admit/deny vs. men/women records are shown in Tables 4-10a and 4-10b.

Tables 4-10a and 4-10b:
Admission records of Departments A and B, Fictitious University

4-10a: Department A			
	Admit	Deny	Row totals
Men	200	200	400
Women	100	100	200
Column totals	300	300	600
			$x^2 = 0.0000$ with 1 *df*

4-10b: Department B			
	Admit	Deny	Row totals
Men	50	100	150
Women	150	300	450
Column totals	200	400	600
			$x^2 = 0.0000$ with 1 *df*

The chi-squared statistics for both departments are *exactly zero*. Thus, there is absolutely *no statistical evidence* of a sex-based bias in admission of either department.

However, this two-table display tells one story for each of the two departments, while an aggregation over the two tables tells another story for the university as a whole. Table 4-10c shows the aggregated records.

Table 4-10c:
Aggregated admission records for Departments A and B of Fictitious University

	Admit	Deny	Row totals
Men	250	300	550
Women	250	400	650
Column totals	500	700	1200
			$x^2 = 5.9940$ with 1 *df*

Now things look considerably different. The large value of the chi-squared statistic, clearly larger than the 3.84 criterion, indicates that there *is* a difference between the admission rates of men and women *when the entire university's admission records are employed.*

For discussion:

- Suppose you are an official at Fictitious University and are faced with these data which have arisen as a result of legal action charging sex bias in admission. What would you do?
- This example of Simpson's paradox is not farfetched. It is based on a real situation; see Bickel, *et al.* (1975) for an extended discussion.

- Don't look for simple explanations when you encounter Simpson's paradox. By definition, paradoxes—whether real in terms of a rigorous philosophic analysis or temporarily confusing in terms of "everything you know"—don't come equipped with simple explanations.

Simpson's paradox: Another example

Fictitious University's admissions example of Simpson's paradox might be dismissed as something that happens only in sociology or among academics practicing some other "soft" science. But it can show up in the "hard" sciences, too.

A new drug is tested on diseased mice. The treatment outcomes are given in Table 4-11a for male mice and in Table 4-11b for female mice.[3]

[3] These data were given to DL by a former student whose identity is lost.

Tables 4-11a and 4-11b:
Male and female mice mortality data

4-11a: Male mice

	Lived	Died	Row totals
Treated	18	12	30
Not treated	7	3	10
Column totals	25	15	40

$x^2 = 0.3200$ with 1 *df*

Survival rates:
Treated - 18/30 = 60%
Not treated - 7/10 = 70%

4-11b: Female mice

	Lived	Died	Row totals
Treated	2	8	10
Not treated	9	21	30
Column totals	11	29	40

$x^2 = 0.3762$ with 1 *df*

Survival rates:
Treated - 2/10 =20%
Not treated - 9/30 = 30%

Now compare the results shown in Tables 4-11a and 4-11b with the aggregated data Table 4-11c.

Table 4-11c:
Aggregated mice mortality

	Lived	**Died**	**Row totals**
Treated	20	20	40
Not treated	16	24	40
Column totals	36	44	80

$x^2 = 0.8081$
with 1 *df*

Survival rates:
Treated - 20/40 = 50%
Not treated - 16/40 = 40%

The inescapable conclusion: This treatment is bad for male mice *and* bad for female mice; the treatment is associated with a smaller proportion of survival, as seen in Table 4-11a (for male mice) and Table 4-11b (for female mice). But wait! When you look at the aggregated data, there's an appearance of a distinct improvement: treated mice survive at a greater rate than untreated mice. Even if it's not statistically significant, the calculated chi-squared statistic is larger for the aggregated data than for the male and female data analyzed separately. The quintessential contingency table paradox! So: what're you gonna do?

The answer lies in recognizing that paradoxes seldom have comforting outcomes. But Bickel, *et al.* (1975) are convincing when they argue that you ought not to use the aggregated tables. The reason is that the aggregated tables do not account for an important factor (namely, the department effect in Tables 4-10a-c and the sex effect in Tables 4-11a-c). The moral: Whenever individual tables are in conflict with an aggregated table, you should report the individual tables.

A protocol for $r \times c$ contingency table analysis

This section is devoted to an 11-step protocol for contingency table analysis. Each step is keyed to ***OSDAR***'s elements (Chapter 1) as indicated to the left.

Objective

Step 1. Define the task: test the hypothesis that two classification criteria (rows and columns) are independent.

Scope

Step 2. Decide on the risk you are willing to take in claiming that the rows and the columns are related when in reality they are not. This risk, written as a probability (or as a percentage, when multiplied by 100) is denoted as α. Following the practice discussed in Chapter 3, $\alpha = 0.05$ is used in most applications in this book.

Step 3. Determine n, the number of observations to be used in the study.

Data Collection

Step 4. Collect the data. Tally the observations in a two-way table. Let r denote the number of non-empty rows and c denote the number of non-empty columns.

Step 5. Be absolutely sure that every individual in the sample is counted and that no individual is counted more than once.

Analysis

Step 6. Complete the table with its marginal totals as shown in the schematic $r \times c$ contingency table of Table 4-12, in which O_{ij} denotes the *observed* number of sample members who fall into the cell at the intersection of the ith row and the jth column.

Table 4-12:
A schematic $r \times c$ contingency table

	Column 1	Column 2	. . .	Column c	Row totals
Row 1	O_{11}	O_{12}	. . .	O_{1c}	$\sum_{j=1}^{c} O_{1j}$
Row 2	O_{21}	O_{22}	. . .	O_{2c}	$\sum_{j=1}^{c} O_{2j}$
.	.	.	.	.	.
.	.	.	.	.	.
.	.	.	.	.	.
Row r	O_{r1}	O_{r2}	. . .	O_{rc}	$\sum_{j=1}^{c} O_{rj}$
Column totals	$\sum_{i=1}^{r} O_{i1}$	$\sum_{i=1}^{r} O_{i2}$	. . .	$\sum_{i=1}^{r} O_{ic}$	$\sum_{i=1}^{r}\sum_{j=1}^{c} O_{ij}$ = Grand Total = n

Step 7. For each of the rc cells, determine the observed count, O_{ij}, and the corresponding expected count, E_{ij}. The expected count is computed as the product of the total of the ith row and the total of the jth column, divided by the Grand Total.

Step 8. Complete the following three calculations

$(O_{ij} - E_{ij})$, $(O_{ij} - E_{ij})^2$, and $(O_{ij} - E_{ij})^2/E_{ij}$

for each of the rc cells.

Step 9. Add the values calculated for the third expression in Step 8. The sum thus obtained is the calculated chi-squared statistic. Mathematically, this sum can be expressed in two equivalent formulas:

$$x^2 = \sum_{i=1}^{c}\sum_{j=1}^{r} \frac{(O_{ij} - E_{ij})^2}{E_{ij}} = \sum_{i=1}^{c}\sum_{j=1}^{r} \frac{O_{ij}^2}{E_{ij}} - n,$$

where n is the Grand Total. You may find the second formula faster to use and less subject to rounding errors.

Step 10. Determine the associated degrees of freedom as

$df = (r - 1)(c - 1)$.

Step 11: Consult Table T-2 in this book and find the value from the chi-squared distribution for the row associated with df and the column associated with α. This tabled value is called the *critical value* of the test. Note that interpolation may be necessary.

Report

Step 12: If the chi-squared statistic exceeds the critical value, you conclude that the rows and the columns are not independent. Otherwise, *at best*, you can say that you have insufficient evidence that the rows and the columns are related.

This has been just a beginning ...

categorical data analysis

... of the study of *categorical data analysis*. The extension to k dimensions is not a trivial matter. Until the general availability of substantial and inexpensive computing resources in the 1970s and 1980s, researchers were limited to particular extensions of the two-dimensional $r \times c$ contingency table to perhaps three or four dimensions. However, research and procedures for dealing with k-dimensional tables are developed and being refined. Two references are Agresti (1990) and Upton (1978).

What to remember about Chapter 4

Chapter 4 introduced and explored an especially useful statistical tool: *$r \times c$ contingency tables*. Among the related ideas developed in the chapter were

- *2×2 contingency tables*
- *chi-squared statistic*
- *general two-dimensional contingency tables*
- *some sample-size considerations*
- *McNemar's test statistic*
- *Simpson's paradox.*

The chapter concluded with a general protocol for two-dimensional contingency table analysis and made a connection with OSDAR's guidelines from Chapter 1.

5

Descriptive statistics

What to look for in Chapter 5

Chapter 5 defines, illustrates, and differentiates among a variety of *descriptive statistics* by focusing on data of the continuous type (measured on an interval or a ratio scale).[1] Topics include:

- *measures of central value,* including the *mean*, the *median*, the *mode*, the *midrange*, the *trimmed mean*, the *Winsorized mean*, and the *weighted mean*

- *measures of variability*, including the *range*, the *percentiles*, the *quartiles*, the *quantiles*, the *variance*, and the *standard deviation*

[1] Discrete data are considered in greater detail in Chapters 16-18.

- *coefficient of variation*
- *effects of variable coding.*

The chapter concludes with two handy shortcuts useful in dealing with data:

- *an empirical rule that provides a convenient way of summarizing the portion of a dataset that lies in an interval*
- *a method for estimating the standard deviation of a dataset from its range.*

Why descriptive statistics?

descriptive statistics

Consider a dataset with a single variable. Numeric values extracted from a data set with the intention of characterizing the behavior of the variable are called *descriptive statistics*. If the dataset is a sample, the descriptive statistics often are the basis for making inferences about the population from which the sample is believed to have been drawn.

You are exposed to—and quite likely use—descriptive statistics of some form in your daily life. To name a few:

> What were yesterday's high and low temperatures for your region?
>
> What is your "share" of the national debt?
>
> What was your average annual income over the last 10 years?

Many commonly used descriptive statistics are easily obtained using an electronic calculator and often as a byproduct of a computer package. Because descriptive statistics are so readily available and because the casual, even thoughtless, use of them may have serious consequences, it is incumbent upon each of us, as users of these conveniences, to make sure that only appropriate statistics are used in our work.

For discussion:

- You may encounter the term *data reduction* in reference to descriptive statistics. What does "data reduction" suggest to you? What, indeed, is being reduced?

Measures of central value

measures of central value

Measures of central value (or *location*) are numbers designed to "represent" an entire set of values that a variable might achieve. An often-used alternate term, "central tendency," describes the proclivity of a set of values to cluster around a single value, but the term itself doesn't really help you "to locate" the set of values. Consequently, you seek a measure of a variable's central value to help you position the variable on a scale or on an axis, often with respect to other variables.

In order to make specific quantitative comparisons in the following material, consider a specific dataset composed of 6 values:

{1.7, 3.2, 3.2, 4.6, 1.4, 2.8}.

The *mean* of n observations is the sum of the observations divided by n. For the 6 observations {1.7, 3.2, 3.2, 4.6, 1.4, 2.8}, the mean is calculated as

$$\begin{aligned}&(1.7 + 3.2 + 3.2 + 4.6 + 1.4 + 2.8)/6 \\ &= 16.9/6 \\ &= 2.82.\end{aligned}$$

Here, consistent with venerable statistical practice, the mean is reported to one more decimal place than reported in the data themselves.

Other terms synonymous with the *mean* are the *arithmetic mean*, the *arithmetic average*, and sometimes simply *average*.

The *median* for an *odd* number of observations is the middle observation when the observations are arranged in order of their magnitude (i.e., size). The median for an *even* number of observations is the mean of the two middle observations when the observations are arranged in order of magnitude. The ordered set of 6 observations becomes {1.4, 1.7, 2.8, 3.2, 3.2, 4.6}. Because 6 is an even number, the median is calculated as

$$(2.8 + 3.2)/2 = 3.00.$$

Recall that the median is an important measure used in the construction of box charts (Chapter 2).

The *mode* of a set of observations is the measurement that occurs most often in the set. For the set of observations {1.7, 3.2, 3.2, 4.6, 1.4, 2.8}, the mode is seen to be the value

3.2.

Some datasets may have more than one mode, a condition you may find described by the word *multi-modal*.

The *midrange* of a set of observations is the mean of the smallest and the largest values in the set. For the set of observations {1.7, 3.2, 3.2, 4.6, 1.4, 2.8}, the smallest value is 1.4 and the largest is 4.6, so that the midrange is calculated as

$$(1.4 + 4.6)/2 = 3.00.$$

The *trimmed mean* of a set of observations is the average of the set after the smallest and the largest observations have been removed. For the set of observations {1.7, 3.2, 3.2, 4.6, 1.4, 2.8}, removal of the smallest (1.4) and the largest (4.6) gives the trimmed mean as

$$(1.7 + 3.2 + 3.2 + 2.8) = 10.9/4 = 2.72.$$

You may encounter trimmed means calculated after a certain proportion of the dataset's smaller and larger numbers are removed. For example, a 10%-trimmed mean is computed by removing 5% of the smaller and 5% of the larger values.

The *Winsorized mean* is related to the trimmed mean. If you replace the largest value in the dataset with the next-to-largest value and, symmetrically, the smallest value with the next-to-smallest value, the mean of the modified dataset is called the *first-level Winsorized mean*. (The adjective *first-level* refers to the replacement of only one value at each extreme of the data.) For the set of observations {1.7, 3.2, 3.2, 4.6, 1.4, 2.8}, you create the modified set {1.7, 3.2, 3.2, 3.2, 1.7, 2.8}. Then the first-level Winsorized mean is calculated as

$$(1.7 + 3.2 + 3.2 + 3.2 + 1.7 + 2.8) = 2.63.$$

For a fuller discussion of trimmed and Winsorized means, see Dixon and Massey (1983, pp. 380-382).

For discussion:

- Other measures of central value are found in the literature, some of them developed for specific situations. For example, the *geometric mean* and the *harmonic mean* are described by Kendall and Buckland (1975). How do you feel about the *golden mean* (found in almost any English dictionary)?

- Have you ever encountered terms like *mean median* or *average mode*? What do you suppose is meant by them?

Mathematical representations of the mean

Denote a discrete finite population's size by N and the numerical value associated with the ith element in the population by Y_i. Then the *population mean* is defined by

$$\mu = \frac{Y_1 + Y_2 + \ldots + Y_N}{N}$$

$$= \frac{1}{N}\sum_{i=1}^{N} Y_i;$$

that is, the mean of a population is the arithmetic average of the values of the N elements in the population. Thus, to compute the mean, you add all the values and divide the sum by the number of values in the population.

When there is no ambiguity, you may see the expression for μ written with fewer details, such as

$$\frac{1}{N}\sum_{i}^{N} Y_i \text{ or } \frac{1}{N}\sum_{i} Y_i \text{ or } \frac{1}{N}\sum Y_i \text{ or } \frac{1}{N}\sum Y.$$

When the population is discrete and infinite, similar expressions are used. But now, because the values are not necessarily equally probable, the probability of each value Y_i must be brought into the picture. To this end, let the probability associated with Y_i be $\Pr\{Y_i\}$. Then the mean is given by

$$\mu = \sum Y_i \, Pr\{Y_i\};$$

that is, you sum the products shown here over all the values of Y_i.

The mean of a continuous population is a bit complicated to describe; it requires the integration of the product of each possible value of y multiplied by an expression of its *density function*, which is a function $f(y)$ with two particular properties:

density function

$$f(y) \geq 0$$

and

$$\int_{-\infty}^{\infty} f(y)\,dy = 1.$$

In this case, designate the variable of interest by the unsubscripted y and its density function by $f(y)$. Then the mean is given by the expression

$$\mu = \int y f(y)\,dy,$$

where the integration is understood to be for all values of the variable y. You will find more about density functions in Chapter 7.

sample mean The *sample mean* is defined generally in a similar fashion. Denote the sample size by n and the individual observations by $\{Y_1, Y_2, \ldots, Y_n\}$. Then you can write the mean as

$$\bar{Y} = \frac{1}{n}\sum_{i=1}^{n} Y_i.$$

which, when there is no ambiguity, may be written as

$$\frac{1}{n}\sum Y_i \quad or \quad \frac{1}{n}\sum Y.$$

For discussion:

- What similarities do you see in the notations of the means of different types (finite, discrete, continuous) of populations? What differences?
- Engineers and physicists will recognize the expression for μ as the center of gravity of a system. What does "system" mean to you? In what sense are the mean of a variable and the center of gravity of a system alike?

The weighted mean

The *weighted mean* arises in circumstances in which certain values in a dataset are considered to be "more valuable" than others; that is, some are given more "weight" than others when the data are summarized. In this sense, it is akin to the mean of a population in which the variable's values have differing probabilities of occurring. The *weighted mean* is a statistic calculated in such a fashion as to accommodate these different weights. Note the special notation, $\bar{Y}_w$, used here to designate the weighted mean.

The situation is this: you have K values Y_i; each one is associated with its own weight W_i. The weighted mean is calculated as

$$\bar{Y}_w = \frac{\sum_{i=1}^{K} W_i Y_i}{\sum_{i=1}^{K} W_i}.$$

You may encounter some texts and formulations of the weighted mean that use n_i (usually when Y_i is a summary value based upon n_i individual values), rather than W_i.

Consider now some typical cases where the weighted mean might be used:

Case 1. Specific values are repeated several times in the data. The number of repetitions of a score is the weight of the score.

- Here is an example involving the years-to-first-rust of metal containers; it has $n = 9$ observations with $K = 6$ distinct values. The mean of the observations is 5.30.

  Data: 2.3, 3.4, 3.4, 3.4, 5.2, 5.2, 6.6, 7.8, 10.4; $n = 9$
  Values: $y_1 = 2.3$, $y_2 = 3.4$, $y_3 = 5.2$, $y_4 = 6.6$, $y_5 = 7.8$, $y_6 = 10.4$; $K = 6$

Weights: $W_1 = 1$, $W_2 = 3$, $W_3 = 2$, $W_4 = 1$,
$W_5=1$, $W_6 = 1$

$\Sigma W_i = 9$

$\Sigma W_i y_i = 47.7$

$\bar{y}_w = 47.7/9 = 5.30.$

Case 2. Data are grouped into intervals. The number of responses falling into an interval is the weight associated with that interval. Thus, each value in the ith interval assumes the value of the interval's midpoint; that midpoint is designated by Y_i. The mean of the observations is 5.42.

- Here is an example involving the measured forces required to break 11 bolts, reported in kilopounds (kips):

Data: 2.1, 2.6, 2.7, 3.4, 5.2, 5.5, 5.9, 6.6, 7.4, 7.8, 10.4; $n = 11$

Values: The values are grouped into 9 intervals, each one kip wide, starting at 2 and ending at 11, yielding these 9 midintervals:
$y_1 = 2.5$, $y_2 = 3.5$, $y_3 = 4.5$, $y_4 = 5.5$,
$y_5 = 6.5$, $y_6 = 7.5$, $y_7 = 8.5$, $y_8 = 9.5$,
$y_9 = 10.5$

Weights: $W_1 = 3$, $w_2 = 1$, $W_3 = 0$, $W_4 = 3$,
$W_5 = 1$, $W_6 = 2$, $W_7 = 0$, $W_8 = 0$,
$W_9 = 1$

$\Sigma W_i = 11$

$\Sigma W_i y_i = 59.5$

$\bar{y}_w = 59.5/11 = 5.41.$

Case 3. Responses have different degrees of "importance." According to some authority or tradition, certain responses are more important than others, and the more important responses carry heavier (i.e., larger) weights than those that are less important.

- Here is an example involving the assignment of weights to three classroom examinations: first (20%), second (20%), and final (60%). The mean of the student's three scores is 75.0.

Data: 70, 60, 95 (scores for one student); $n = 3$

Values: $y_1 = 70, \; y_2 = 60, \; y_3 = 95$

Weights: $W_1 = 20, \; W_2 = 20, \; W_3 = 60$

$\Sigma \, W_i = 100$

$\Sigma \, W_i y_i = [(20)(70) + (20)(60) + (60)(95)]/100$

$= 8300$

$\bar{y}_w = 8300/100 = 83.0.$

For discussion:

- In *Case 1*, in the discussion of the weighted mean, the mean of the data (47.7/9 = 5.30) is identical to the weighted mean. Is this a coincidence?

- In *Case 2*, the mean of the data (59.6/11 = 5.42) is *not* identical to the weighted mean (59.5/11 = 5.41). Thus, some detail is lost when you assign the values to the midintervals. Find at least one observation whose value is changed by this process.

- In *Case 3*, weights of 20, 20, and 60 are used. However, you get the same result when you use weights of 2, 2, and 6, or of 1, 1, and 3. Thus, it is the ratios of the elements of the set of weights, not their specific values, that are important in this type of application. How might you demonstrate (i.e., prove) this statement? Is there something special about using 20, 20, and 60 as weights in the example?

Measures of variability

A measure of central value—be it the mean, the median, the mode, or whatever—generally is considered to be the most important parameter/statistic associated with a set of observations. However, a measure of location does not tell the whole story about a variable. To be able to make inferences about a variable, some measure of variability or uncertainty or "noise" of the data is necessary. Recall the box plot concept from Chapter 2 and how it illustrates more about a set of data than merely its location.

Several different measures of variability are described in this section: the range, the percentile and the quartile—both of which are special cases of the quantile—the variance, and the standard deviation. Calculational details are illustrated by the same set of six data values, {1.7, 3.2, 3.2, 4.6, 1.4, 2.8}, used to illustrate the various measures of central tendency.

The *range* is the difference between the largest value and the smallest value in a set of numbers. In the set {1.7, 3.2, 3.2, 4.6, 1.4, 4.8}, the largest value is 4.8 and the smallest is 1.4, so that the range is

$$4.8 - 1.4 = 3.4.$$

For discussion:

- Ranges can be associated with a population of values as well as with samples. However, for some populations and samples, the range cannot be determined. Can you think of cases where the range can be determined

  - □ for neither the population nor the sample?
  - □ for the sample but not for the population?
  - □ for the population but not for the sample?

- The sample range cannot be larger than the population range; indeed, seldom are the two equal.

The *percentiles* can be obtained for any set of values on an ordinal scale or on an interval scale. They are based upon the division of a set of data into 100 equal parts. The kth percentile is a value such that $k\%$ of the measurements are smaller than that value and $(100 - k)\%$ are larger than that value.

The *quartiles* are three particular percentiles of special interest in describing sets of data:

The *first quartile* is the *25th percentile* of a dataset.
The *second quartile* is the *50th percentile* of a dataset.
(From Chapter 2, recall that the *second quartile* also is called the *median*.)
The *third quartile* is the *75th percentile* of a dataset.

Thus, the quartiles divide the dataset into 4 equal parts. Related to the quartiles are *the quintiles* (5 equal parts) and *the deciles* (10 equal parts).

The ***quantiles*** are measures of variability which are, according to Chambers, *et al.* (1983, pp.11-12),

> ... closely connected to the familiar concept of percentile[s]. When we say that a student's college board exam score is at the 85th percentile, we mean that 85 percent of all college board scores fall below that student's score, and that 15 percent of them fall above. Similarly, we define the .85 quantile of a set of data to be a number on the scale of the data that divides the data into two groups, so that a fraction .85 of the observations fall below and a fraction .15 fall above. We will call this value $Q(.85)$. The only difference between percentile and quantile is that percentile refers to a percent of the set of data and quantile refers to a fraction of the set of data....

Thus, a quantile can refer to any portion of a set of data while percentiles necessarily refer only to portions measured in hundredths.

Figure 5-1 depicts the 25 water impurities data given in Table 2-3 plotted along a number line with $Q(.85)$ indicated. Notice especially that none of the observations serves to set a value for $Q(.85)$ and some arbitrariness cannot be avoided. Here, $Q(.85)$ lies between the 21st and 22nd ordered value.

Figure 5-1:
Number line for impurity measurements from Table 2-3

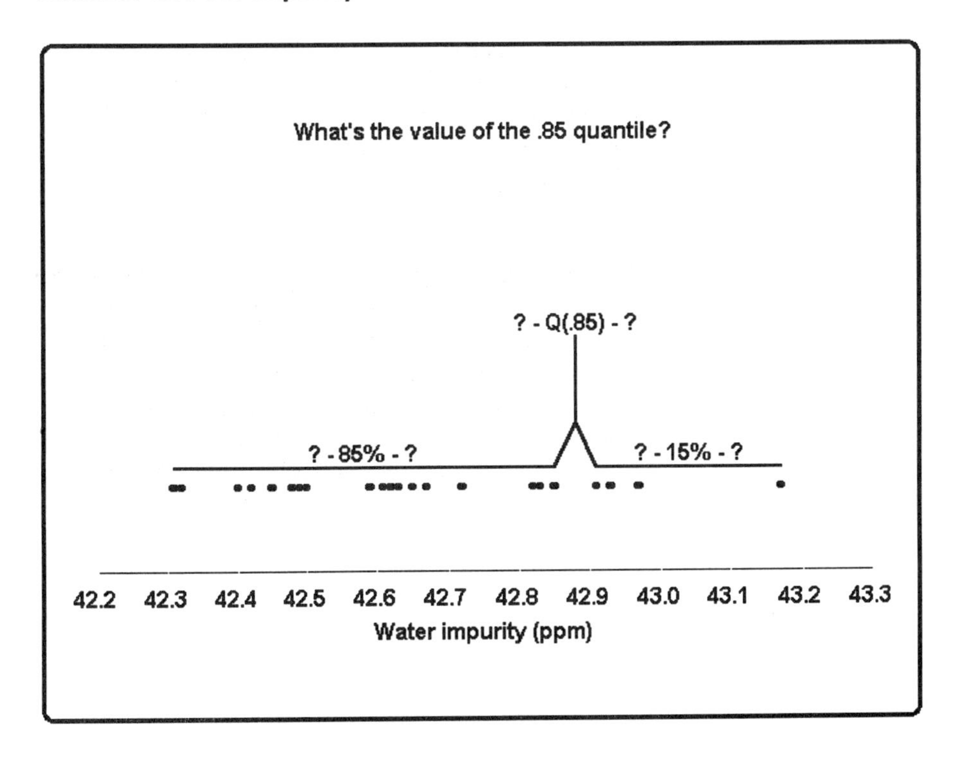

As Chambers, *et al.* (1983, pp. 12-14) go on to explain, this definition

> ... runs into complications when we actually try to compute quantiles from a set of data. For instance, if we want to compute the .27 quantile from 10 data values, we find that each observation is 10 percent of the whole set, so we can split off a fraction of .2 or .3 of the data, but there is no value that will split off a fraction of exactly .27. Also, if we were to put the split point exactly at an observation, we would not know whether to count that observation in the lower or upper part.

To overcome these difficulties, we construct a convenient operational definition of quantile. Starting with a set of raw data y_i, for $i = 1$ to n, we order the data from smallest to largest, obtaining the sorted data $y_{(i)}$, for $i = 1$ to n.[2] Letting p represent any fraction between 0 and 1, we begin by defining the quantile $Q(p)$ corresponding to the fraction p as follows: Take $Q(p)$ to be $y_{(i)}$ whenever p is one of the fractions $p_i = (i - 5)/n$, for $i = 1$ to n.

Thus the quantiles $Q(p_i)$ of the data are just the ordered data values themselves, $y_{(i)}$. The quantile plot in ... [Figure 5-2 is a plot of p_i against $Q(p_i)$ of the water impurities] data. The ... [vertical] scale shows the fractions p_i and goes from 0 to 1. The ... [horizontal] scale is the scale of the original data. Except for the way the ... [vertical] axis is labeled, this plot would look identical to a plot of ... [i against $y_{(i)}$].

[2] This notation for the "ordered statistics" of a set of data is widely used. Thus, $y_{(1)}$ is the smallest member of the dataset, $y_{(2)}$ is the second smallest, ..., and $y_{(n)}$ is the largest.

Figure 5-2:
Quantile plot for impurity measurements from Table 2-3

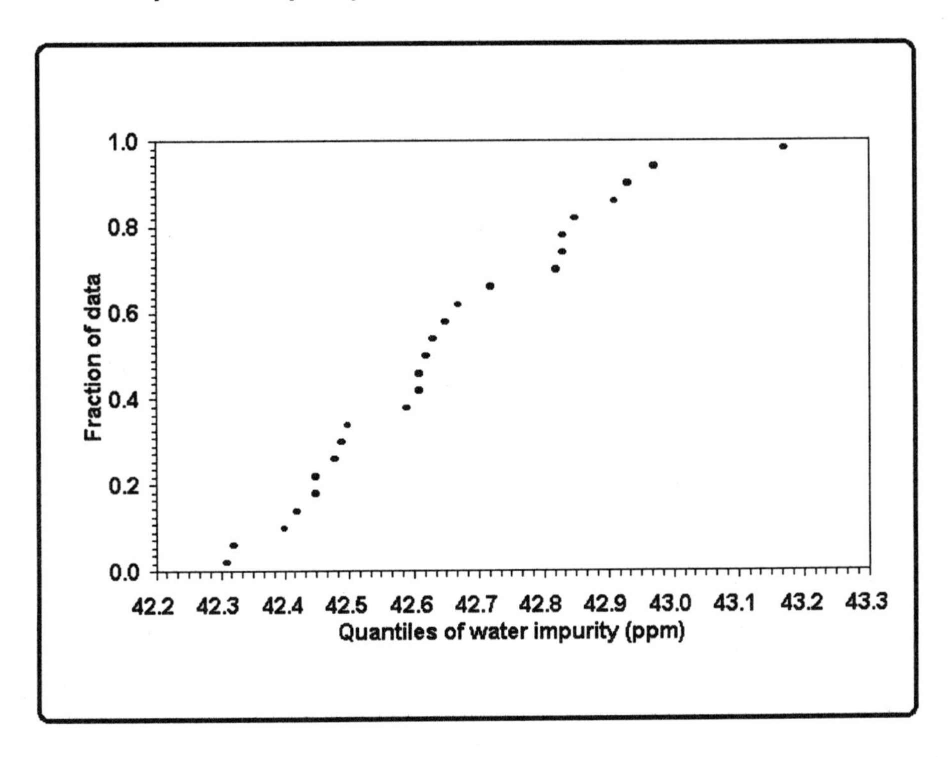

So far, we have only defined the quantile function $Q(p)$ for certain discrete values of p, namely p_i. Often this is all we need; in other cases, we extend the definition of $Q(p)$ within the range of the data by simple interpolation. In ... [Figure 5-2] this means connecting consecutive points with straight line segments, leading to ... [Figure 5-3]. In symbols, if p is a fraction f of way from p_i to p_{i+1}, then $Q(p)$ is defined to be

$$Q(p) = (1 - f)Q(p_i) + f\,Q(p_{i+1}).$$

Figure 5-3:
Interpolated quantile plot for impurity measurements from Table 2-3

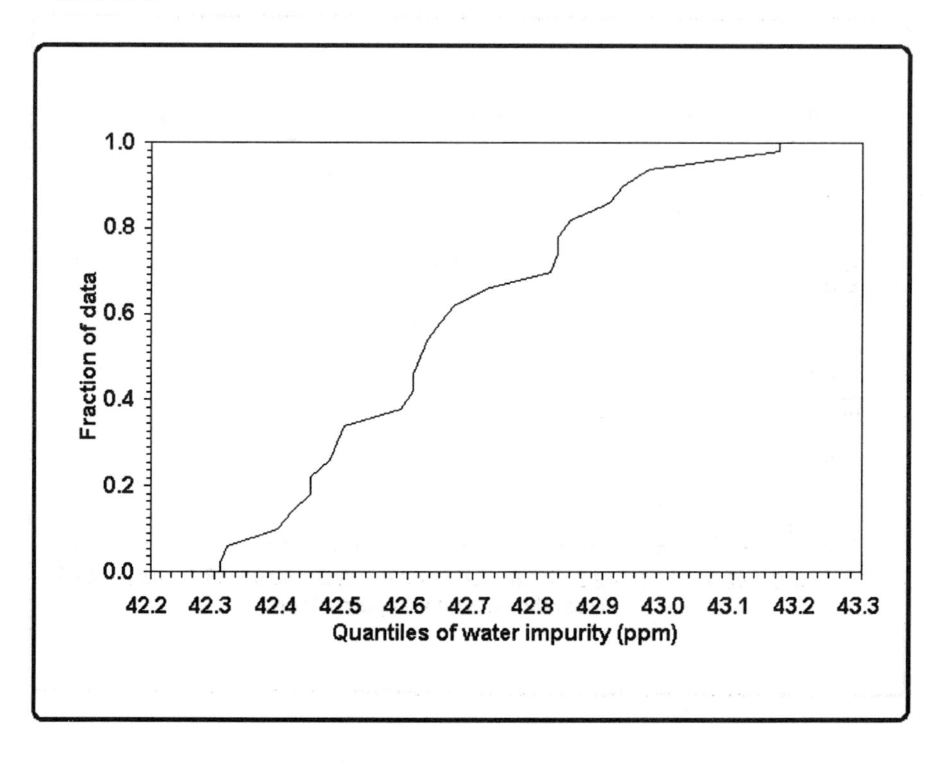

We cannot use this formula to define $Q(p)$ outside the range of the data, where p is smaller than $.5/n$ or larger than $1 - .5/n$. Extrapolation is a tricky business; if we must extrapolate we will play safe and define $Q(p) = y_{(1)}$ for $p < p_1$ and $Q(p) = y_{(n)}$ for $p > p_n$, which produces the short ... [vertical] segments at the beginning and the end of ... [Figure 5-3].

Why do we take p_i to be $(i - .5)/n$ and not, say i/n? There are several reasons, most of which we will not go into here, since this is a minor technical issue. (Several other choices are reasonable, but we would be hard pressed to see a difference in any of our plots.) We will mention only that when we separate the ordered observations into two groups by splitting exactly on an observation, the use of $(i - .5)/n$ means that the observation is counted as being half in the lower group and half in the upper group.

The variance and the standard deviation for populations are commonly used statistical tools for measuring variability. For a *finite population* of size N, with values $\{Y_i, i = 1, \ldots, N\}$, the variance is defined by

$$\sigma^2 = \frac{\sum_{i=1}^{N} (Y_i - \mu)^2}{N}.$$

For a continuous population with density function f(y), the variance is defined by the integral

$$\sigma^2 = \int_{-\infty}^{\infty} (Y - \mu)^2 f(Y)\, dY.$$

In either case, the standard deviation of the population is the positive square root of the variance; i.e.,

$$\sigma = \sqrt{\sigma^2}\ .$$

For discussion:

- "The population variance measures the fluctuation of the data around the population mean." How do you interpret that remark?
- The value of σ^2 is rarely known. Can you think of situations in which it is known? Indeed, what does *to know* σ^2 really mean?
- Engineers may recognize σ^2 as the moment of inertia of a system around its center of gravity. What does the term "system" mean?

What does the term "moment of inertia" mean? How do you link them to the statistical issues discussed in this chapter?

- The variance is sometimes called the Mean Square Error. Why?

- The concept of standard deviation is generally easier to conceptualize than that of the variance because the standard deviation is expressed in the same units of measurement as the original data.

The ***variance and standard deviation for samples***, like those for populations, also are commonly used statistical tools for measuring variability.

For a sample of size n, the sample variance is defined by one of two expressions, according to whether you *know* the value of the population's mean:

- On those rare occasions when μ is known, the sample variance is

$$S^2 = \frac{\sum_{i=1}^{n} (Y_i - \mu)^2}{n}.$$

- When μ is not known and must be estimated from the sample itself by $\bar{Y}$, the sample variance is

$$S^2 = \frac{\sum_{i=1}^{n} (Y_i - \bar{Y})^2}{n-1}.$$

The second expression is the *definition formula* for the sample variance. An alternative formula, called the *working formula*, can be written in several forms, two of which are:

$$s^2 = \frac{\sum_{i=1}^{n} y_i^2 - \frac{(\sum_{i=1}^{n} y_i)^2}{n}}{n - 1} = \frac{n\sum_{i=1}^{n} y_i^2 - (\sum_{i=1}^{n} y)^2}{n(n - 1)}.$$

In either case, the standard deviation of the sample is the positive square root of the sample variance; i.e.,

$$s = \sqrt{s^2}\,.$$

For discussion:

- S^2, the variance of the sample, is an *unbiased estimator* of the population variance, σ^2. This means that S^2 may be (and usually is!) different from σ^2, but that *on the average* S^2 equals σ^2. What does *on the average* mean in this context?

- The denominator of S^2, either n or $(n - 1)$ according to whether you know the mean, in the formula for the sample variance is called the *degrees of freedom*. What does *degrees of freedom* mean in this context, compared to its use in the analysis of contingency tables (Chapter 4)?

- Some personal calculators use σ_n and σ_{n-1} to denote the standard deviations for the population and the sample, respectively. Some calculators use σ and s to make the distinction. Still other calculators fail to differentiate between the two in any way. To see what your calculator does, enter the values 1, 2, and 3 and press the standard deviation key. If the display shows exactly 1, your calculator uses $(n - 1)$ in the denominator. If the display shows a number in the neighborhood of 0.8165, your calculator uses n. If some other number appears in your display, check your operator's manual or buy another calculator (no, it's not the batteries gone sour!). What does your calculator do?

- If you use spreadsheet software, be aware that some built-in functions (e.g., Lotus 1-2-3's @VAR(*list*) and @STD(*list*)) are incorrect for samples in that they use n, rather than $(n - 1)$, in their denominators. If you must use these functions, multiply @VAR(*list*) by $n/(n - 1)$ and @STD(*list*) by the square root of $n/(n - 1)$ to get sample estimates of σ^2 and σ, respectively. You may find other interesting choices among the functions in various spreadsheet packages. Check your software's manual(s).

Nine steps to computing a sample standard deviation

Even though modern computers and calculators have taken much of the fear, loathing, anger, anxiety, inaccuracy, and pain out of variance and standard deviation computations, this section is designed to help you become comfortable with various ideas and nomenclatures that you may encounter elsewhere. It is a nine-step procedure:

Step 1: Denote the n observations in your sample as the set

$$\{Y_1, Y_2, \ldots, Y_n\}.$$

Step 2: Denote the sample sum by one of the forms:

$$\sum_{i=1}^{n} Y_i, \quad \sum_{i}^{n} Y_i, \quad \sum Y_i, \quad \sum Y, \quad Y_+, \quad or \quad Y.$$

Note: You will find the last of these forms, i.e., $Y.$ (called y-dot), used extensively in statistical literature.

Step 3: Calculate the sample mean:

$$\overline{Y} = \frac{\sum_{i=1}^{n} Y_i}{n}.$$

Step 4: Calculate the *unadjusted sum of squares*:

$$SS_{(unadj)} = \sum Y_i^2.$$

Step 5: Calculate the *correction term*:

$$CT = \frac{(\sum Y_i)^2}{n} = n(\bar{Y})^2.$$

Step 6: Calculate the *sum of squared deviations* (sometimes called the *adjusted sum of squares*):

$$\begin{aligned} SSD &= \sum (Y_i - \bar{Y})^2 \\ &= \sum (Y_i^2) - \frac{(\sum Y_i)^2}{n} \\ &= SS_{(unadj)} - (CT). \end{aligned}$$

Step 7: Calculate the degrees of freedom:

$$df = (n-1).$$

Step 8: Calculate the sample variance:

$$S^2 = \frac{SSD}{df}.$$

Step 9: Calculate the sample standard deviation:

$$S = \sqrt{S^2}.$$

For discussion:

- Is it obvious that the relation

  $$\sum (Y_i - \bar{Y}) = 0$$

  holds for any set of values $\{Y_1, Y_2, \ldots, Y_n\}$? How would you prove it? Can you prove it using "simple algebra"? Aside from the algebraic demonstration, why might you expect this relation to hold true?

- The equivalence of the definition formula and the working formula for the variance can be verified by algebra. Try your hand at showing this.

- If your hand-held calculator has statistical functions, it calculates the variance using the working formula. Why should this be the case?

- The working formula, generally, is easier to apply than the definition formula. However, the working formula may cause an error if the data entries are large enough to cause a computer/calculator overflow.

- Would you rather use the definition formula or the working formula? Can you suggest cases where the definition formula is superior? What is the advantage of the working formula?

- As noted, the two methods may disagree due to rounding errors. For most datasets, the working formula is recommended; it requires fewer operations. For numbers with many significant figures, the definition formula is recommended; it is less sensitive to rounding or truncation of intermediate results.

- A variance cannot be negative. If you see a negative variance, you know something is wrong. What might that be?

- The variance *can* be identically zero. Under what conditions? Give an example.

The mean and standard deviation of coded variables

Consider a set of values $\{Y_i, i = 1, \ldots, n\}$ with a mean $\bar{Y}$ and a standard deviation S_Y. Suppose you multiply each value by a constant K and then add another constant C. What happens to the mean and the standard deviation?

Algebraically, you have created a new set of values by coding (i.e., applying a *linear transformation*[3]) the Y_i to create a new set of values, call them X_i, where

$X_i = KY_i + C, \quad i = 1, \ldots, n.$

The mean of the X_i becomes $\bar{X} = (K\bar{Y} + C)$, and the standard deviation of the X_i becomes $S_X = KS_Y$.

Because any linear transformation can be written in terms of the constants K and C, you have full knowledge of the results without actually calculating them. For example, if you wish to add 10 units to each Y_i in a dataset, simply set $K = 1$ and $C = 10$. Then you know immediately that $\bar{X} = \bar{Y} + 10$ and that $S_X = S_Y$, no matter what the actual values of $\bar{Y}$ and S_Y.

[3] A linear transformation is a procedure in which a variable is multiplied by a constant K, immediately followed by the addition of a constant C.

For discussion:

- Consider this set of 10 values: {98, 99, 101, 97, 100, 98, 95, 101, 99, 99}. Compute the mean, variance, and standard deviation.
- Add 10 to each of the values and repeat the calculations. Are they as the formulas specify?
- Subtract 10 from each of the original values. What will be the mean, the variance, and the standard deviation? Do they agree with your expectation?
- Multiply each of the original values by 100. What will be the mean, the variance, and the standard deviation? Do they agree with your expectation?
- Divide each of the original values by 5. What will be the mean, the variance, and the standard deviation? Do they agree with your expectation?
- Subtract 32 from each of the values and then multiply those values by 5/9. What will be the mean, the variance, and the standard deviation? Do they turn out to be as you predict?
- Why would you even consider coding a set of data in the first place? What are the advantages in coding data?
- You may have occasion to apply other, but non-linear, transformations to a set of data. Proceed with some caution because the results are not necessarily predictable in terms of the descriptive statistics. For example, suppose you apply the square root transformation to the 10 values given here. Then what are the mean and the standard deviation? Is the square root of the original mean the same as the mean of the square root values?

The coefficient of variation

The *coefficient of variation*—also known as the *coefficient of variability* and/or the *relative standard deviation*—is widely used in a number of disciplines. It is defined as the standard deviation divided by the mean. Interestingly, the coefficient of variation has not acquired a widely used single symbol. Thus, in this book, for a population, you have

$$CV_p = \frac{\sigma}{\mu}.$$

In a similar fashion, in this book, for a sample, you have

$$CV_s = \frac{S}{\bar{Y}}.$$

For discussion:

- Compute the coefficient of variation for each of the transformations you made in the previous For discussion: section. Draw your conclusions. Why do you suppose you need the coefficient of variation? What are its units? What problems do you see with using and reporting the coefficient of variation?

An "Empirical Rule"

Many sets of measurements, when rendered as a histogram, yield "mound-shaped" images. That is, when all the measurements are plotted as a histogram, where the horizontal axis represents the measurements and the vertical axis gives the frequencies (or the relative frequencies) of the measurements, you obtain a relatively symmetric, relatively smooth curve with tapering tails to the left and to the right of a center "mound."

The histogram of such measurements is seldom truly symmetric. Nevertheless, for many mound-shaped sets of numbers with mean $\bar{Y}$ and standard deviation S, Ott and Mendenhall (1984, pp. 76-82) offer the following "Empirical Rule":

> For a histogram that is mound-shaped, the interval
>
> $$\text{from } (\bar{Y} - S) \text{ to } (\bar{Y} + S)$$
>
> contains approximately 68% of the measurements;
>
> $$\text{from } (\bar{Y} - 2S) \text{ to } (\bar{Y} + 2S)$$
>
> contains approximately 95% of the measurements; and
>
> $$\text{from } (\bar{Y} - 3S) \text{ to } (\bar{Y} + 3S)$$
>
> contains all or nearly all of the measurements.

This empirical rule supports convenient summaries regarding the nature of various portions of a set of data. For example, consider the 150 uranium ingot weights (Bowen and Bennett, 1988, pp. 12-15) displayed in Table 5-1. The mound shape derived from these weights is shown in Figure 5-4. The mean of these 150 weights is $\bar{y} = 426.40$, the standard deviation is $s = 2.83$, and the range is 15.7.

Table 5-1:
Weights of 150 uranium ingots

425.0	426.7	423.3	429.4	427.9	425.9	422.1
422.5	424.4	427.8	428.9	425.3	425.4	427.3
424.9	422.4	426.1	424.8	432.4	427.9	421.9
431.7	432.2	422.4	427.3	427.3	423.7	425.7
426.3	427.8	424.8	428.0	426.3	428.6	425.9
424.5	431.3	431.2	427.3	418.5	428.8	431.6
426.1	425.8	429.8	429.5	425.3	424.5	424.6
423.1	426.8	430.9	423.9	421.9	425.1	421.8
428.3	424.8	427.0	425.1	425.2	424.4	432.3
423.2	423.6	427.9	427.9	428.5	424.7	428.8
428.2	424.8	421.0	423.6	428.0	427.7	425.7
429.1	429.7	419.6	421.3	426.8	421.2	425.2
424.2	430.3	424.6	430.0	423.5	427.2	430.0
429.7	423.2	428.8	425.4	427.5	429.4	424.9
424.8	431.0	427.9	423.6	421.7	425.9	426.6
427.2	428.0	428.0	429.7	427.4	426.6	426.2
428.3	426.6	428.4	427.1	427.5	425.5	426.2
429.3	425.4	423.1	426.9	425.7	429.2	434.2
421.8	427.3	425.2	427.3	425.7	426.5	420.4
424.0	426.0	424.9	430.5	426.3	426.3	424.9
428.0	423.3	431.1	426.4	429.0	429.9	423.3
427.4	424.2	428.2				

Figure 5-4:
A mound-shape derived from 150 uranium ingot weights given in Table 5-1

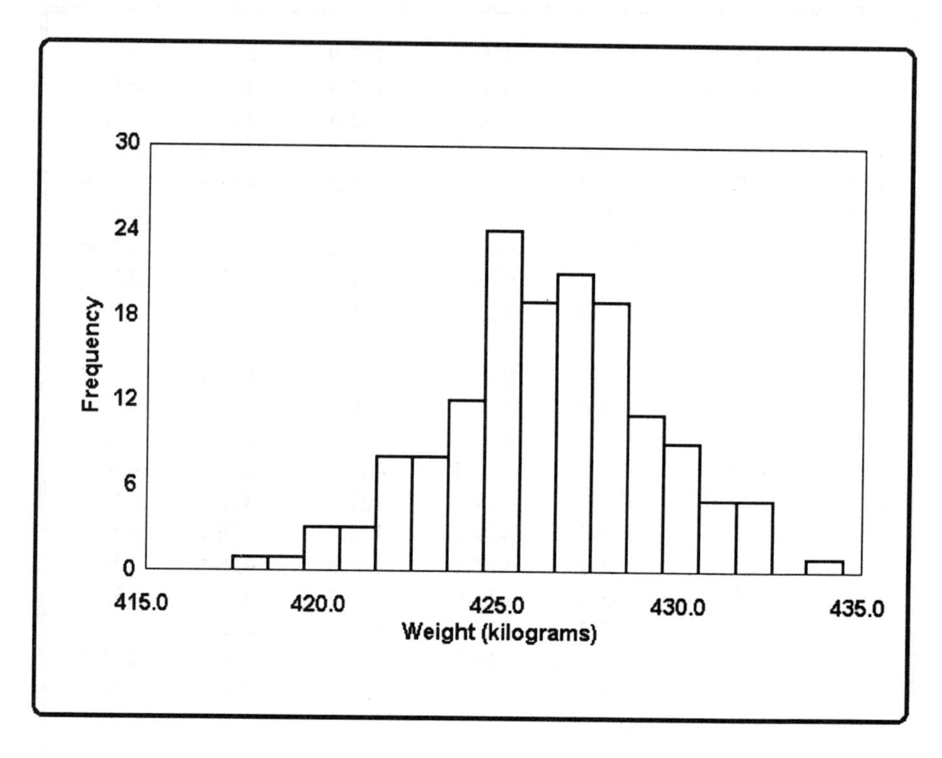

These intervals and their corresponding percentages for the uranium ingot data are displayed in Table 5-2.

Table 5-2:
Comparison of expected and actual contents of intervals based upon the "Empirical Rule" and the "mound-shaped" uranium ingot weight data in Table 5-1 and Figure 5-4

The interval defined by	which evaluates as	is expected to contain	but actually contains
$(\bar{Y} - S), (\bar{Y} + S)$	(423.58, 429.23)	68%	69.3%
$(\bar{Y} - 2S), (\bar{Y} + 2S)$	(420.75, 432.06)	95%	95.3%
$(\bar{Y} - 3S), (\bar{Y} + 3S)$	(417.93, 434.88)	100%	100%

Estimating the standard deviation from the range

Related to the "mound-shaped" conditions that support the Empirical Rule is an inequality involving the ratio of the range and the standard deviation for almost every set of data you will ever encounter:

$$3 < \frac{\textit{range}}{\textit{standard deviation}} < 6.$$

You may use this relationship to check on the calculation of the standard deviation. For example, if the ratio of the range to the standard deviation is 15, you get a hint that something is either unusual or just plain wrong. Furthermore, it can be shown mathematically that no matter what the shape of the distribution is, the ratio of the range to the standard deviation cannot be smaller than 1.4142 (the square root of 2).

You can use the following rule to obtain rough estimates of the standard deviation from the range as:

> range/4 for a *small* set of measurements (say, 20 measurements or fewer),
>
> range/5 for a *large* set of measurements (say, between 20 and 100 measurements), and
>
> range/6 for a *very large* set of measurements (say, more than 100 measurements).

Because $n = 150$ and the range = 15.7 for the uranium ingot data of Figure 5-4, you could use 15.7/6 = 2.62 as a rough estimate of the standard deviation. Recall that the "real" standard deviation is 2.83.

These two empirical results provide useful, albeit informal, computational and/or "eyeball" checks in many statistical investigations. However, because they do not always apply to every set of data (not even to those that are "mound-shaped"), some care is needed in their use. They should never be used as substitutes for exact calculations in formal studies.

What to remember about Chapter 5

Chapter 5 defined, illustrated, and differentiated among a variety of *descriptive statistics* by focusing on data of the continuous type (measured on an interval or a ratio scale). Topics included:

- *measures of central value,* including the *mean*, the *median*, the *mode*, the *midrange*, the *trimmed mean*, the *Winsorized mean*, and the *weighted mean*

- *measures of variability*, including the *range*, the *percentiles*, the *quartiles*, the *quantiles*, the *variance*, and the *standard deviation*

- *coefficient of variation*

- *effects of variable coding*.

The chapter concluded with two handy shortcuts useful in dealing with data:

- *an empirical rule that provides a convenient way of summarizing the portion of a dataset that lies in an interval*

- *a method estimating the standard deviation of a dataset from its range.*

Errors, errors, ... everywhere

What to look for in Chapter 6

Chapter 6 invites you to consider methods of characterizing and analyzing *errors*. You will encounter such special ideas as:

- *measurement system*
- *error*
- *accuracy*
- *bias*
- *precision*
- *uncertainty*.

These ideas pave the way for Chapter 7's examination of the *normal distribution* whose nature and wide applicability provide the touchstone for all modern statistical procedures.

Thinking about errors: Some examples that set the scene

Have these things ever happened to you?

You weigh yourself in the morning. You don't like the number the scale gives you. You rationalize that the scale is in error. Further, you decide that this dubious measurement could be the result of one or more of the following:

- Instrument error
- Resolution error
- Calibration error
- Replication error
- Sampling error
- Accuracy error
- Design error
- Day-of-the-week error
- Time-of-the-day error
- Seasonal variation error
- Random error
- Rounding off error
- "Something I ate" error
- "After/before" shower error
- Your spouse, your children, and your dog make you nervous.

But which error is it that gets your morning off to such a start? Or which set of errors is the culprit? Whatever it is, you cannot cope with it at the moment because now it is time to go to work.

Just as you get settled at your work station, your manager brings you the report you recently submitted so confidently—and well ahead of your deadline. Some computer some place has contradicted your "bottom line" figure by an order of magnitude. You immediately attribute the error to one or more of the following:

- Program error
- Programmer error
- Counting error
- Laboratory error
- Technician error
- Recording error
- Transcription error
- Modelling error
- Significant figure error
- "Wrong column" error
- Definition error
- Printer error
- Transmission error
- Erasure error
- Your boss, your colleagues, and your deadline make you nervous.

But which error is it that causes your professional discomfort? Or which set of errors?

For discussion:

- Were any errors left off either list? What are they?
- Which errors in these lists are redundant?
- What is the difference between an *error* and a *mistake*?
- If, as Alexander Pope observed, "To err is human," then what's this fuss all about?
- Pick a process from the work you do. List the potential errors. How could you proceed to quantify those errors?

Characterizing errors: Accuracy and precision and uncertainty

metrology

Measurement is fundamental to scientific study. But measurements are made, not born. *Metrology* is the science that studies and codifies the measurement process. Coleman (1985) says:

measurement system

> The first concept to grasp about metrology is that any measurement comes from a measurement *system*, not from measurement equipment alone. In other words, the weight shown on your bathroom scale is the result of more than the mere existence of you and the scale....

Eisenhart, in Ku (1969, p. 23-163), offers this perspective:

> Measurement is the assignment of numbers to material things to represent the relations existing among them with respect to particular properties. The number assigned to some particular property serves to represent the relative amount of this property associated with the object concerned.
>
> Measurement always pertains to properties of things, not to the things themselves. Thus we cannot measure a meter bar, but can and usually do, measure its length; and we could also measure its mass, its density, and perhaps, also its hardness....
>
> As Walter A. Shewhart ... has remarked:
>
> > "It is important to realize ... that there are two aspects of an operation of measurement; one is quantitative and the other qualitative. One consists of *numbers* or pointer readings such as the observed lengths in *n* measurements of the length of a line, and the other consists of the *physical manipulations* of physical things by *someone* in accord with instructions that we assume to be describable in words constituting a text." [Shewhart 1939, p. 130]

More specifically, the qualitative factors involved in the measurement of a quantity are: the *apparatus* and *auxiliary equipment* (e.g., reagents, batteries or other source of electrical energy, etc.) employed; the *operators* and *observers*, if any, involved; the *operations* performed, together with the *sequence* in which, and the *conditions* under which, they are respectively carried out.

the truth

Now consider a particular system of measurement. It is designed to measure *the truth* of something. That truth is designated in this discussion by the Greek letter τ (tau). Let Y be a measurement produced by the system. That measurement may or may not be "close" to the truth τ. The difference between Y and τ is called an *error*.

error

To formalize these matters, let the English letter E[1] be the error associated with the Y that is produced by the system in its attempt to determine τ. The three symbols are linked by the expression

$$E = Y - \tau.$$

Now consider *a series* of such measurements produced by the system, $\{Y_1, Y_2, \ldots, Y_i, \ldots\}$, and the series of their corresponding errors, $\{E_1, E_2, \ldots, E_i, \ldots\}$. The concepts of accuracy and *precision* are quantifiable in terms of the behavior of the errors shown in the sequence $\{E_1, E_2, \ldots, E_i, \ldots\}$.

accuracy

Each measuring system carries its own *accuracy* with it. If the errors in the measurements average to zero *in the*

[1] To designate the error, we elect to use the upper-case English letter E. This is consistent with the notation established in Chapter 3, but you will find other designations in other literature. Many, if not most, writers in this area use the lower-case Greek letter epsilon, ϵ, perhaps because—by convention or tradition or strong training in calculus—ϵ suggests "smallness," a quality most of us would like our errors to posses.

accurate

inaccurate

long run, then the system is said to be *accurate*. If the errors in the measurements average to something other than zero, the system is said to be *inaccurate*.

For the most part, accuracy is a relative matter that is based upon the comparison of two or more measuring systems. Consider a second measurement system, also aimed at determining the truth τ. If—again *in the long run*—the average of the values of Y of the first system are closer to τ than the average of the values of Y for the second system, then the first system is said to be *more accurate* than the second system. Of course, all of this presumes that the truth is known—calling for another discourse altogether.

The concept of accuracy is directly illustrated by rifle target practice. Even if you aim carefully at the target, sometimes you shoot to the left and sometimes to the right, sometimes high and sometimes low. If, on the average, you score a bull's-eye, the system (you, the rifle, the weather, the shooting range, *inter alia*) is considered accurate. Note that the system can be accurate, even if you never hit the bull's-eye.

bias

unbiased

A system is said to have a *bias* (or to be *biased*) if it is inaccurate. Whether the magnitude of the bias is critical depends upon the system and the use to which it is applied. In a similar fashion, an accurate system is *unbiased*.

precision

Irrespective of a system's accuracy, it carries its own *precision* with it. The idea behind precision is how "close" the elements of the series of measurements are to each other. The "closeness" (i.e., the precision) of a series is measured by its variance. One system of measurements is said to be more *precise* than a second system if it has a smaller variance than the second system.

As with accuracy, the concept of precision is directly illustrated by rifle target practice. Regardless of where your bullets hit, the system is precise if all the hits are in exactly the same place. Clearly, if you and your rifle and your shooting range are simultaneously accurate and precise, you have a statistically superb system.

Figure 6-1 displays four combinations of good/poor accuracy and good/poor precision in a one-dimensional system.

Figure 6-1:
Combinations of good and poor accuracy and good and poor precision

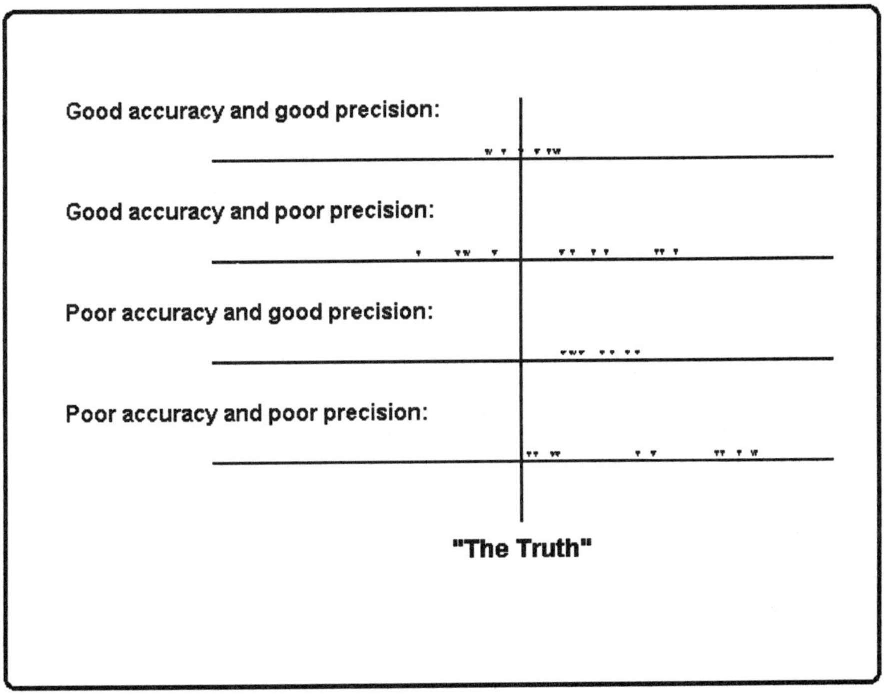

Good accuracy and good precision:

Good accuracy and poor precision:

Poor accuracy and good precision:

Poor accuracy and poor precision:

"The Truth"

uncertainty Attention must be paid to the concept of *uncertainty*. This broadly used term refers to the "incorrectness" in a set of data arising from inaccuracy, imprecision, or both. The term itself is troublesome because it conveys such a broad spectrum of meaning and suffers from an equally broad spectrum of interpretation. On the whole, it seems best to avoid using "uncertainty" in technical discussions unless it is carefully qualified and, if possible, quantified.

For discussion:

- The phrase *in the long run* arises several times in this chapter. What does it mean to you? Do you have a way of quantifying the associated ideas?

- Re-examine the two lists of errors presented at the beginning of the chapter regarding your morning's weigh-in and your office report. Which errors affect accuracy? Which affect precision? Which affect neither?

- If a measurement system is declared to be *accurate and precise*, what virtues are being claimed for the system?

- Kendall and Buckland (1971, p. 50) have this to say:

> **Error** In general, a mistake ... in the colloquial sense. There may, for example, be a gross error or avoidable mistake; an error of reference, when data concerning one phenomenon are attributed to another; copying errors; an error of interpretation. In a more limited sense the word "error" is used in statistics to denote the difference between an occurring value and its "true" ... value. There is no imputation of mistake on the part of a human agent; the deviation is a chance effect....

Does this statement clarify things for you?

- Your old watch stopped working years ago. How good is it as an instrument for measuring time? When you look at it (at random moments, of course) it always reads 8:07. Thus, it is an accurate instrument: half of its readings overstate the true time and half understate it, and thus their errors average to zero. It also is a very precise instrument: all of its readings are identical, and thus their variance is exactly zero. With such an accurate *and* precise instrument for measuring time, why should you ever consider buying a new watch?

What to remember about Chapter 6

Chapter 6 focused on methods of characterizing and analyzing *errors*. You found definitions of:

- *measurement system*
- *error*
- *accuracy*
- *bias*
- *precision*
- *uncertainty*.

These ideas lead directly to Chapter 7's examination of the *normal distribution*. Indeed, you will encounter them in a variety of special statistical topics as you move through Chapters 8-21.

The normal distribution

What to look for in Chapter 7

Chapter 7 explores the single most important concept in modern statistics—the *normal distribution*. In addition to its mathematical tractability, the normal distribution provides a more-than-adequate model for many natural phenomena and supports a wide variety of decision-making processes. As you work through this chapter, you will encounter such important ideas as

- *distribution function*
- *density function*
- *the normal distribution function*
- *the normal density function*
- *the standardized normal distribution*
- *testing data for normality*
- *the Central Limit Theorem.*

Finally, you will learn how to use a single-page tabulation (Table T-1) to quantify nearly everything you need to know about the normal distribution.

An experiment

When you weigh an object, "chances are" that your reading is incorrect, no matter what the quality of your scale or its manufacturer's claims. For one thing, no scale is blessed with an infinite number of significant figures. As discussed in Chapter 6, your reading most certainly is in error. Whether you can live with that error depends upon many factors. The discipline of statistics provides methods for addressing, assessing, and coping with such errors.

If a single weighing of an object contains an error, then a reasonable strategy is to weigh the object a number of times and "average out" the errors. Thus, you expect to find comfort in a procedure that balances positive and negative errors to produce a value that is "closer to the truth" than any single value.

To illustrate, suppose you place an object (weighing, say, in the neighborhood of 12 kilograms) on a scale many times during the same day (say, every 15 minutes between 8 a.m. and 5 p.m.), removing the object from the scale after each weighing. With a good scale (say, correct to the nearest decigram), you should not expect to get identical readings each time. Rather you should expect them to vary. More than that, you hope that you obtain random (i.e., unpatterned) readings.

You record all your readings (say, a total of 32 because you take time for lunch), and you then construct a histogram for these values. You expect to see a "peak" frequency in the center with reduced frequencies to the left and the right. The more readings you have and the narrower the histogram's class interval width, the more you might expect your histogram to look symmetric and "bell-shaped," two of the telling characteristics of a mathematical-statistical concept which we affectionately call the *normal distribution*.

For discussion:

- In the object-weighing experiment, what is a reasonable interval width to use in the construction of the histogram? Do you have all the facts to make this judgement now?
- What will you gain (or lose) if the scale is correct to the nearest gram rather than the nearest decigram?
- Why would you bother to remove the item from the scale between weighings?
- What do you gain by taking measurements both in the morning and in the afternoon?
- What do you do if the histogram is not as you expect; i.e., *not* "peaked" in the middle with reduced frequencies to the left and to the right? Most importantly, do not despair. Sometimes it's just a matter of transforming your measurements. Sometimes it requires extended modeling. Sometimes it means recognizing that you're into new "stuff" and need some help. Making measurements is by no means a trivial or casual activity.[1]

---

[1] One place to start your inquiry is with Coleman (1985) who defines metrology as "the science of studying and understanding measurements." In his concluding paragraph, Coleman reminds us that:

> Measurements enter all phases of work and life. However, hardly any of us ever measure measurements. Problems with measurements are commonly thought to belong to spectometrists who try to identify minute chemical quantities, or ... operators ... who inspect incoming material. Actually, problems with measurements are epidemic. They cloud understanding, and they reduce quality and productivity.... We can take safeguards by:
>
> (1) Recognizing that measurements come from measurement systems.
> (2) Deliberately conducting measurement capability studies according to statistically designed experiments.
> (3) Analyzing the data correctly afterwards.

Some necessary and useful mathematics and details

The modern mathematical formulation of the normal distribution derives essentially from the function developed by Carl Friedrich Gauss (1777-1855) in his 1809 book, *Theoria Motus Corporum Coelestium in Sectionibus Conicis Solem Ambientium* (Theory of the Motion of the Heavenly Bodies Revolving Around the Sun in Conic Sections). The importance of the normal distribution in statistical analysis and interpretation cannot be overstated. Nor is there any substitute for your biting the "math-stat bullet" for a bit as you enter into these matters. It will benefit you in all future discussions involving data-based decisions.

Consider a continuous random variable, say Y. In order for Y to be declared *normally distributed*, certain conditions must be met. These conditions are summarized in the normal variable's *distribution function* and its corresponding *density function*.[2] Both of these functions are expressed in terms of the normal distribution's two *parameters*: μ (the mean) and σ (the standard deviation). Here are the basics of these ideas.

distribution function

The *distribution function*, indicated by $F(y) = Pr\{Y \leq y\}$, of a random variable Y is the probability of obtaining a value for the random variable that is less than or equal to a specific value y.

density function

The corresponding *density function*, indicated by $f(y)$, specifies the way in which the frequencies of possible values that can be taken by the random variable are dispersed among Y's possible values; i.e., mnemonically, the density function tells you how "densely packed" you will find those possible values. Mathematically, the

[2] In adopting this terminology, which is applicable to all random variables, we follow the leads of Feller (1957, p. 168) and Mood, *et al.* (1974, pp. 57-84)

density function is the first derivative of the distribution function with respect to y.

The mathematical expression for the normal distribution function associated with a random variable Y is given by

$$F(y) = Pr\{Y \leq y\} = N(y)$$

$$= \int_{-\infty}^{y} \frac{1}{\sqrt{2\pi}\,\sigma} e^{-\frac{1}{2}\frac{(s-\mu)^2}{\sigma^2}} ds, \qquad -\infty < s < +\infty,$$

where $N(y)$ is a special notation used to indicate the normal distribution function. The symbol e denotes is the base of the natural logarithms (approximately 2.71828); the symbol π demotes the basic mathematical constant (approximately 3.14159).

ogive

When you plot the normal distribution function, $F(y)$, against y, you get a curve called an *ogive* (an elongated S-shape).

The corresponding normal density function is given by

$$f(y) = n(y) = \frac{1}{\sigma\sqrt{2\pi}} e^{-\frac{1}{2}\frac{(y-\mu)^2}{\sigma^2}}, \qquad -\infty < y < +\infty,$$

where $n(y)$ is a special notation used to indicate the normal density function.

bell-shape

When you plot the normal density function, $f(y)$, against y, you get the familiar symmetric *bell-shape*.

A widely used notation for the normal distribution is $N(y;\, \mu,\, \sigma^2)$ where μ and σ^2 are the mean and variance, respectively. Thus, $N(y;\, 100,\, 200)$ denotes a normal distribution with $\mu = 100$ and $\sigma^2 = 200$. Another common notation for the normal distribution function omits the y from the parentheses, thus writing $N(\mu, \sigma^2)$. Some authors prefer to employ σ, the standard deviation, in place of σ^2 in this notation—so be alert!

standardized

When $\mu = 0$ and $\sigma^2 = \sigma = 1$, the normal distribution is said to be *standardized*. Figure 7-1 contains plots of both the standardized normal distribution function and the standardized normal density function.

Figure 7-1:
The standardized normal distribution function, *F(y)*, and density function, *f(y)*

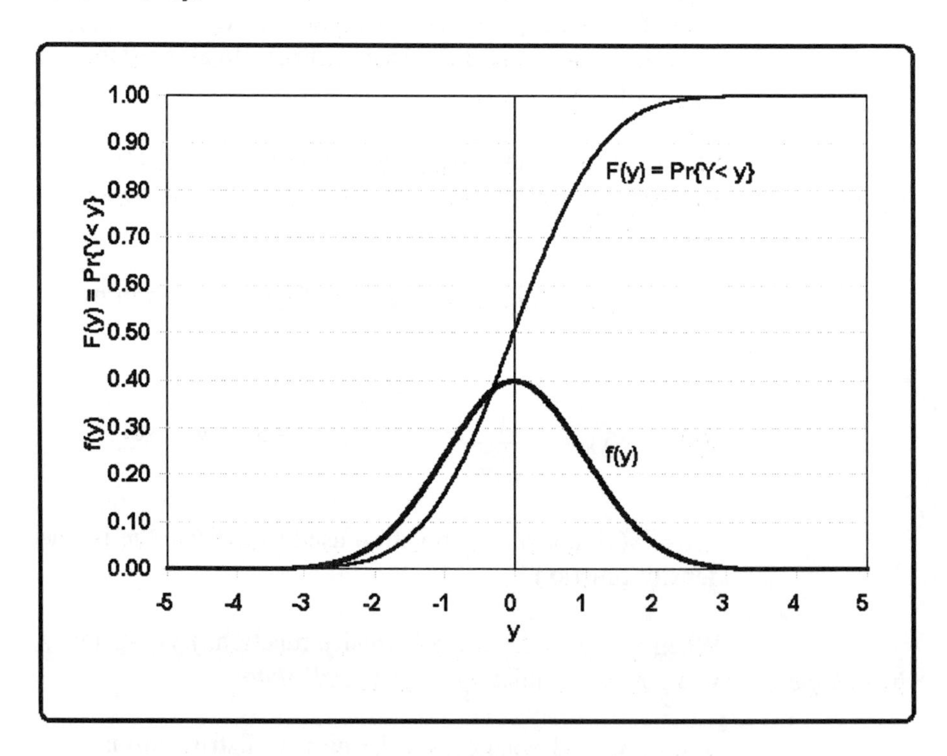

The shape of the density function in Figure 7-1 shows that the normal values are more closely packed in the center than they are near the left and right ends; that is, a random normal variable is more likely to be in the center where the values are "denser" than it is to be in the left or right tails where the values are "less dense."

Recall from Chapter 6 that a difference between an observation, say Y, and the mean, say μ, written $Y - \mu$, is called an error. When a set of errors is said to be normally distributed, you conclude that those errors have *at least* the following two properties:

- Positive and negative errors are equally likely to occur.
- Large errors are less likely to occur than small errors.

In many investigations about the numerical characteristics of populations, the populations are assumed to be "normal." For many measurements of interest, this assumption is reasonable, proper, and consistent with what is known about the process under study. However, systems that have errors with these properties are *not necessarily* producing *normally distributed* errors. Thus, at times, you may have reasons to challenge the normality assumption. Accordingly, you need some tools to investigate that assumption.

For discussion:

- There are several distributions which are bell-shaped but which are not normal. One such distribution is Student's *T* distribution which you will meet in detail in Chapter 9 and see its applications in later chapters.

- Are there datasets generated from your work with variables that are usually characterized—or at least regarded and accepted—as normal? How do you know that they *are* normal?

Testing data for normality

Numerous methods for testing normality may be found in the statistical literature. A commonly used "eye ball" method requires the use of special "normal probability paper." This technique, called *probability plotting* and described by Hahn and Shapiro (1967, pp. 261-294), tends to be both limited and subjective.

Among analytical methods (those involving objective data-based hypothesis-testing in contrast to subjective "eye-ball" techniques), some are superior to others in their sensitivity to departures from normality under different scenarios. Four of the more widely applied analytical methods are listed here:

- The *chi-squared goodness-of-fit test* compares a histogram of sample values with the frequencies of a related normal density function. For a development of the chi-squared goodness-of-fit test, see Dixon and Massey (1983, pp. 64-66).

- The *Anderson-Darling statistic* compares the sample distribution function with a normal distribution function. Two cases are considered: (1) both μ and σ have known values and (2) neither μ nor σ have known values. For development of the Anderson-Darling statistic, see Pearson and Hartley (1976, pp. 117-119).

- The *Lilliefors test for normality* compares the sample standardized distribution function with a standardized normal distribution function. For a discussion of the

Lilliefors test for normality, see Bowen and Bennett (1988, pp. 531-532).

- The *W test* is applicable when neither μ nor σ have known values. The calculations are lengthy but straightforward; their tedium is relieved with the use of a computer. For a development of the *W* test and procedures when $n \leq 50$, see Hahn and Shapiro (1967, pp. 295-298); if $n > 50$, see Madansky (1988, pp. 20-29). The *W* test is called an "omnibus test for normality" because of its applicability to sample sizes as small as 3 and its superiority to other procedures over a wide range of problems and data-connected conditions requiring the investigation of the assumption of normality. A perhaps unexpected feature of the *W* test is its use of lower quartiles as critical values. This is shown in here, based on the six steps given by Hahn and Shapiro and the example given by Bowen and Bennett (1987, pp. 532-535).

Example 7-1: Testing for normality with the W test

The following data are the percent uranium value for 17 cans of ammonium diuranate (ADU) scrap:

35.5	79.4	35.2	40.1	25.0	78.5
78.2	37.1	48.4	28.6	75.5	34.3
29.4	29.8	28.4	23.4	77.0.	

You wish to test these data for normality with $\alpha = 0.05$.

The *W* test is detailed here as an eleven-step procedure.

Step 1. Arrange the *n* sample observations in *ascending* order. Using established convention, for a sample $\{Y_1, Y_2, \ldots, Y_i, \ldots, Y_n\}$ of size *n*, let $Y_{(1)}$ denote the smallest observation, $Y_{(2)}$ the second smallest observation, ..., $Y_{(i)}$ the ith ordered observation, ..., and $Y_{(n)}$ the largest observation. Thus, $Y_{(1)} \leq Y_2 \leq \ldots \leq Y_i \leq \ldots \leq Y_n$. The ordered observations are given in the second column of Table 7-1, wherein the *rank* of an observation (in the first column) refers to its numeric-order assignment.

rank

Table 7-1:
The *W* test for normality applied to ADU scrap

Rank (i)	Ascending ordered data $y_{(i)}$	Descending ordered data $y_{(n-i+1)}$	Difference $y_{(n-i+1)} - y_{(i)}$	Table T-6a coefficients for $k = 8$ a_i	$a_i(y_{(n-i+1)} - y_{(i)})$
1	23.4	79.4	56.0	0.4968	27.8208
2	25.0	78.5	53.5	0.3273	17.5106
3	28.4	78.2	49.8	0.2540	12.6492
4	28.6	77.0	48.4	0.1988	9.6219
5	29.4	75.5	46.1	0.1524	7.0256
6	29.8	48.4	18.6	0.1109	2.0627
7	34.3	40.1	5.8	0.0725	0.4205
8	35.2	37.1	1.9	0.0359	0.0682
9	35.5				
10	37.1				
11	40.1				
12	48.4				
13	75.5				
14	77.0				
15	78.2				
16	78.5				
17	79.4				
					b = 77.1796

Step 2. If n is even, set $k = n/2$; if n is odd, set $k = (n - 1)/2$. For Example 7-1, where $n = 17$, set $k = 8$.

Step 3. Rearrange the observations in *descending* order of magnitude, and enter the first k of them as shown in the third column of Table 7-1.

Step 4. Calculate the differences between the corresponding entries of the third and the second columns of Table 7-1 and enter those differences in the fourth column.

Step 5. From Table T-6a, copy the k coefficients $\{a_1, a_2, \ldots, a_i, \ldots, a_k\}$ associated with sample size n into the fifth column of Table 7-1.

Step 6. Multiply the associated elements of the fourth and the fifth columns of Table 7-1. Enter the corresponding products in the sixth column of Table 7-1.

Step 7. Sum the sixth column of Table 7-1. Denote the sum by B. In Example 7-1, the sample value is b = 77.1796.

Step 8. Calculate S^2, the sample variance of the n observations. In Example 7-1, the sample value is $s^2 = 476.5968$.

Step 9. Calculate the test statistic, W, where

$$W = \frac{B^2}{(n-1)S^2}.$$

In Example 7-1, the sample value is

$$w = \frac{(77.1796)^2}{(16)(476.5968)} = 0.7811.$$

Step 10. From Table T-6b, obtain the critical point $w_\alpha(n)$ for the corresponding sample size and the appropriate level of significance. In Example 7-1, $\alpha = 0.05$, $n = 17$, and you find $w_{0.05}(17) = 0.892$.

Step 11. Compare w from **Step 9** to w_α in **Step 10**. If w is *smaller* than $w_\alpha(n)$, the hypothesis of normality is rejected. In Example 7-1, $w = 0.7811$ is less than $w_{0.05}(17) = 0.892$; thus, the data in this example yield sufficient evidence to reject normality.

For discussion:

- What would your conclusion be for Example 7.1 if $w > w_{0.05}$?
- In calculating the variance, s^2, for the sample, would you use the original data $\{y_1, y_2, \ldots y_n\}$ or the ordered data $\{y_{(1)}, y_{(2)}, \ldots, y_{(n)}\}$. Does it make a difference?
- Suppose you conduct a test and do not reject the hypothesis of normality. What can you say about the population from which the sample was drawn? What can you say about the sample itself? What can you say about the *next* sample you draw from that population? How comfortable are you with your answers? What would you write in a single-paragraph statement to convince your supervisor that you have a sample from a normal population?
- Measurements of some natural phenomena (like wind velocity) are believed to follow the *lognormal* distribution, in that the *logarithm* of the observations is distributed normally. Show how you would use the W test to test whether a set of observations is consistent with a lognormal distribution.

The Central Limit Theorem

The Central Limit Theorem is fundamental to many statistical procedures because it links many commonly encountered situations to the normal distribution. When its conditions are met, it provides powerful ways of answering a variety of statistical questions. This section leads you into the Central Limit Theorem through a series of plausibility arguments.

In preparation for the ensuing discussion which involves the tossing of several coins, we elect to call the value associated with a single toss a "mean." This idea leads to no ambiguity because the mean of a sample consisting of a single number is the number itself.

First, consider a coin-tossing experiment similar to those set out in Chapter 3. This time, you record your responses, using 1 if a head appears and 0 if a tail shows up. With repeated tossing of an unbiased coin, you would expect to score a "1" for half of the tosses and a "0" for the other half. These two means can be indicated by the set {0/1, 1/1}. In graphical form, this expectation is represented as a density function as shown in Figure 7-2. Thus, Figure 7-2 shows a density function that displays the probability of each possible value.

Figure 7-2:
Density function for the mean of a one-coin toss

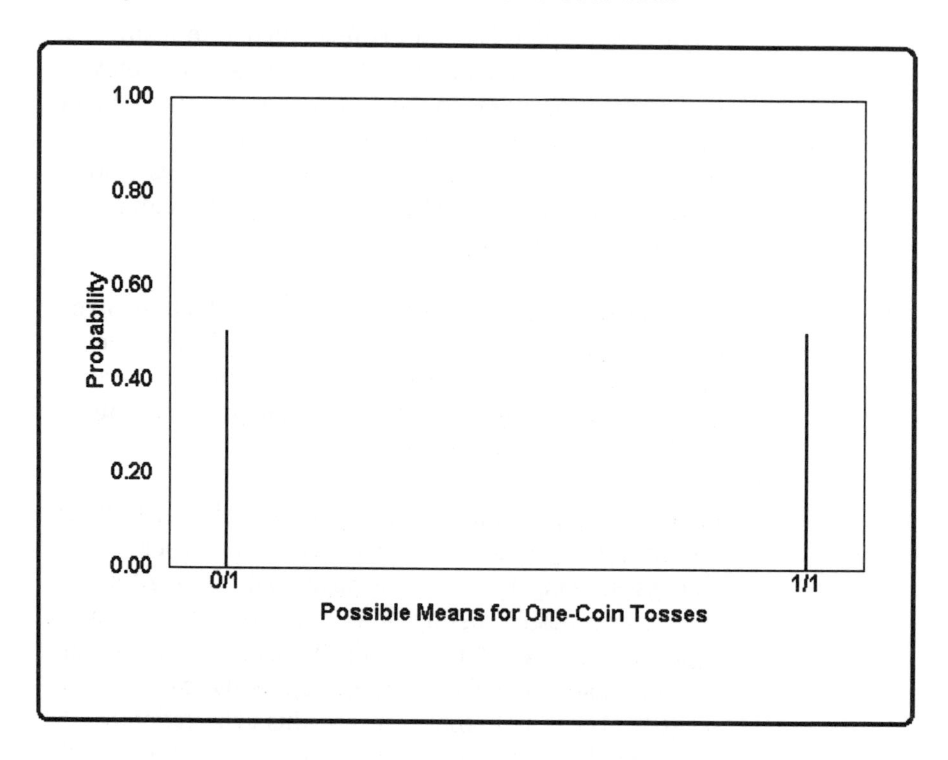

Next, suppose you toss three unbiased coins, each yielding a score of 0 or 1. The mean of these three scores can be characterized as belonging to the set {0/3, 1/3, 2/3, 3/3}. If you toss the three coins many times, each time calculating the described mean, those means will "pile up" in a pattern similar to the one given in Figure 7-3.

Figure 7-3:
Density function for the mean of a three-coin toss

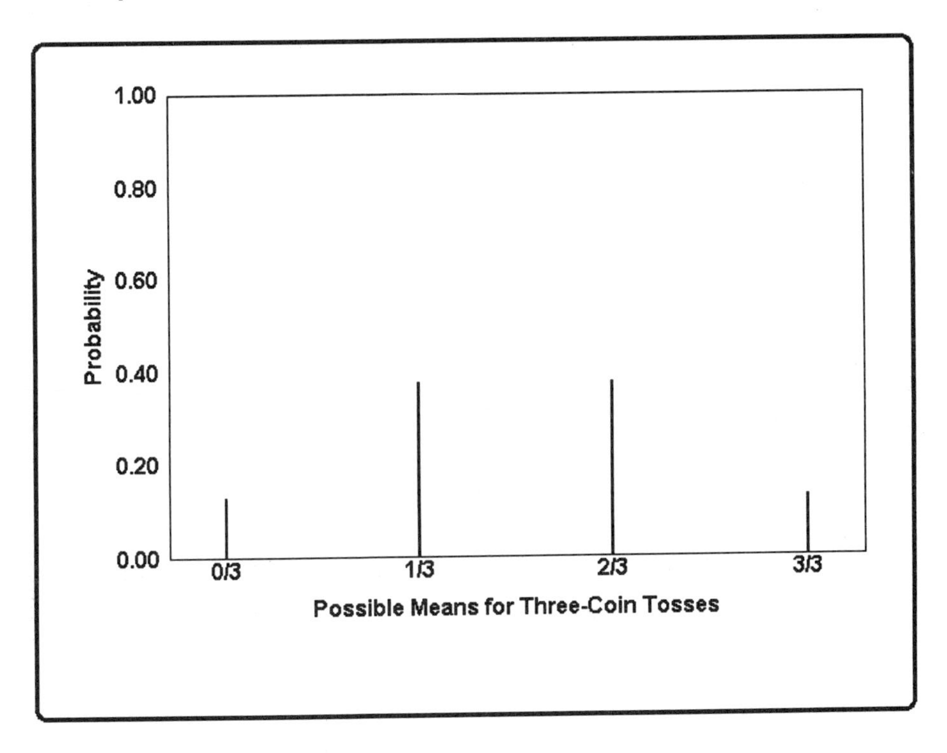

The pattern (i.e., the "density") in Figure 7-3 has several characteristics: It is centered around 0.5 (not surprisingly, the mean of 0 and 1); it is symmetric about 0.5; it peaks around the center; and it tapers off symmetrically from the center.

Next, suppose that you toss ten coins. The possible means are the set {0/10, 1/10, 2/10, 3/10, 4/10, 5/10, 6/10, 7/10, 8/10, 9/10, 10/10}. The characteristics that are displayed in Figure 7-3 are even more pronounced in a ten-coin toss, depicted in Figure 7-4.

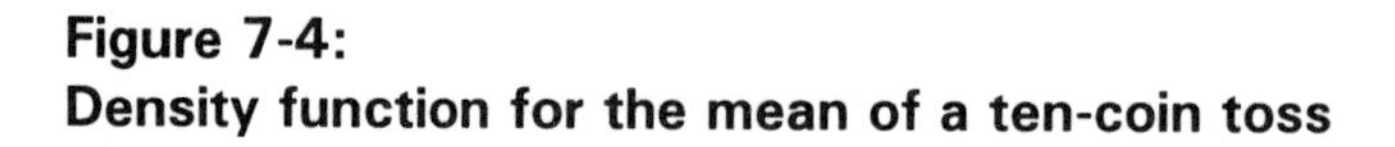

Figure 7-4:
Density function for the mean of a ten-coin toss

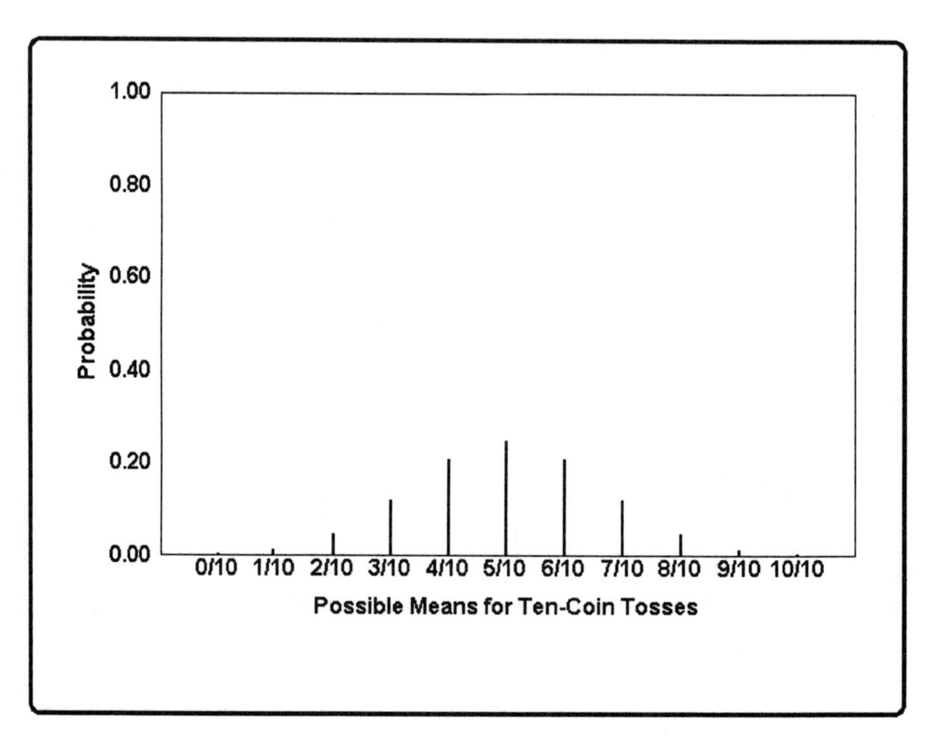

For discussion:

- Referring to Figures 7-2 through 7-4, which show results for small samples *from an obviously non-normal distribution*, you observe that, as the sample size increases:

  - ☐ The density function of the sample means tends toward a "bell-shaped" curve.
  - ☐ The mean of the sample means remains equal to the mean of the "parent" distribution.
  - ☐ As the sample size increases, the sample means "hug" their "parent" mean more tightly.

- Suppose you were to perform this experiment with 100 coins. What are the possible values of the sample means? What would you expect the "mean of the sample means" to be? Sketch the resulting density function. Did you use the same scale as that in Figures 7-2, 7-3, and 7-4? Any comment?

- In order to define a density function on a finite set of discrete points, the probabilities must add to one. How might you go about proving that this coin-tossing experiment yields a proper density function, no matter how many coins you use?

One formulation of the Central Limit Theorem and some interpretations

The Central Limit Theorem holds a singularly important place in the development of modern statistics because it ties the normal distribution to a vast variety of inferential methods. Here is one version of this important theorem:

***Suppose that random samples, each containing a fixed number,* n*, of measurements, are repeatedly drawn from a population with a mean* μ *and a standard deviation* σ*. If* n *is sufficiently large, the sample means will have a distribution that is approximately normal with a mean equal to* μ *and a standard deviation equal to* $\sigma/\sqrt{n}$.**

Three interpretations, of varying degrees of mathematical-statistical complexity, of the Central Limit Theorem are offered here. But first, your experimental situation must be established:

- You collect a sample of size n from some population of measurements. You do not have to know the parameters of this distribution.

- Whatever this population is, assume that it has a mean (call it μ) and a standard deviation (call it σ).

- The sample mean, $\overline{Y}_n$, is a random variable. Note: the subscript in the symbol $\overline{Y}_n$ serves as reminder that the mean is function of the sample size n.

- The sample size, n, is "large."

Now for the three interpretations of the Central Limit Theorem:

Interpretation 1: The sample mean, $\overline{Y}_n$, is distributed approximately as a normal variable having a mean μ and a standard deviation $\sigma/\sqrt{n}$.

Interpretation 2: The more observations, n, you have in the sample the better, because the sample mean, $\overline{Y}_n$, has a standard deviation of $\sigma/\sqrt{n}$, which becomes smaller as n becomes larger. Moreover, even if the original distribution is not known, thus making certain other inferences difficult if not impossible, the distribution of $\overline{Y}_n$ can be approximated by the normal distribution which has manageable characteristics.

Interpretation 3: $Y \sim ?(\mu, \sigma^2) \Longrightarrow \overline{Y}_n \sim N(\mu, \sigma^2/n)$, where the symbol "$\sim$" is read as "is distributed as," the ? indicates an unspecified distribution, the letter N designates the normal distribution, and the symbol $\Longrightarrow$ indicates "tends to."

The standard error of the mean

Let Y be a random variable with mean μ and standard deviation σ.

Let $\overline{Y}_n = (\Sigma Y)/n$, where all n of the Y values come from the same distribution as Y. Obviously, $\overline{Y}_n$ also is a random variable. Like any "decent" random variable, $\overline{Y}_n$ has a mean and a standard deviation. How do these random variables, Y and $\overline{Y}_n$, compare? Look at Table 7-2.

Table 7-2:
Comparing a single random variable with the mean of a sample of size *n*

Variable	Sample Size	Mean	Standard Deviation
Y	1	μ	σ
$\overline{Y}_n$	n	μ	$\sigma/\sqrt{n}$

standard error (of the mean)

The standard deviation of a *sample* mean often is called the *standard error of the mean*; when there is no ambiguity, it is simply the *standard error.*

The term *standardized variable* denotes a special transformation. This transformation is brought about by subtracting the variable's mean from the variable and dividing the result by the variable's standard deviation.

This is an important point:

> Regardless of the distribution of the random variable, the mean and variance of its standardized variable are *always* 0 and 1, respectively.

Applying the transformation to the sample mean yields:

$$Z_n = (\overline{Y}_n - \mu)/(\sigma/\sqrt{n}).$$

By the Central Limit Theorem, then, the standardized variable for the mean of a sample of size n, Z_n, is approximately distributed normally with mean 0 and standard deviation 1. The approximation improves with increasing sample size. In words, you say that "Z_n approaches $N(0, 1)$ as n approaches infinity."

A sampling exercise

This exercise is designed to reinforce some of the concepts associated with the Central Limit Theorem. The strategy for the exercise is to first generate several random samples of a fixed size from a distribution with known population mean (μ) and standard deviation (σ). The next step is to calculate the mean and the variance for each of these samples and to compare those statistics to the known population parameters. Finally, we investigate the variability of the sample means to show that the average of the means is quite close to μ and that the standard deviation of the sample means is approximately $\sigma / \sqrt{n}$, as suggested by the Central Limit Theorem. As you recognize, this exercise doesn't *prove* the Central Limit Theorem; but we are confident that, with a sufficiently large number of samples, the flavor of the theorem will come through.

For this exercise we ask that you pick 10 numbers, each of which is made of three random digits between 0 and 9. Hence, each of the 10 numbers you pick will be a random number between, and including, 000 and 999. One way of picking those digits is to "throw darts" at Table T-12, or you might follow the procedure suggested at the explanatory notes to that table. Or you may use a calculator with a built-in random number generator. Such a generator typically generates numbers between 0 and 1. The function key for this operation is usually labeled "RND" or "RAND." In order to get three-digit random integers from 000 to 999, multiply the number obtained from your calculator by 1000 and discard the fractional part. Thus, if the number generated is 0.12345, multiply that number by 1000 to get 123.45, discard the fraction (.45), and record 123.

Now write down your ten 3-digit numbers:

1. _______ 2. _______ 3. _______

4. _______ 5. _______ 6. _______

7. _______ 8. _______ 9. _______

10. _______

Determine and record the following statistics for your 10 values:

mean _________

minimum (the smallest number) __________

maximum (the largest number) __________

range _______________

variance ________________

standard deviation ____________

Now that you have completed the exercise, here is some information about the population you just sampled. From mathematical considerations, the mean of that population is $\mu = 499.5$, the variance is $\sigma^2 = 83{,}166.75$, and the standard deviation is $\sigma = 288.39$. Your own sample statistics ought not to be too far from these parameter values.

If you did this sampling as a part of a class exercise, give your results to your instructor. This will enable you and your colleagues to give credence to the statements that

- the average of serves $\overline{Y}_n$ is μ
- the average of S^2 is σ^2
- the variance of $\overline{Y}_n$ is $\sigma/\sqrt{n}$
- $\overline{Y}_n$ is distributed normally.

For discussion:

- In performing the sampling exercise, we collect several means, each of which is based on 10 observations from a "flat" distribution— usually called a *uniform distribution* because each of the possible responses (000 through 999) is equally likely to be selected. In the following bulleted items, you will investigate the behavior of these means and attempt to see if they support the Central Limit Theorem and if they agreement with the Empirical Rule introduced in Chapter 5. For the purpose of discussion, we denote the number of participants in this exercise by *k*.

- Examine the *k* sample means generated in the sampling experiment. Whereas individual observations can be very small (say, less than 100) or very large (say, more than 900), you should be surprised if a mean of 10 observations is less than 100 or larger than 900. Are such means possible?

- Denote the average of the individual means by $\overline{\overline{Y}}$. Since each mean is constructed for the same number of observations, that average equals the mean of all the 10k observations collected in class and is called the *grand mean*. (Does that imply that the mean of means is the meanest of them all?). The grand mean may not necessarily be closer to μ than any one of the individual sample means, but you may be surprised at how close it gets. In fact, the Central Limit Theorem, coupled with other statistical tools, can give you a reasonable assurance that the grand mean will not differ from μ by more than $179/\sqrt{k}$, where k is the number of participants in the class. Thus, with 25 students in the class, we are "reasonably assured" that the samples' grand mean does not differ from μ by more than 36.

- What is meant by the term "reasonably assured" in the previous bullet?

- Note how close the standard deviation for each sample is to $\sigma = 288.39$. Average those standard deviation values, and determine how close this average is to 288.39.

- Sketch a histogram of the k sample means. You are likely to get a graph that would remind you of the normal distribution. The degree of similarity to the normal distribution would increase if sample size were larger.

- The histogram thus constructed gives you an empirical distribution of the mean of a sample of size 10 from a specified distribution. You can now estimate measures of location and dispersion for this distribution. This sampling approach to estimation of population parameters is one—an extremely simple one—application of the *Monte Carlo Method.* As the name suggests, Monte Carlo methodology was developed in connection with gambling, where the distribution of possible outcomes was determined by multiple runs of a roulette wheel. The method gained wide currency during World War II when it was applied to such important problems as bomb-sight accuracy and nuclear fission control. Suggested references are Hammersly and Hanscomb (1964) and Shreider (1966).

Using Table T-1: *The cumulative standardized normal distribution and selected quantiles*

The purpose of any normal distribution table is to support and simplify the calculations of probabilities associated with a random variable Y which is distributed normally with mean μ and a standard deviation σ. Since there are infinite combinations of values that μ and σ may assume, you could, in principle, have an infinite number of tables to consider.

Alternatively, and much less dauntingly, you can standardize Y by subtracting μ from Y—yielding $(Y - \mu)$ which is a variable with zero mean—and then dividing $(Y - \mu)$ by σ—yielding $(Y - \mu)/\sigma$, which is a variable with unit variance. As a result, you produce a variable with mean 0 (zero) and standard deviation 1 (one). The variable $Z = (Y - \mu)/\sigma$ is called a standardized normal variable. Using the convention introduced earlier, you can write $Z \sim N(0,1)$.

Remember: In this convention, the second number denotes the variance, *not* the standard deviation. Of course, apart from the units of measurement, the variance and the standard deviation of the standardized normal are numerically the same, i.e., $\sigma = \sigma^2 = 1$.

The standardized normal distribution, displayed in Figure 7-1, is a probability distribution. As such, it has several properties:

- The area under the density function is 1.

- The probability that Z is less than or equal to a specified value, say z, $Pr\{Z \leq z\}$ is the total area under the curve for the region where $-\infty < Z < z$. This "cumulative probability" is denoted by

  $$F(z) = Pr\{Z \leq z\}.$$

- You may also write

$$F(z) = Pr\{Z < z\}$$

because

$$F(z) = Pr\{Z \leq z\} = Pr\{Z < z\} + Pr\{Z = z\}$$

and $Pr\{Z = z\}$ is identically 0 for continuous distributions.

- The probability that Z is between two specified values, say a and b, where $a < b$, is

$$Pr\{a < Z < b) = F(b) - F(a) = Pr\{a \leq Z \leq b).$$

To make these ideas even more concrete, turn now to Table T-1: *The cumulative standardized normal distribution and selected quantiles*. Notice that, although a standardized normal extends from $-\infty$ to $+\infty$, only positive values of z are available in the table. However, this is not a problem because the normal density function is symmetric around its mean; i.e., the area to the left of the mean is "mirrored" in the area to the right of the mean. Symbolically, for the standardized normal distribution, $F(z) = 1 - F(-z)$. The use of the Table T-1 for calculating probabilities is illustrated with three examples:

Example 7-2:
Find the probability that a random variable Y from a normal distribution with mean $\mu = 75$ and standard deviation $\sigma = 8$ will not be larger than 91

You will find it valuable to sketch a rough drawing of the problem as stated and the corresponding standardized normal distribution. You don't have to be an artist—indeed, as you see from our drawings, we're not artists either. Your sole purpose in making such a sketch is to emphasize the area, or probability, of interest (as indicated in Figure 7-5).

Figure 7-5:
Sketch for Example 7-2

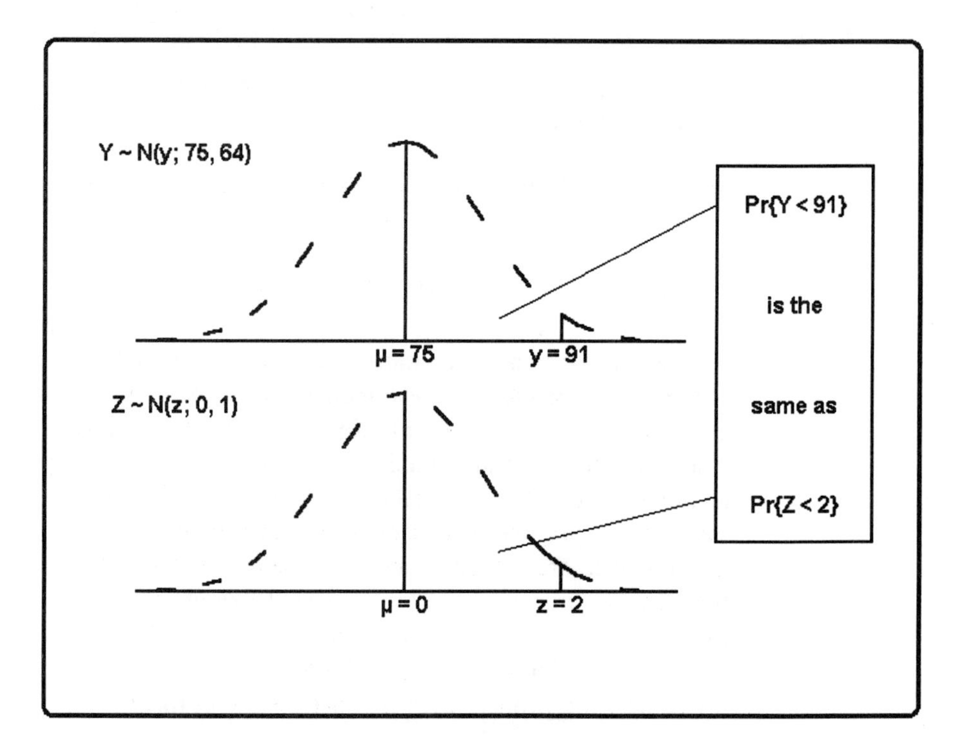

Now you can transform the problem mathematically:

$$Pr\{Y \leq 91\} = Pr\{Y < 91\}$$
$$= Pr\{\frac{Y - \mu}{\sigma} < \frac{91 - 75}{8}\}$$
$$= Pr\{Z < 2\}.$$

Next, consult Table T-1. The left margin lists values of z from 0.0 to 3.4 in increments of 0.1, whereas the column headings list numbers from 0.00 to 0.09. For $z = 2.00$, find the intersection of the row beginning with 2.0 and the column headed by 0.00, where the table entry is 0.9772. This means that the probability of Z being less than or equal to 2.0 is 0.9772. Therefore, in answer to the posed problem, $Pr\{Y \leq 91\} = 0.9772$.

Example 7-3:
If Y ~ N(100,16), find Pr(Y > 105)

Start with a sketch of the problem, similar to that displayed in Figure 7-6.

Figure 7-6:
Sketch for Example 7-3

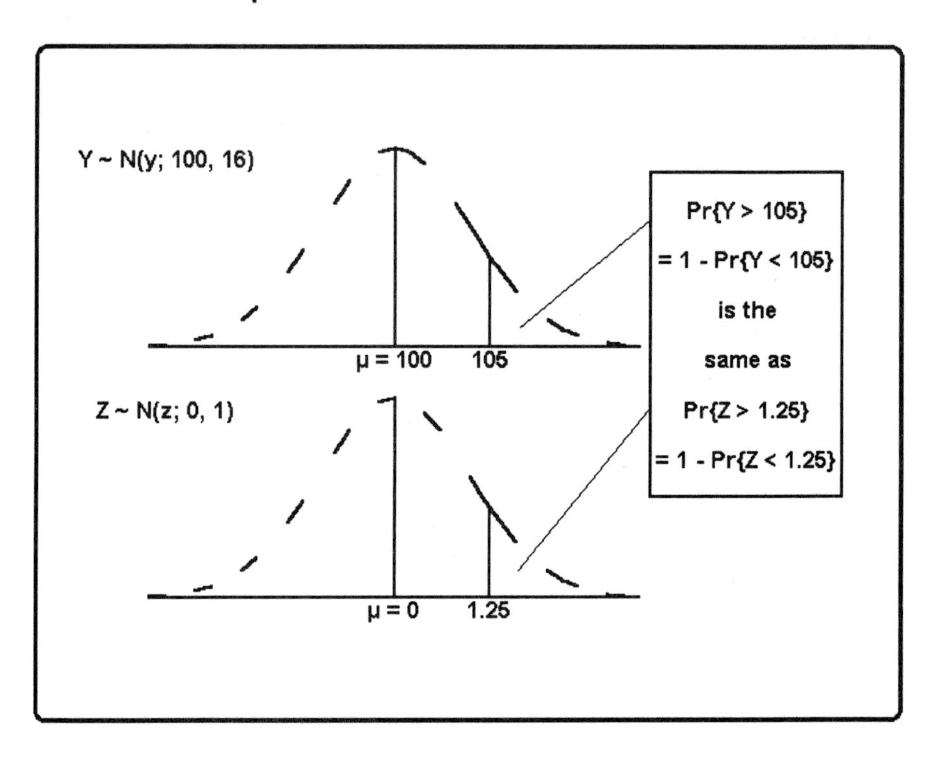

The mathematical transformation from Y to Z is:

$$Pr\{Y > 105\} = Pr\{\frac{Y - \mu}{\sigma} > \frac{105 - 100}{4}\}$$

$$= Pr\{z > 1.25\}.$$

Consult Table T-1. At the intersection of the row beginning with 1.2 and the column of 0.05, you will find the table entry of 0.8944. This number is the probability that Z is less than or equal to 1.25. Hence the probability that Z is larger than 1.25 is 1.0000 - 0.8944 = 0.1056. Hence $Pr\{Y > 105\}$ is also 0.1056.

Example 7-4:

Let $\bar{Y}$ be the mean of 5 observations from* N*(86.8, 1.21). Find the probability that $\bar{Y}$ will be less than or equal to 86.0

Start your analysis with a sketch of the desired probability as shown in Figure 7-7.

Figure 7-7:
Sketch for Example 7-4

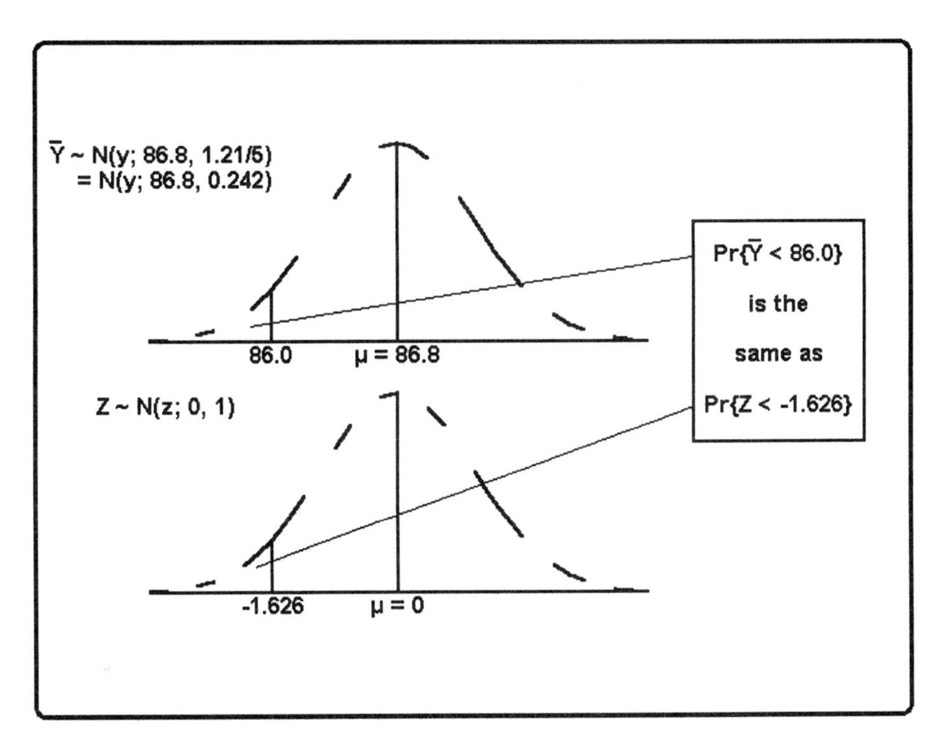

The standard deviation of Y is 1.10. The standard deviation of $\bar{Y}$ is $1.10/\sqrt{5}$. The requested probability is therefore written as

$$Pr\{\bar{Y} \leq 86.0\} = Pr\{\bar{Y} < 86.0\}$$

$$= Pr\{\frac{\bar{Y} - \mu}{\sigma}\sqrt{n}\} < \frac{86.0 - 86.8}{1.1/\sqrt{5}}\}$$

$$= Pr\{Z < -1.626\}$$

There are no negative values for Z in Table T-1. You resort to the symmetry of the normal distribution and soon realize that $Pr\{Z < -1.626\} = Pr\{Z > 1.626\}$ $= 1 - Pr\{Z \leq 1.626\}$.

But Table T-1 does not give you $\{(Z \leq 1.626\}$ directly, so some table interpolation is required. From the Table T-1 you find $Pr\{Z \leq 1.62\} = 0.9474$ and $Pr\{Z \leq 1.63\} = 0.9484$. Consequently, $Pr\{Z \leq 1.626\}$ must lie somewhere between 0.9474 and 0.9484, and a little closer to 0.9484 than to 0.9574. Interpolating linearly between these two numbers yields $Pr\{Z \leq 1.626\} = 0.9480$. Putting it all together, you have: $Pr\{\bar{Y} \leq 86.0\} = Pr\{\bar{Y} < 86.0\}$ $= Pr\{Z < -1.626\} = Pr\{Z > 1.626)$ $= 1 - Pr\{Z \leq 1.626\} = 1 - 0.9480 = 0.0520$.

The selected quantiles. Several specific quantiles of the standardized normal distribution are common in statistical investigations. You often will encounter two of them, the 0.95 and the 0.975 quantiles, in this text; they are included at the bottom of Table T-1 as 1.645 and 1.960, respectively.

What to remember about Chapter 7

Chapter 7 explored the single most important concept in modern statistics—the *normal distribution*. Specific topics included:

- *the normal distribution and density functions*
- *the standardized normal distribution*
- *testing data for normality*
- *the standard error of the mean*
- *the Central Limit Theorem.*

The chapter concluded with a demonstration of a remarkable feature of the normal distribution: a single-page tabulation (Table T-1) can be used to quantify nearly everything you need to know about the normal distribution.

Statistical estimation

What to look for in Chapter 8

Chapter 8 examines the general process of *statistical estimation*; that is, it examines the question: "How do you put a number on it?" Among the special terms employed here, you will find:

- *parameter*
- *estimator*
- *estimate*
- *unbiased estimators*
- *minimum variance*
- *point estimators*
- *interval estimators*.

The chapter concludes with discussions of :

- *confidence intervals for a mean*
- *tolerance limits for a normal distribution*
- *confidence intervals for a variance*

and illustrates simple cases of sample size determination for normally distributed data.

On the concepts of statistical estimation and inference

The principal objectives of statistics are often characterized as *estimation* and *inference*.[1] These two objectives are strongly interrelated, a condition that often leads to their being confused. In their practical aspects, they both involve the collection, analysis, and interpretation of data; in their theoretical bases, they both stem from mathematical formulations of statistical issues. But they *must* be distinguished to bring out their special individual powers.

Consider these two statements:

Estimation

Estimation involves the processes brought to bear on a problem of quantification; i.e., "How do you put a number on it?"

Inference

Inference involves the processes brought to bear on a problem of decision; i.e., "Based upon these data, should you or shouldn't you take a particular action?"

This chapter is devoted to estimation processes; inferential processes are described in Chapter 9. Once established, both concepts are explored in a variety of application areas in the remainder of the book.

[1] The more you work with statistical concepts, the more you will find that two distinct approaches to statistical estimation and inference dominate the literature: *classical statistics* and *Bayesian statistics*. Because the distinction is so basic and because, at the time of this writing, no fully satisfactory ecumenical bridges exist between these two schools of statistical thought, we choose to focus this book on classical statistical processes. You will find presentations of Bayesian processes, for example, in Press (1988).

Getting started with statistical estimation

Statistical estimation is based upon three principal elements:

A *parameter* is a quantity you wish to estimate.

An example of a parameter is a population mean, denoted in many formulations by the symbol μ. (This book designates parameters by Greek letters wherever possible.)

An *estimator* is a rule (usually expressed as a formula) for obtaining a numerical value to estimate a parameter of interest.

An example of an estimator is the sample mean, $\bar{Y} = (\sum Y)/n$, which may be used to estimate the parameter μ. (This book designates estimators by upper-case English letters wherever possible.)

An *estimate* is the numerical value obtained by applying the estimator to the sample data.

An example of an estimate is the *calculated* sample mean. Thus, for the three sample values {3, 7, 2}, the estimate of the parameter μ is $\bar{y} = (\sum y)/n = (3 + 7 + 2)/3 = 4$. (This book designates estimates by lower-case English letters wherever possible.)

An estimator is a tool used in an inference about a population parameter. An estimator is a random variable because it is a function of random variables. Thus, an inference about the parameter is not absolute. However, from statistical theory, you may know something about the behavior of estimators like the one you are using. It is only in the context of the estimator's distribution (i.e., how the estimator behaves when it is used over and over again) that you can make any proper statistical inference.

There are two major categories of classical statistical estimators: *point estimators* and *interval estimators*.

Point estimators

A ***point estimator*** is a rule that produces a single number to be used as an estimate of a population parameter. Consider these examples:

- A sample mean is a point estimator of a population mean.
- A sample standard deviation is a point estimator of a population standard deviation.
- The sample proportion of defects is a point estimator of the population proportion of defects.

Statistical researchers have developed a number of desirable properties of estimators and use these properties to decide among several candidate estimators. This discussion is restricted to two important properties and their combination: *unbiased*, *minimum variance*, and *unbiased minimum variance*.

unbiased

A point estimator is said to be *unbiased* if, on the average, it equals the parameter which it estimates.

minimum variance

A point estimator is said to have *minimum variance* if it has a smaller variance (or, equivalently, a smaller standard deviation) than other estimators of the same class.

minimum variance unbiased

A point estimator is said to be *minimum variance unbiased* if it has both properties; that is, it is both *unbiased* and has *minimum variance*.

For discussion:

- The properties of unbiasedness and minimum variance are applied to an estimator, not to an estimate. Why?

- The statistic S^2 can be shown to be an unbiased estimator of a population's variance σ^2. What, therefore, can you say about the long-run average of S^2?

- The statistic S can be shown to be a biased estimator of a population's standard deviation σ. What, therefore, can you say about the long-run average of S?

- Is an unbiased estimator accurate? Is it precise?

- A tricky issue: Is a minimum variance unbiased estimator accurate? Is it precise?

- In technical discussions of varying degrees of seriousness, you may hear someone make a statement something like this: "This is the best estimate of so-and-so." What do you suppose the speaker means by this phrase? Kendall and Stuart (1971, pp. 13-14) give this advice:

> The estimation of population parameters from information provided by the sample raises the question whether there is a "best" estimator. The answer depends mainly on the criteria which are laid down as to the "goodness" of the estimator. If there is a criterion which distinguishes one of two estimators as better than the other and if there exists an estimator which is better than any other, it is said to be the best.
>
> Various criteria [for "best estimator"] have been suggested..... It is not always true that a "best" estimator exists.

Interval estimators

An *interval estimator* is a rule that produces numerical bounds on a population parameter (or on a function of the population parameters). These bounds usually are reported as an interval determined by two numbers, neither of which is necessarily finite. Two such intervals are discussed: *confidence intervals* and *tolerance intervals*.

Confidence intervals are perhaps the most widely used type of interval estimators. They are formed according to a set of basic statistical principles so that they focus on a parameter of interest *and* a probabilistic statement that conveys a measure of assurance that the interval contains that parameter of interest. This probabilistic measure of assurance is called the *confidence coefficient*.

confidence coefficient

Note that synonyms for the term "confidence coefficient" abound in statistical literature, including *confidence level* and *level of confidence*. *Confidence coefficient* and *confidence level* are used interchangeably in this book.

confidence level

Note further that various writers, and we are no exception, use a variety of notations and expressions in quantifying the confidence coefficient. For instance, you sometimes will see the phrase "95% confidence" and sometimes the phrase "0.95 confidence." Sometimes you will see tests of significance (introduced in Chapter 9) conducted at some level of confidence.

Confidence intervals support the following types of statements:

> We are 95% confident that the population mean is between 8.4 and 10.1 kilograms.

We are 75% confident that the mean-time-before-failure (MTBF) of a population of electric motors is at least 8,000 hours.

We are 99.99% confident that the percentage of plutonium, by weight, in a batch of PuO_2 (plutonium oxide) is between 87.16 and 88.02.

We are 91% confident that the proportion of cables traced to control unit A is between 7% and 11.5%.

Statistical tolerance intervals serve an entirely different purpose than confidence limits. Statistical tolerance intervals are constructed when you wish to place a bound (or bounds) on a specified portion of a population. Such bounds have important application, as, for example, when you want to find limits within which the "bulk" of the population lies and you need assurance that only a small fraction of the population falls outside those limits.

Look at it this way: If you have the entire population at your disposal, your problem is trivial; for limits that contain the middle 90% of the population, sort the population in ascending order and find the 0.05 and 0.95 quantiles (5th and 95th percentile, respectively). The quantiles thus found contain the middle 90% of the population with 100% confidence.

But you rarely have the entire population, and you settle for a sample. From that sample you calculate the sample mean and the standard deviation, and then proceed to calculate *statistical tolerance limits* that you believe are likely to contain the stated proportion of the population. Armed with these limits, you will be able to make statements with the following flavors:

We are 95% confident that 80% of the population lies between 3.52 cm and 3.84 cm.

We are 90% confident that 99% of our employees earn between $39,014 and $57,755 per annum.

> We are 99% confident that the annual per-house savings from residential energy conservation work is between 156 and 1,942 kilowatt-hours.

These examples are clearly two-sided statistical tolerance limits. One-sided statistical tolerance limits have a similar flavor. Although many two- and one-sided statistical tolerance limits are built around a mean and a standard deviation calculated from a sample, you may have occasion to use nonparametric tolerance limits. Here is an all-encompassing definition given by Kendall and Buckland (1971, p. 145):

> **Statistical Tolerance Limit** An upper, and lower, value of a [random variable] following a given distribution form between which, it is asserted with confidence β, a proportion γ will lie.

For discussion:

- In technical discussions of varying degrees of seriousness, you may hear someone say something like this: "I'm 99% confident that so-and-so is true." What do you suppose is intended by this assertion?

- Hahn and Meeker (1991) use the generic term *statistical intervals* to cover interval estimation processes. To confidence intervals and statistical tolerance intervals, they add a third category: prediction intervals. All three categories are important. When you face interval-related problems, be sure you use the correct type. Hahn and Meeker (1991) will make your task easier. To help keep these intervals straight, compare these (slightly modified for consistency) definitions by Kendall and Buckland (1971, p.30, p. 145, and p. 117):

> **Confidence Interval** If it is possible to define two statistics t_1 and t_2 (functions of sample values only) such that, θ being a parameter under estimate, $Pr(t_1 \leq \theta \leq t_2) = 1 - \alpha$, where α is some fixed probability, the interval between t_1 and t_2 is called a confidence interval. The assertion that θ lies in this interval will

be true, on the average, in a proportion (1 - α) of the cases when the assertion is made.

Statistical Tolerance Limit An upper, and lower, value of a [random variable] following a given distribution form between which, it is asserted with confidence β, a proportion γ will lie.

Prediction Interval The interval between the upper and lower limits attached to a predicted value to show, on a probability basis, its range of error.

Confidence intervals for a mean

An Easy Case (with Strong Assumptions)

A confidence interval about the mean of a normal population is developed under the following three assumptions:

- The confidence coefficient is 0.95.
- A two-sided confidence interval about the population mean, μ, is desired.
- The population variance, σ^2, is known.

Consult the standardized normal table, Table T-1. Observe from the table that 95% of the standardized normal distribution lies between -1.96 and +1.96. Alternatively, the number +1.96 cuts off 2.5% of the area under the curve to the right and the number -1.96 cuts off 2.5% of the area to the left. This means that

$$Pr\{-1.96 < Z < 1.96\} = 0.95,$$

where Z is a standardized variable; i.e., $Z = (Y - \mu)/\sigma$. For a sample of size $n = 1$, this yields:

$$Pr\{-1.96 < \frac{Y - \mu}{\sigma} < 1.96\} = 0.95$$

while, for a sample of size n, the following expression holds:

$$Pr\{-1.96 < \frac{\bar{Y} - \mu}{\sigma/\sqrt{n}} < 1.96\} = 0.95.$$

Now manipulate the inequality in the brackets to obtain this sequence:

$$-1.96\sigma/\sqrt{n} < \bar{Y} - \mu < 1.96\sigma/\sqrt{n}$$

$$1.96\sigma/\sqrt{n} > \mu - \bar{Y} > -1.96\sigma/\sqrt{n}$$

$$\bar{Y} + 1.96\sigma/\sqrt{n} > \mu > \bar{Y} - 1.96\sigma/\sqrt{n}.$$

One more reversal of the inequalities yields the conventional form of a confidence interval (i.e., with the smaller end at the left and the larger end at the right):

$$\bar{Y} - 1.96\sigma/\sqrt{n} < \mu < \bar{Y} + 1.96\sigma/\sqrt{n}.$$

Hence, a 95% confidence interval for the mean of a normal distribution, μ, based upon a sample of size n when the standard deviation, σ, is the interval

$(\bar{Y} - 1.96\sigma/\sqrt{n}\,,\ \bar{Y} + 1.96\sigma/\sqrt{n})$, which sometimes is written as $\bar{Y} \pm 1.96\sigma/\sqrt{n}$.

Example 8-1:
Beam momentum measurements from Frodesen,* et al. *(1979, p. 141)

Measurements on the momentum of monoenergetic beam tracks on bubble chamber pictures have led to the following sequence of 10 readings in units of GeV/c: 18.87, 19.55, 19.32, 18.70, 19.41, 19.37, 18.84, 19.40, 18.78, and 18.76. Assume that this sample originated from a normal distribution with a known standard deviation of σ = 0.3 GeV/c. Find a 95% confidence interval for the beam momentum, μ.

From the sample values, you calculate the mean $\bar{y}$ = 19.100. Since σ is given, there is no need (nor license) to calculate and use the sample standard deviation. The 97.5th percentile of the normal distribution (from Table T-1) is 1.960. The lower and upper 95% confidence limits for μ, respectively, are therefore calculated as:

$$19.100 - 1.960(0.300/\sqrt{10}\,) = 18.914$$
$$19.100 + 1.960(0.300/\sqrt{10}\,) = 19.286.$$

Notice especially that the resulting interval, (18.914, 19.286), has a fixed length, no matter what the sample values are. Only the center of the interval changes from sample to sample.

The next easy cases—relaxing the assumptions

Selecting the confidence coefficient. The first assumption calls for a 95% confidence coefficient. As mentioned earlier, your confidence coefficient can be any value between, and including, 0% and 100%. Thus, if your confidence coefficient is 100(1 - α)%, consult a standard normal table to find $z_{1-\alpha/2}$ and then compute $\bar{Y} \pm z_{(1-\alpha/2)}\sigma/\sqrt{n}$ to get the endpoints of the corresponding interval for the mean of the population. For example, for a confidence coefficient of 0.99, $z_{(1 - 01/2)} = z_{0.995} = 2.576$; this results in a 99% confidence interval for the mean, μ, being constructed from $\bar{Y} \pm 2.576\sigma/\sqrt{n}$. Similarly, for a 0.90 confidence coefficient, use $z_{1 - 0.10/2} = z_{0.95}$ $= 1.645$, resulting in a 90% confidence interval for the mean, μ, being constructed from $\bar{Y} \pm 1.645\sigma/\sqrt{n}$.

Selecting a one-sided confidence interval. The second assumption calls for a two-sided confidence interval. To construct a one-sided confidence interval, use $z_{(1 - \alpha)}$, the value from Table T-1 to determine the 100(1-α) quantile of the distribution. Note the switch here from $z_{(1 - \alpha/2)}$ to $z_{(1 - \alpha)}$. Thus, you use $z_{(1 - \alpha)}$ if an upper one-sided confidence bound is sought and $-z_{(1 - \alpha)}$ if a lower one-sided confidence bound is desired. For example, $z_{\alpha} = 1.645$ is the multiplier of $\sigma/\sqrt{n}$ for a 95% upper one-sided confidence limit about the mean, and $z_{\alpha} = -1.645$ is the multiplier of $\sigma/\sqrt{n}$ for a 95% lower one-sided confidence limit about the mean. Accordingly, a one-sided *upper* 95% confidence limit is $\bar{Y} + 1.645\sigma/\sqrt{n}$, and a one-sided *lower* 95% confidence limit is $\bar{Y} - 1.645\sigma/\sqrt{n}$.

Selecting the value of the standard deviation. The third assumption states that the value for σ is known. Lacking that knowledge, you use the sample standard deviation, S, which is based on $df = (n - 1)$ degrees of freedom. However, instead of

Student's T *distribution*

on $df = (n - 1)$ degrees of freedom. However, instead of using the standard normal table's quantiles, you now use a table of *Student's* T *distribution*,[2] whose quantiles are given in Table T-3. Look in Table T-3 for the cell at the intersection of the row corresponding to *df*, associated with *S* and the column with the desired confidence coefficient, $(1 - \alpha)$, to find the required value, $t_{(1-\alpha)}(df)$. Thus, for a one-sided upper 95% confidence interval on μ when $n = 12$ ($df = 11$), find $t_{0.95}(11) = 1.806$ and compute $\bar{Y} + 1.80s/\sqrt{n}$. For a *two*-sided 80% confidence interval for μ when $n = 15$ ($df = 14$), use $t_{(1-\alpha/2)}(df) = t_{0.90}(14) = 1.35$ and compute $\bar{Y} + 1.35s/\sqrt{n}$.

Example 8-2:
Reconsidering Example 8-1: The beam momentum measurements example from Frodesen, et al. _(1979, p. 141), this time when σ is not known and must be estimated_

Calculate the sample standard deviation and obtain $s = 0.335$. From Table T-3, obtain the associated 97.5th percentile for 9 degrees of freedom: $t_{0.975}(9) = 2.26$. The respective lower and upper confidence limits for μ are

$$19.100 - 2.26\,(0.335/\sqrt{10}) = 18.861$$

and

$$19.100 + 2.26\,(0.335/\sqrt{10}) = 19.339.$$

In contrast to the case in which the standard deviation is known, intervals constructed by this process have lengths that depend upon the data.

---

[2] Developed by W. S. Gossett (1876-1937), an English experimenter/brewmeister with Guinness in Dublin, who published under the pen name "Student."

Interpreting confidence intervals

Suppose you collect a sample of size n and construct a 50% two-sided confidence interval about the mean. If σ is known, then the desired interval is constructed by

$\bar{Y} \pm 0.674\sigma/\sqrt{n}$. Since $\bar{Y}$ is the only term that can change from sample to sample in this expression, the constructed confidence interval is necessarily a random interval of fixed length, namely $(2)(0.674)\sigma/\sqrt{n}$. But any given interval may or may not contain μ. What, then, are you to make of the behavior of such random intervals?

From theoretical considerations, we know that if we were to repeat the experiment many times under identical conditions, each time constructing another confidence interval, we would expect 50% of those intervals to contain μ. This is shown schematically in Figure 8-1 for 100 samples, each of which is produced as a 50% confidence interval on the unknown mean.

Figure 8-1:
Schematic display of 100 individual 50% confidence intervals covering μ when σ is known

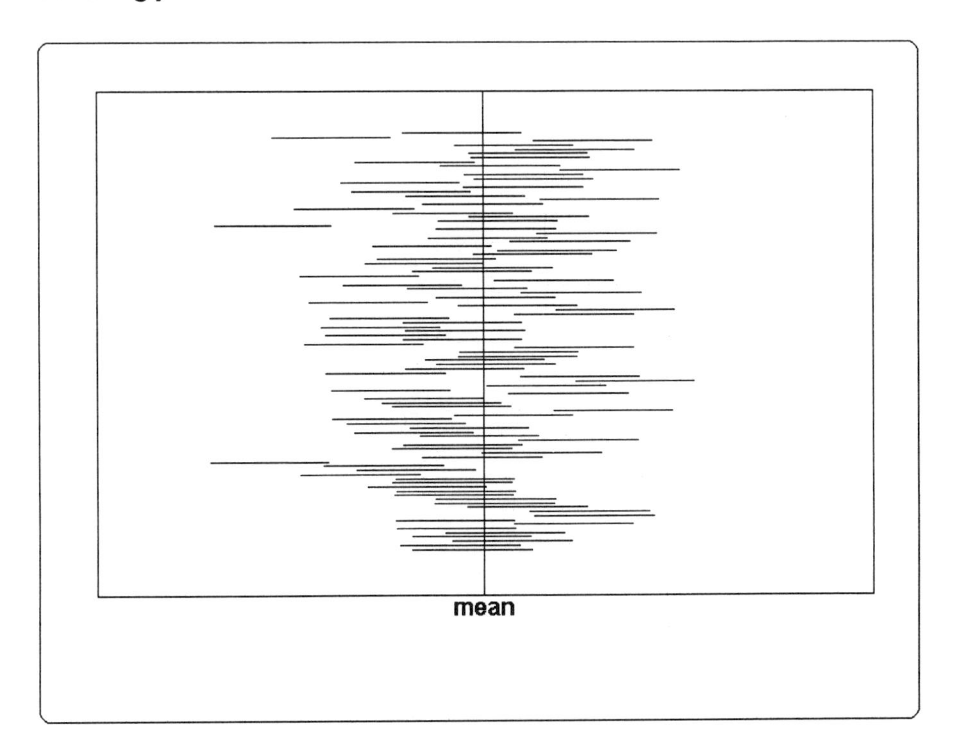

If σ is not known, then the 50% confidence interval is given by $\bar{Y} + tS/\sqrt{n}$, where $t = t_{0.50}(df)$ is obtained from Table T-3 with $df = n - 1$ degrees of freedom. Now both $\bar{Y}$ and S are random variables, yielding confidence intervals of different lengths. A typical replication of such intervals may look like Figure 8-2.

Figure 8-2:
Schematic display of 100 individual 50% confidence intervals covering μ when σ is unknown

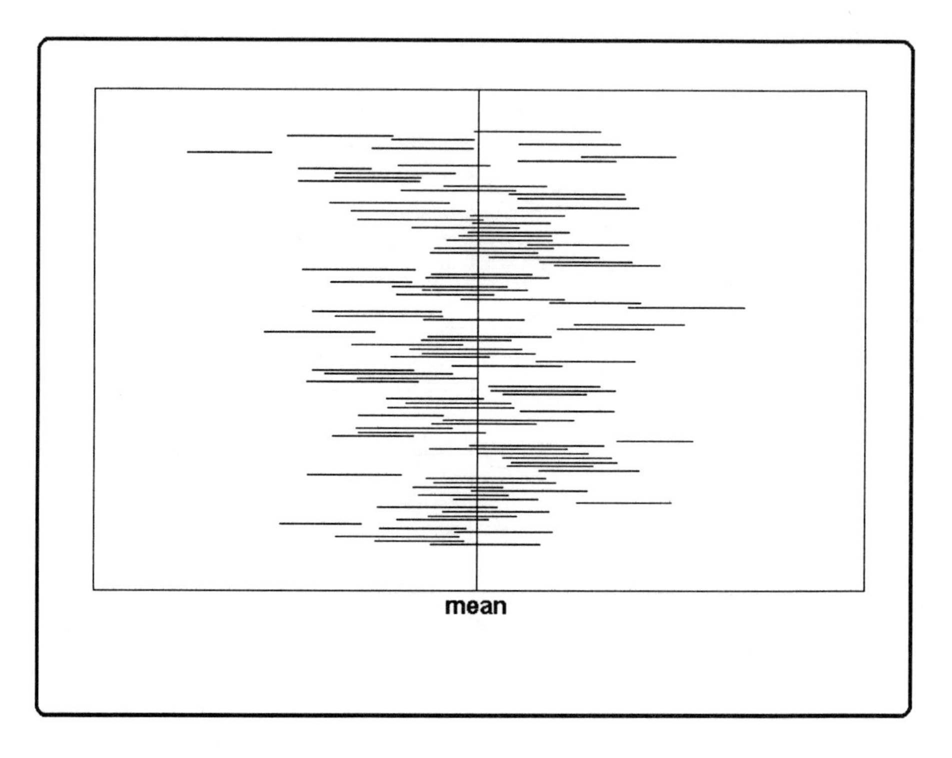

One more time: theoretical considerations dictate that, in the long run, 50% of similarly constructed intervals contain the true mean, μ.

For discussion:

- Once you construct a confidence interval, you must articulate your conclusions correctly. The following statements, made with respect to the mean momentum measurements, μ, may sound alike. But which statements are correct, which are not, and which are trivial? Assume $\sigma = 0.300$ GeV/c.

  We are very sure that the interval (18.914 GeV/c, 19.286 GeV/c) contains μ.

  The true mean lies between 18.914 GeV/c and 19.286 GeV/c with 95% confidence.

  The probability that the interval (18.914 GeV/c, 19.286 GeV/c) does not contain the true measurement mean is 0.05.

  95% of future measurements will be between 19.286 GeV/c and 18.914 GeV/c.

  We are 95% sure that 95% of future measurements will be between 18.914 GeV/c and 19.286 GeV/c.

  The sample mean lies between 18.914 GeV/c and 19.286 GeV/c with 95% confidence.

  We are 95% confident that if a similar sample were drawn tomorrow, under identical conditions, and a confidence interval were constructed, the confidence interval would include μ.

  We are willing to bet 19 to 1 that μ is somewhere between 18.914 GeV/c and 19.286 GeV/c.

- Consider Figure 8-1. How many of the 100 intervals contain μ? How many intervals do you *expect* to contain μ? Is there a difference? If so, how do you explain it? Does it throw doubt on the efficacy of confidence intervals?

Statistical tolerance limits for a normal population

Given a sample of size n from the normal distribution with a mean $\bar{Y}$ and a standard deviation S, the construction of *two-sided statistical tolerance limits* is much like the construction of a two-sided confidence interval about the mean. The form of the tolerance limits is $\bar{Y} \pm kS$ where the factor k—a function of γ, the level of confidence; π, the proportion of the population you wish to bound; and n, the sample size—is obtained from Table T-11a. Note that, unlike the confidence interval for the mean, the standard deviation is not divided by $\sqrt{n}$.

Statistical tolerance limits are sometimes written in a shorthand notation as γ/π. Thus, limits constructed to include, with 90% confidence, 95% of the population may be written as 90/95 tolerance limits.

As an example, suppose that a sample of $n = 10$ from a normal population yields the statistics $\bar{y} = 120.5$ and $s = 8.4$. For 95% confidence that these limits include 90% of the population, consult Table T-11a, under $\gamma = 0.95$, $\pi = 0.90$, and $n = 10$ to find $k = 2.856$. The desired tolerance limits are therefore given by $120.5 \pm (2.856)(8.8)$, or (95.4, 145.6).

One-sided tolerance limits have the structure of either $\bar{Y} + kS$ or $\bar{Y} - kS$. However, now you must obtain the factor k from Table T-11b, constructed specifically for one-sided tolerance limits. Thus, in the last example, a one-sided 95% upper confidence upper tolerance limit for 90% of the population is $120.5 + (2.355)(8.8) = 141.2$.

For discussion:

- Why would you want tolerance limits when you can have a confidence interval for the mean?
- Compare the form of the tolerance limits to that of the confidence interval for the mean. The latter has a square root of n in the formula, while the former does not. Any explanation?
- It is suggested that tolerance limits are nothing but confidence intervals on quantiles. Do you agree?

Confidence interval for a variance

As stated earlier, the ratio $(n - 1)S^2/\sigma^2$ is distributed as a chi-squared statistic with $(n - 1)$ degrees of freedom. You thus have the following relation that occurs with probability of $1-\alpha$:

$$\chi^2_{\alpha/2}(n-1) < \frac{(n-1)S^2}{\sigma^2} < \chi^2_{1-\alpha/2}(n-1).$$

From this relation, solve for σ^2 to obtain two limits that bound σ^2:

$$\frac{(n-1)S^2}{\chi^2_{1-\alpha/2}(n-1)}, \quad \frac{(n-1)S^2}{\chi^2_{\alpha/2}(n-1)}.$$

These two limits provide a $100(1-\alpha)\%$ confidence interval for σ^2. If a one-sided confidence interval for σ^2 is required (say an upper limit) you need calculate

$$(n-1)S^2/\chi^2_{(1-\alpha/2)}(n-1).$$

As an example, suppose a sample of size 10 yields a standard deviation of 1.47, and we wish a 95% upper confidence limit for σ^2 and for σ. First, from Table T-2 we find the 0.05 quantile for the chi-square distribution with 9 degrees of freedom to be 3.33. Next the desired limit is calculated as (9)(1.47)/3.33 = 3.972. The upper limit for σ is typically calculated as $\sqrt{3.973} = 1.993$.

The interpretation of the confidence interval follows the same reasoning we used in the construction of confidence intervals for the mean. Thus, if we were to repeat the experiment many times, each time calculating a 95% one- or two-sided (as needed) confidence interval for σ^2, then 95% of those confidence intervals would include the true σ^2.

For discussion:

- Using simple algebra derive the confidence limits for σ^2.
- Construct a diagram similar to Figure 8-1 to aid in the interpretation of the confidence interval for σ^2.

A note on the determination of sample size

Perhaps the most common question asked of a statistician is: "How large a sample do I need?" (Irrespective of its frequency, it certainly is the most poignant.)

Put simply, the answer is not simple. It depends on numerous factors and assumptions. This section considers a situation in which your problem is the estimation of the mean of a variable measured on an interval scale. Furthermore, you seek "reasonable assurance" that the resulting estimate of the mean does not deviate from the true mean by more than a specified amount.

To make the problem a bit more specific, suppose you wish to estimate the mean μ of a population of interest and require $100(1\ \alpha)\%$ assurance that the estimate will be within d units of μ. Your protocol now follows one of two cases:

Case I: the population standard deviation σ is known.

Case II: the population standard deviation σ is unknown.

Each case is pursued separately.

Case I: You know the standard deviation σ and seek a sample size such that you are at least $100(1 - \alpha)\%$ confident that the sample estimate, $\bar{Y}$, will not differ from μ by more than a fixed distance d. This requirement can be expressed mathematically as

$$Pr(|\bar{Y} - \mu| \leq d) \geq 1 - \alpha.$$

After some manipulation, the sample size is determined from

$$n = (z_{(1-\alpha/2)}\sigma/d)^2.$$

To derive the expression for n with 95% assurance, recall that a 95% two-sided confidence interval for the mean is constructed from $\bar{Y} \pm 1.96\sigma/\sqrt{n}$. You set the quantity $1.96\sigma/\sqrt{n}$ equal to d and solve for n (remember: σ is known). Thus, the sample size is derived from

$$n = [(1.96\,\sigma)/d]^2.$$

Example 8-3:
Estimating the average use of residential electricity

A review of the September 1991 statements sent to all residential customers of the Salem (Oregon) Electric Cooperative showed that the average use of electricity per household in the month was 818.18 kilowatt-hours (kWh), with an associated standard deviation of 567.11 kWh. How do you use this information to determine the sample size to estimate the average monthly residential electricity use a year later to within a 95% "error bound" of 100 kWh?

Assuming that the standard deviation of 567.11 kWh still is applicable a year later, you have

$$n = \left[\frac{(1.96)(567.11)}{(100)}\right]^2 = 123.55.$$

Rounding up the last number yields the sample size: n = 124. Note that, if you wish this estimate to have a 95% error bound of 100 kWh, the sample size would be considerably larger, namely, 12,356.

Case II: You *do not know* the standard deviation σ, but you have the same goals as in Case I. You discover a two-stage method called *Stein's procedure*, discussed and slightly modified by Desu and Raghavarao (1990, pp. 4-5). As before, d is the error bound and $100(1 - \alpha)\%$ is the confidence coefficient. The sample size determination is made in four steps:

(1) Take an initial sample of size n_1 of at least 3 observations. Calculate the initial sample's variance, S_1^2.

(2) Use Table T-3 to find $t_{(1-\alpha/2)}(n_1 - 1)$. Determine n_0 from the expression

$$n_0 = [s_1 t_{(1-\alpha/2)}(n_1 - 1)/d]^2.$$

(3) If $n_2 = n_0 - n_1 > 0$, take n_2 additional observations. If $n_2 \leq 0$, no additional observations are required.

(4) Let $\bar{Y}$ be the mean of all $n_1 + n_2$ observations. Then $\bar{Y}$ satisfies the stated requirement.

Example 8-4:
Illustrating Stein's procedure with the problem of estimating the use of residential electricity

Begin with the supposition that σ is not known and you want $d = 100$ kWh. Then the four-step process provides that:

(1) You take an initial sample of size, say, $n_1 = 15$ observations. Suppose the sample yields $\bar{y} = 745.53$ and $s_1^2 = 176{,}526.0225$ (i.e., $s_1 = 420.15$).

(2) Use Table T-3 to find that the 0.975 quantile for 14 degrees of freedom is $t_{0.975}(14) = 2.14$. Determine

$n_0 = (2.14 \text{ x } 420.15/100)^2$

$= 80.98$.

(3) Since $n_2 = n_0 - n_1 = 80.98 - 15 = 65.98$ is positive, you take 66 additional observations which yields, say, $\bar{y}_2 = 811.12$.

(4) You compute $\bar{y}$, the mean of all $n_1 + n_2$ observations, from

$$\bar{y} = [(15)(745.53) + (66)(811.12]/(15 + 66)$$
$$= 789.23,$$

which satisfies the stated requirement that the length of the 95% confidence interval be less than $d = 100$ kWh.

See Desu and Raghavarao (1990) for an extended treatment of sample-size determination.

For discussion:

- In determining sample size, why do you round *up* the sample size? Why not round to the nearest integer?

What to remember about Chapter 8

Chapter 8 examined the general process of *statistical estimation*; that is, it looks at the question: "How do you put a number on it?" Among the special terms employed here, you will find:

- *parameter*
- *estimator*
- *estimate*.

A distinction is made between:

- *point estimators*
- *interval estimators*.

The chapter concluded with discussions of :

- *confidence intervals for a mean*
- *tolerance limits for a normal distribution*
- *confidence intervals for a variance*

and illustrated simple cases of sample size determination for normally distributed data.

Testing statistical hypotheses: one mean

What to look for in Chapter 9

Chapter 9 formalizes the process of *statistical inference* about the mean of a population of interest. The presentation identifies the circumstances when such inference is valid and points out conditions that invalidate the inference. The chapter thus lays a foundation for specific material developed in Chapters 10, 11, and 12. Among the special terms employed in the discussion of statistical inference are:

- *null hypothesis*
- *alternative hypothesis*
- *test statistic*
- *critical region*
- *non-critical region*
- *critical value (critical point)*
- *power of a test*.

You will encounter two types of errors in hypothesis-testing: *Type I* and *Type II*. The consequences of each type must be considered in any hypothesis-testing.

The chapter concludes with a formal protocol for testing a hypothesis about a single mean, whether or not you know the variance of the underlying random variable.

Testing statistical hypotheses: Setting the stage

The logic of testing statistical hypotheses is motivated and explained by two seemingly unrelated examples. The first involves traffic violations, while the second examines response times to security violations.

Example 9-1
A traffic-violation example (Innocent until found guilty?)

You are in your car on your way to your statistics class when a nice police officer pulls you over and suggests that you just ran a stop sign. Adamantly, but gently, you explain that you did no such thing. Just as adamantly, but somewhat less gently, the officer holds an alternate opinion and invites you to traffic court to resolve the matter.

How do you deal with this problem? Indeed, what is the problem? Is there a hypothesis to be examined? What does it have to do with statistics anyhow?

Example 9-2:
A response-time example (When is quick quick enough?)

The Director of Security at a power station claims that the station's electronic surveillance system can pinpoint the location of an intrusion with an average response time that is no longer than 8.1 seconds, give or take a standard deviation of 2.2 seconds. The system is tested by numerous daily random provocations. At each such provocation, the system's response time is recorded. A random sample of 100 response times is collected, and the average response time, $\bar{y}$, is made available for your statistical analysis.

How do you deal with this problem? Indeed, what is the problem? What do you think " ... give or take a standard deviation of 2.2. seconds ... " means? And what do you think Examples 9-1 and 9-2 have in common?

Some basic hypothesis-testing concepts: Two examples compared

Consider the following parallel elements that are designed to compare traffic court processes with those of statistical hypothesis-testing:

Traffic courts in the United States work under a process by which evidence is examined to see whether it contradicts the accused's claim of innocence.... *What process do you face as you attempt to defend yourself against the charge of running the stop sign in Example 9-1?*

Statistical hypothesis-testing is a process by which a set of data is examined to see whether the data contradict some idea or perception about the population from which the data were drawn.... *What manipulations will you apply to the 100 response times in Example 9-2?*

The claim of innocence is a hypothesis about the accused.... *You are innocent until you are proved guilty to a high degree of certainty; i.e., "beyond a reasonable doubt."*

The idea or perception is a hypothesis about the population.... *The contention that the average response time is 8.1 seconds is maintained until data fail to support it.*

The mechanics of examining evidence in traffic courts are well defined; if applied properly, they set out an orderly sequence of rational and defensible steps leading to a finding of guilty or not-guilty.... *You are assured that you can face your accuser, can question evidence, and can present counter-evidence during your hearing.*

The mechanics of statistical hypothesis-testing are well defined; if applied properly, they set out an orderly sequence of rational and defensible steps leading to a rejection of the hypothesis or to no rejection.... *Data must be selected according to certain rules to avoid bias, they must be analyzed according to a prescribed protocol, and the results must be properly interpreted.*

This parallel structure reveals our motivation. You will encounter numerous similarities between traffic-court procedures in which material is examined as evidence of violation and the process of hypothesis-testing in which data are examined as evidence of the falseness of some idea about a population.

Some terminology and processes

Statistical hypothesis-testing is a procedure by which inference about a population is made. The procedure requires that the idea or ideas about the population be articulated succinctly and carefully.

A *statistical hypothesis* is a statement about a population. Once you have data, you are in a position to ask whether the data contain information that contradicts a stated hypothesis.

The *null hypothesis (symbolized by H_0)* is a hypothesis about a population's parameters, a function of those parameters, and/or the structure of the population. In Example 9-2 (response-time), the null hypothesis conventionally is written as H_0: $\mu = 8.1$.

The null hypothesis usually is written without explicitly stating the units of measurement in order to focus on the essence of the statistical process. But you must never forget that data carry units with them, and those units must be restored when you conclude and report your analysis.

Note: There is a class of null hypotheses that focuses on a range of values. However, this book follows the practice of stating each null hypothesis as an equality. Thus, although the claim is that response time is equal to or less than 8.1 seconds, you need only to write H_0: $\mu = 8.1$ because, if a mean of 8.1 seconds is acceptable, so would be any value of μ less than 8.1 seconds.

The subscript 0 (read and pronounced "zero," "not," or "naught") in H_0 implies "no difference," "no change," or "status quo." Indeed, the statistical hypothesis assumes neutrality and retains such neutrality unless the data prove otherwise.

The *alternative hypothesis (symbolized by H_1)* is the hypothesis you accept (the conclusion you draw) about the population if you have sufficient evidence to reject the null hypothesis. H_1 is usually stated as an interval; e.g., if the null hypothesis is H_0: $\mu = 8.1$, then the alternative hypothesis is H_1: $\mu > 8.1$ or H_1:$\mu < 8.1$ or H_1: $\mu \neq 8.1$.

Some writers denote the alternative hypothesis by H_A.

right-sided, left-sided, and one-sided hypotheses

An alternative hypothesis written with the inequality symbol > (e.g., $\mu > 8.1$) is called a *right-sided* hypothesis. An alternative hypothesis written with the inequality symbol < (e.g., $\mu < 8.1$) is a *left-sided* hypothesis. Left-sided and right-sided hypotheses are both *one-sided.*

two-sided hypothesis An alternative hypothesis written with the not-equal symbol $\neq$ (e.g., $\mu \neq 8.1$) is called a *two-sided hypothesis*.

The *test statistic* is a function of the data and the null hypothesis; it is used as an indicator of whether or not to reject the null hypothesis. In Example 9-2, the test statistic is the standardized normal variable $z = (\bar{y} - \mu_0)/(\sigma/\sqrt{n})$, where μ_0 is the value of μ specified by H_0.

The *level of significance (symbolized by α)* is an expression of the risk, quantified as a probability, that you are willing to take in rejecting the null hypothesis *if the null hypothesis is correct*. In various contexts, you will find the level of significance called *the probability of Type I error*, *the false alarm rate*, *the false-positive rate*, *the producer's risk*, and others.

The *critical region* is the set of values such that realization of any one of them by the test statistic leads to the rejection of the null hypothesis, H_0. The critical region is sometimes called the *rejection region*.

The *non-critical region* is the set of values such that realization of any one of them *does not* lead to the rejection of H_0; i.e., it is the complement of the critical region.

Some writers call this the "acceptance region." However, there is a bit of a problem with this terminology, so be careful about using it. Here is the crux of the issue: If you *reject* H_0, you do so because the evidence is *against* its being true and you know α, the probability of your being in error; on the other hand, if you *"accept* H_0", you rarely know the probability that you made the wrong decision because you *don't know* the *true value* of the alternative hypothesis.

A *critical value* (also called ***a critical point***) is any value (point) that separates the critical region from the non-critical region. A test of a hypothesis may involve more than one critical point. Thus, neither the critical region nor the non-critical region is necessarily contiguous.

A *Type I error* is an error committed when you *reject* the null hypothesis as false when, in fact, it is true.

The *probability of a Type I error, denoted by* α, is the probability of committing a Type I error.

A *Type II error* is an error committed when you *do not reject* the null hypothesis as false when, in fact, it is false.

The *probability of a Type II error, denoted by* β, is the probability of committing a Type II error.

The *power of a test* is the probability that the null hypothesis will be rejected. The power of the test depends upon

- the choice of α
- the sample size
- above all, how far the null hypothesis is from the "truth."

You want your statistical test to be especially effective in telling you when your null hypothesis, H_0, is not true. Thus, you want the test to be "powerful" (i.e., sensitive) to the falseness of the null hypothesis.

For discussion:

- With respect to the response-time example at the beginning of this chapter:

  What is the null hypothesis? How is it expressed in symbols?

What is the alternative hypothesis? How is it expressed in symbols?

What test statistic do you propose to use? How is it expressed in symbols?

What level of significance do you propose to use? How is it expressed in symbols?

What is the rejection region? What is the non-rejection region? Produce a sketch that shows both of them on one chart.

If you were to test only the hypothesis that the standard deviation of the response time is indeed 2.2 seconds, as claimed by the Director of Security, how would you write the null hypothesis?

- Try responding to the questions above with respect to the traffic-court example.

Examining hypothesis-testing with a "truth table"

A null hypothesis is either correct or it is incorrect. You collect data, conduct an analysis, and draw a conclusion about that null hypothesis. However, even when you use objective statistical procedures, you may or you may not reject the null hypothesis, irrespective of its "truth." No matter how much you hope that your decision is correct, you may never know.

To organize your thinking about this decision-making process, consider the "truth table" displayed in Table 9-1.

Table 9-1:
A truth table for statistical hypothesis-testing

	Reality	
Statistical decision	**Null hypothesis is true**	**Null hypothesis is false**
Do not reject null hypothesis	Correct decision	Wrong decision (*Type II error*)
Reject the null hypothesis	Wrong decision (*Type I error*)	Correct decision

The truth table sets out the possibilities inherent in a hypothesis-testing situation. Your hypothesis-testing mechanism may coincide with *reality*; that is, (1) you do not reject the null hypothesis when it is true *or* (2) you reject the null hypothesis when it is false. In either of these cases, your decision is correct.

On the other hand, if you reject the null hypothesis when it is true, you commit an error. Similarly, if you do not reject the null hypothesis when it is false, you commit an error. At least, they are two different kinds of errors. But how are they characterized, and what are their consequences?

For discussion:

- What is the genesis of the term "truth table"?
- Set out the truth table specific to the response-time example; i.e, remove all the general notations from Table 9.1 and replace them with

terminology and numerical values that are directly related to the response-time question.

Consider the consequences: What happens *after* the hypothesis is tested?

It is not too early in the presentation of these matters to reflect on some of the basic notions, principles, and concepts of testing statistical hypotheses. Although some of these principles repeat ideas already encountered, you need not fear overexposure.

The following display compares and contrasts the implications and consequences of statistical hypothesis-testing. The first column focuses on rejecting the null hypothesis, while the second column examines *not* rejecting it.

After the hypothesis is tested ...	
... if the test statistic falls into the critical region, you reject the null hypothesis H_0.	**... if the test statistic *does not* fall into the critical region, you *do not* reject the null hypothesis H_0.**
The "story told by the data" is *not* consistent with H_0.	The "story told by the data" is not *in*consistent with H_0.
You may have made an error in rejecting H_0.	You may have made an error in not rejecting H_0.
If the null hypothesis is true, you committed a *Type I error*.	If the null hypothesis is false, you committed a *Type II error*.
The *probability of committing a Type I error* is denoted by α (the lower-case Greek letter *alpha*). You were free to select the value of α to suit yourself; conventionally, its value is "small."	The *probability of committing a Type II error* is denoted by β (the lower-case Greek letter *beta*). It's a fact of statistical life that β usually is a complicated function of the null hypothesis (H_0), the sample size (usually n), the level of significance (α), and the variability of the test statistic.
In rejecting H_0, one of the following eventualities must have happened: (1) H_0 is incorrect and rejected. (2) H_0 is correct but rejected.	In not rejecting H_0, one of the following eventualities must have happened: (1) H_0 is correct and not rejected. (2) H_0 is incorrect but not rejected.
To protect yourself from telling lies (and from lawsuits), you qualify your statement of rejection. This qualification places a bound on the probability of your being wrong. Your qualification is usually written in parenthesis like "($p < .05$)."	Your first impulse simply to accept H_0 is understandable. Be aware that you *must* learn to control this impulse. You do not *know* the truth or falseness of the null hypothesis. (If you did, you wouldn't bother to perform a statistical test, would you?) Therefore, you don't know β. So, if you claim that H_0 is correct, you cannot augment your statement with an error qualification.
Consult the Truth Table and note that when you reject H_0 the only inferential error you could have made is Type I, for which the associated probability is known.	However, you can—and should—examine the size of β for a range of possible alternative hypotheses.

Another look at Example 9-1 (traffic-court justice)

To repeat: Testing a statistical hypothesis is analogous to trying a case in an American traffic court where the defendant is accused of, say, running a stop sign. The null hypothesis presumes the defendant is innocent. The alternative hypothesis states that the defendant is guilty. The test statistic is the presentation of evidence.

If the evidence is strong enough to suggest guilt, the null hypothesis is rejected (i.e., the defendant is declared guilty). On the other hand, if the evidence is not convincing, the null hypothesis is not rejected (i.e., the defendant is declared not guilty), and the case is dismissed. *Notice a subtlety here: Not guilty is not the same as innocent.*

You may be bothered by the fact that falseness/guilt can be demonstrated (by rejecting H_0) whereas truth/innocence cannot. You may further surmise that there is no symmetry of "justice" between the two types of decisions. Well, you are right; no symmetry is intended!

A closer look at Example 9-2 (response time)

A schematic display of the test of a hypothesis always is instructive. Let's return to the response-time example, where the null hypothesis H_0: $\mu = 8.1$ is tested against the alternative hypothesis H_1: $\mu > 8.1$, and where $\sigma = 2.2$, $n = 100$, and $\alpha = 0.05$. This clearly is a one-sided (and *right*-sided at that) test; you would expect that only a sample mean larger than 8.1—in a sense yet to be determined—would lead to the rejection of H_0.

Two basic, but equivalent, schematic representations for testing H_0 can be displayed:

- **The first schematic representation** directly indicates the critical value (and, hence, the critical region) in the units of the given problem; this yields

  $$8.1 + (1.645)(2.2)/\sqrt{100} = 8.462,$$

  so that H_0 is rejected when the sample mean, $\bar{y}$, is larger than 8.462.

  In this representation, you leave the hypothesized mean and the sample mean intact and determine the critical region (a function of H_0, n, and α); i.e., those values which result in the rejection of the null hypothesis if the sample mean equals any of them.

- **The second schematic representation** employs the transformation to the standardized normal statistic, z, where

  $$z = \frac{\bar{y} - 8.1}{2.2/\sqrt{100}},$$

  so that H_0 is rejected when z is larger than 1.645.

  In this representation, you leave the standardized critical region intact and calculate the test statistic (a function of H_0, n, and α); i.e., the value which is compared to those in critical region and which results in the rejection of the null hypothesis if it equals any of them.

The equivalence of these two expressions is sketched in Figure 9-1.

Figure 9-1:
Two schematic representations for the testing of the response-time hypothesis H_0: $\mu = 8.1$ against the alternative hypothesis H_1: $\mu > 8.1$

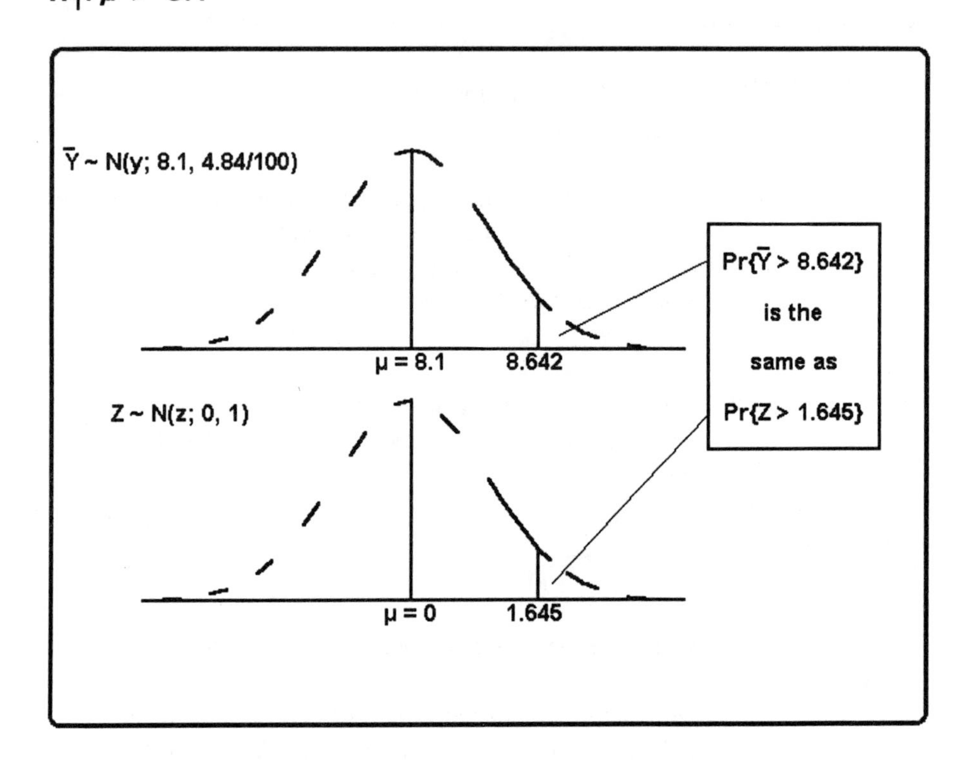

With few exceptions, we prefer the second schematic representation to conceptualize any test of a statistical hypothesis; some test statistics simply are hard to deal with unless a transformation is made, as you will see in Chapters 11, 12, and 14. However, the first schematic representation is especially useful in discussing the power of a test, an idea that's given more attention later in this chapter.

Note especially the compounding of "falseness" and "guilt" and of "truth" and "innocence" in the discussion of Example 9-1. The analogy between traffic-court justice and statistical hypothesis-testing is imperfect; see Large and Michie (1981) for further discussion.

For discussion:

- In Example 9-2, the response-time example, you saw that $z > 1.645$ if and only if $\bar{y} > 8.462$. What does this tell you?
- If you can determine that the hypothesis is correct without using the standardized statistic, why standardize?
- Why was the test in the response-time example characterized as "clearly a one-sided (and *right*-sided at that) test"?
- If, in the response-time example, the standard deviation were larger than 2.2, the critical value (now at 8.462) would change. In which direction would it change? Would you need a larger $\bar{y}$ or a smaller $\bar{y}$ to reject H_0? What do you think are the implications?
- If, in the response-time example, the sample size were smaller than 100, the critical point (now at 8.462) would change. In which direction would it change? Would you need a larger $\bar{y}$ or a smaller $\bar{y}$ to reject H_0? What do you think are the implications?

A view from the other side: What's *left* of the response-time example?

Example 9-2 can be restated to show how the same situation can yield a *left*-sided test of hypothesis. Recall the claim by the Director of Security that their system's average response time is "no longer than 8.1 seconds, give or take a standard deviation of 2.2 seconds." As originally given, we stated the test in such a manner that it tolerates (i.e., does not reject) a $\bar{y}$ of 8.1 seconds; even some $\bar{y}$ values greater than 8.1 are tolerated. As suggested in the last For discussion: section, a smaller sample size and/or a larger standard deviation would cause the test to tolerate a larger $\bar{y}$ without declaring it statistically significant.

Let's modify the problem by suggesting that 8.1 seconds response time is *not* acceptable and ask the Director of Security to demonstrate that the response time is significantly *shorter* than 8.1 seconds. As you may have already sensed, if the Director's claim is valid, it would be to the company's advantage to have a standard deviation as small as possible and a sample as large as practical.

Suppose that you are told that of the 100 trials, 36 were "rehearsed" and cannot be considered random. You therefore disregard the 36 suspect observations and calculate the mean of the remaining 64 observations as $\bar{y} = 7.6$. You still believe that $\sigma = 2.2$. In the new framework, you set the hypothesis-testing into motion as follows:

H_0: $\mu = 8.1$
H_1: $\mu < 8.1$
$\sigma = 2.2$
$n = 64$
$\alpha = 0.05$.

Your strategy is to construct the Z statistic and to reject H_0 if the calculated z is less than -1.645. Schematically, as shown in Figure 9-2, you reject H_0 if z is to the left of -1.645.

Figure 9-2:
Schematics for the z statistic for a left-sided test of H_0: μ = 8.1 against the alternative H_1: μ < 8.1

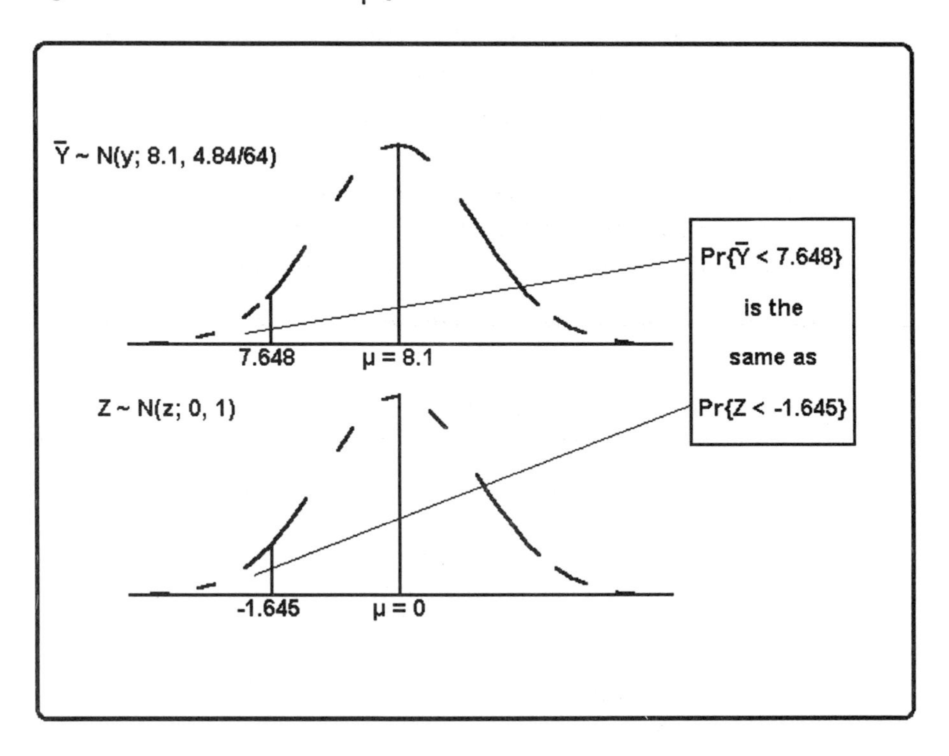

Suppose you find that the mean of the 64 valid observations is $\bar{y}$ = 7.6. Your calculations now yield

$$z = \frac{\bar{y} - 8.1}{2.2/\sqrt{64}} = \frac{7.6 - 8.1}{2.2/\sqrt{64}} = -1.82.$$

Because $z < -1.645$, you have statistical evidence that the Director of Security is credible, assuming the standard deviation is correct, a problem which you will address at a later time.

For discussion:

- Consider the "left-sided" version of Example 9-2. In terms of the first schematic representation, rejecting the hypothesis if $z < -1.645$ translates into $\dfrac{\bar{y} - 8.1}{2.2/\sqrt{64}} < 1.645$ or the equivalent $\bar{y} < 7.648$. Since $\bar{y} = 7.6 < 7.648$, the null hypothesis is rejected, a result consistent with that given in the text and in Figure 9-2. Show analytically the equivalence of these results for this problem.

Summary of hypothesis-testing

Testing a hypothesis calls for the following steps:

- State the null and the alternative hypotheses, paying special attention to the direction of the inequality of the alternative hypothesis.

- Select the level of significance which is acceptable to you.

- State the "givens," such as the sample size and the value of the known standard deviation.

- State explicitly other assumptions, such as data model and structure and independence of data points.

- State the proposed statistical test and sketch the problem in terms of the test statistic and the associated critical region.

Only when you have fully expressed these matters do you examine the data, construct the test statistic, and determine the question of the test statistic's significance.

For discussion:

- In the last presentation of the response-time problem, sketched in Figure 9-2, you reject H_0. What are your conclusions? Could you be wrong? What's the chance of that?

- Suppose $\bar{y} = 8.1$ exactly. Would you reject the Division of Security's claim? Could you be wrong? What is the chance of that?

- Consider the entire problem described in Example 9-2, including the design of the study, and relate it to OSDAR (Chapter 1). Are all of OSDAR's elements addressed?

- Look at your own experience with your organization. Describe "real" examples in which the test of a hypothesis about a mean may be addressed and analyzed.

- Here is an example from another walk of life that shows the wide applicability of hypothesis-testing. A manufacturer of a pain reliever claims that its product relieves bee sting pain (measured in minutes after its application) with a mean $\mu = 6.33$ minutes, and a standard deviation $\sigma = 1.10$. A sample of $n = 12$ is collected, giving a mean of $\bar{y} = 7.12$, and standard deviation $s = 1.20$. Do the data support the manufacturer's claim about the mean? Use a one-sided test with $\alpha = 0.05$.

Power curves for Example 9-2

You have encountered numerous suggestions in this text that acceptance of a null hypothesis usually is ill-advised. Isn't acceptance the other side of rejection? What's going on here?

To fix these ideas, recall Example 9-2, the response-time example, in which $\sigma = 2.2$, H_0: $\mu = 8.1$, H_1: $\mu > 8.1$, and the critical region is set at $\bar{y} > 8.462$. If $\bar{y} > 8.462$, you reject H_0, feeling relatively comfortable because you set the probability of wrongly rejecting the null hypothesis at a reasonably small value, namely 0.05. But there is another nagging question lurking here: What is the probability, denoted by β, that you will commit a Type II error and fail to reject H_0 when H_0 is incorrect?

Because you don't know the true value of μ, you do the next best thing: you calculate β for several different candidate values of μ.

To see how this is accomplished, let $\mu = 9.0$. Calculate $z=(8.46-9.0)/(2.2/\sqrt{100})=-2.455$. Then β (the probability of Type II error) is obtained from the standard normal table as the area to the left of -2.455, or, equivalently, as the area to the right of +2.455, which is about 0.007. Hence, the probability of failing to reject the hypothesis that $\mu = 8.1$ when in fact $\mu = 9.0$ is $\beta = 0.007$. If, in the same manner, you set $\mu = 8.8$, you find that $z = -1.545$ and that $\beta = 0.061$.

You continue to calculate β for other candidate values of μ. The process gains credibility when you think of it as a way of examining an ensemble of alternative values of μ, not just a single μ. You can then build a table like Table 9-2.

Table 9-2:
Probability (β)—and its complement (1 - β)—of failing to reject the null hypothesis for a selection of alternative candidate values of μ

μ	β	1 - β = Power
8.00	0.98	0.02
8.10	0.95	0.05
8.20	0.88	0.12
8.30	0.77	0.23
8.40	0.57	0.43
8.462	0.50	0.50
8.50	0.43	0.57
8.60	0.26	0.74
8.70	0.14	0.86
8.80	0.06	0.94
8.90	0.02	0.98
9.00	0.01	0.99

The power of the test—i.e., the value of (1 - β) for a candidate alternative—is the probability that the null hypothesis will be rejected. Therefore, the power when $\mu = 8.1$ is *exactly* 0.05, the prescribed level of significance.

power curve

If you now plot Power = (1 - β) against μ, you get a curve that relates μ to β for the statistical test used in the response-time example. This curve is called the *power curve* for the test. In addition to being a function of μ, the power curve is a function of H_0, n, α, and σ. For the example worked out in Table 9-2, the power curve is displayed in Figure 9-3.

The main feature of the power curve is that it gives—for each candidate value of μ—the probability of rejecting H_0 if that candidate were the correct mean. These probabilities are plotted to display the test's power curve in Figure 9-3. Thus, if H_0 claims that $\mu = 8.1$ when in reality $\mu = 8.5$, the probability that H_0 would be rejected is about 0.57; similarly, if $\mu = 8.9$, the probability of rejection of H_0 is 0.98. This pattern makes sense: The further μ is from H_0, the greater is the probability that the test will "sense" that H_0 is incorrect. For this reason, the power of the test is sometimes referred to as the *sensitivity* of the test.

Figure 9-3:
Power curve for the test in Table 9-2

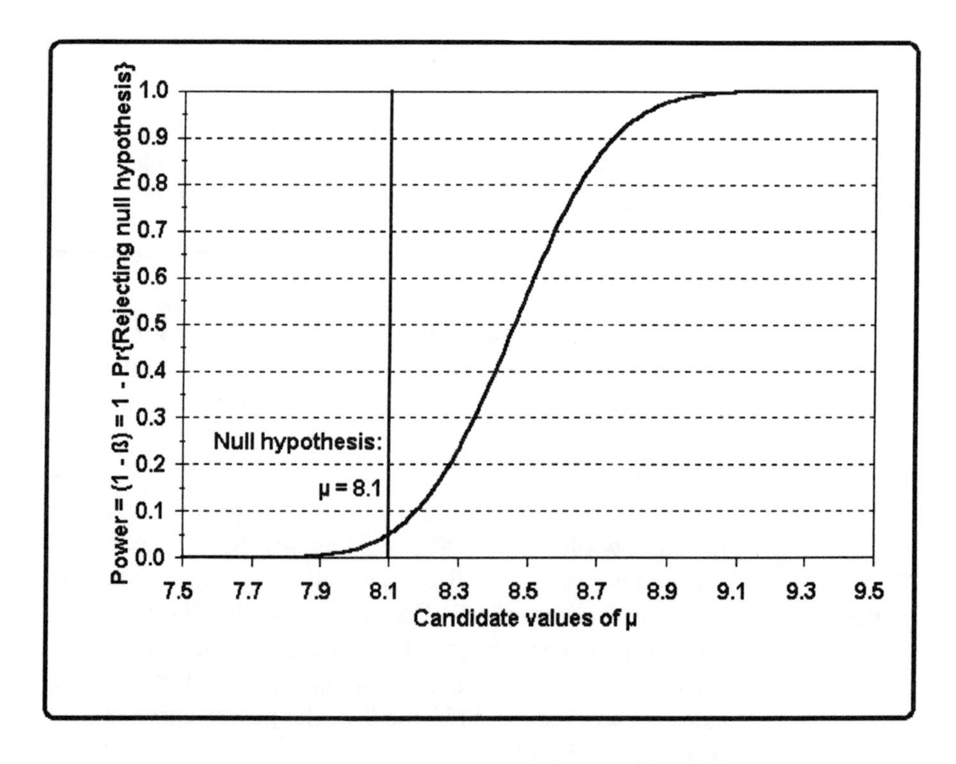

For discussion:

- How will you, from now on, deal with situations in which you are tempted to *accept the null hypothesis*?

- Does the term *power of the test* convey useful concepts about hypothesis-testing?

More power to you

If you are concerned that you may fail to reject a null hypothesis that is false, you must address the power of the test and see where that power can be improved. Reconsider example 9-2 and its associated power curve depicted in Figure 9-3.

What you really want is a test whose power curve is very steep and as close as possible to the vertical line drawn at $\mu = 8.1$. If you could find such a test, you would have no problem deciding whether $\mu = 8.1$ is true or not. But curve balls are part of life—you only know you've been fooled by one *after* the opportunity has passed.

The power reflects the sensitivity of the test to departures from the null hypothesis. The question, then, is what can you do to make the curve steeper and closer to the vertical line at $\mu = 8.1$. Three options are considered here: (1) change the probability of Type I error, (2) change the sample size, and (3) change the standard deviation.

Option 1: *Change the probability of Type I error.* As α increases, β necessarily decreases. It should be obvious—but is it?—that as you increase α, you increase the size of the critical region, giving the test statistic greater opportunity of falling into the critical region and thereby rejecting the null hypothesis. This is shown in Figure 9-4, where power curves are plotted for three values of the level of significance: $\alpha = 0.10$, 0.05, and 0.01.

Figure 9-4:
Power curves for three different levels of significance

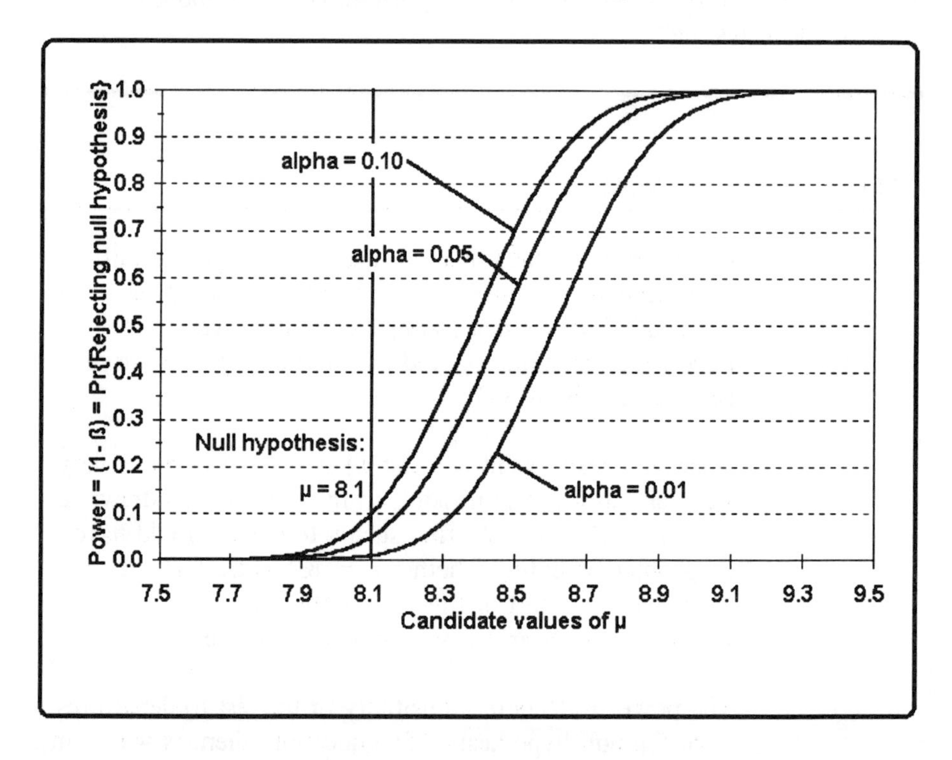

Option 2: *Change the sample size.* As the sample size, n, increases, the standard error of $\overline{Y}$ decreases, and the power of the test increases. This principle is demonstrated in Figure 9-5, where the power of the test is compared for $n = 400$, 100, and 25. The change in power from $n = 100$ to $n = 400$ is quite pronounced, although not as pronounced as the loss of power when you go from $n = 100$ to $n = 25$.

Figure 9-5:
Power curves for three different sample sizes

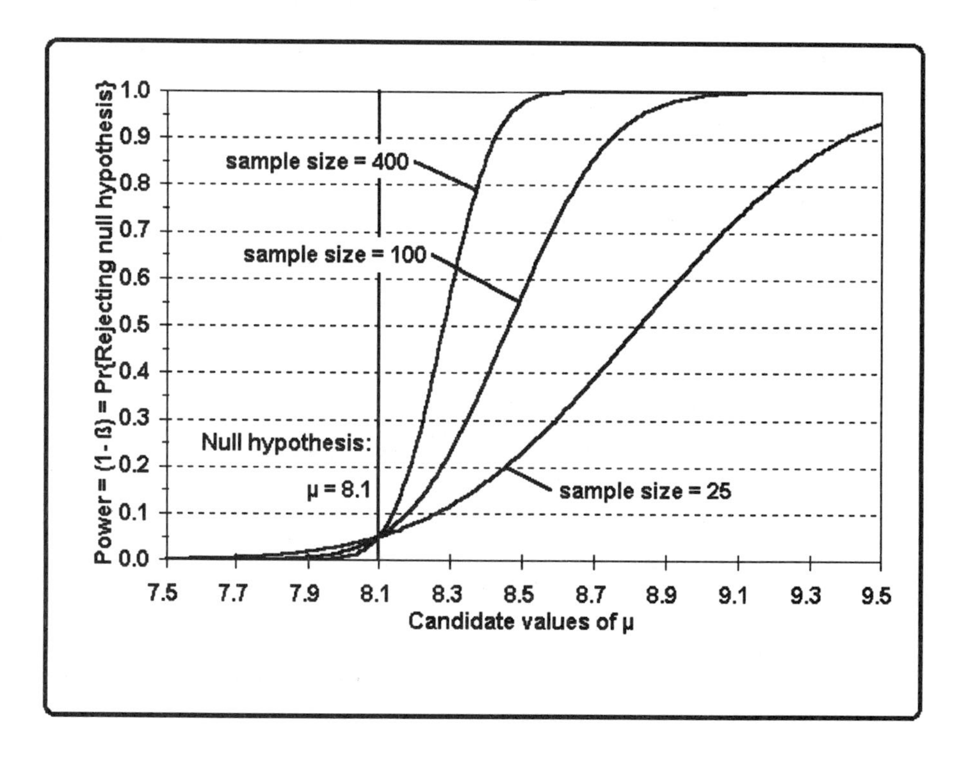

Option 3: *Change the standard deviation.* The effect of changes in σ is apparent in Figure 9-6. Notice especially how much you gain in power by halving the size of σ from 2.2 to 1.1.

Figure 9-6:
Power curves for three different values of the standard deviation

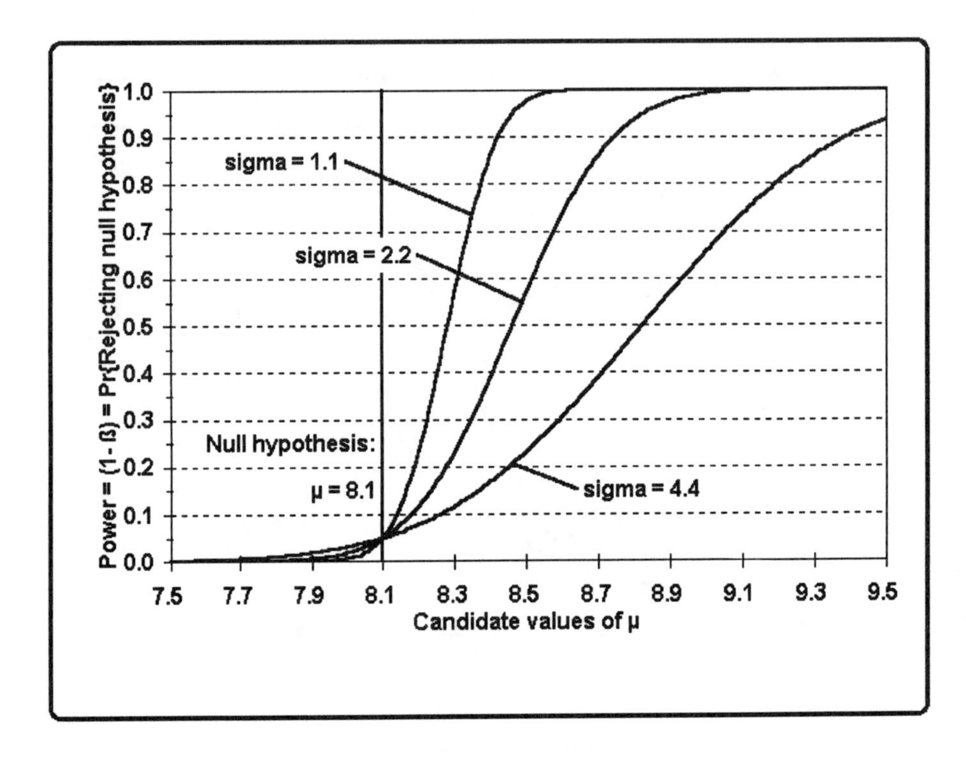

Of course, you may employ more than one of these options and perhaps generate a power curve for some combination of two or more of them. But you need first to determine which of the stated options are viable. And, of course, you need to determine if adopting one or more of these options is warranted, giving due consideration to budget, time, and resource constraints.

This section illustrates a "sensitivity study" conducted with respect to the effects of sample size, level of significance, and variability on a power curve for a particular hypothesis-testing situation. Note especially that, in general, although you and your management may be able to select n and α, you will have no direct control over the actual magnitude of σ. Thus, many studies of this type will require determination of a minimum and maximum value for the standard deviation.

For discussion:

- Consult Figures 9-3, 9-4, 9-5, and 9-6, which are used in testing $H_0 = 8.1$ versus $H_0 > 8.1$, and attempt to answer the following questions. Rough answers and approximations are cheerfully accepted.

  - ☐ Given that $\sigma = 2.2$, and n=100, what is the probability that H_0 would be rejected if $\mu = 8.5$? If $\mu = 8.3$? What would be your answer if $\alpha = 0.10$? If $\alpha = 0.50$?

  - ☐ Since the power is greater for $\alpha = 0.50$ than for $\alpha = 0.10$, why don't you always select a large α?

  - ☐ For a fixed α, show the power is the same when $\sigma = 2.2$ and $n = 100$ as when $\sigma = 1.1$ and n = 25.

What to do when σ is unknown

Rarely will you be fortunate enough to know the true standard deviation, σ, associated with a population that you examine. The steps necessary to test a hypothesis about a population mean, however, are natural extensions of the case where the population variance is known. As you might expect, the population standard deviation is estimated by the sample standard deviation, S, which is based on n observations and $(n - 1)$ degrees of freedom.

Now, reconsider the Z statistic used in the known-variance situation. Your intuition tells you that the σ in the denominator of the Z statistic ought to be replaced by S. The statistic thus constructed, however, is *not* normally distributed; it is a Student's T statistic with $(n - 1)$ degrees of freedom. It looks like this:

$$T = \frac{\bar{Y} - \mu}{S/\sqrt{n}}.$$

Once the sample data are collected, the values of $\bar{y}$, s, and n and the value of μ under the null hypothesis are substituted in the expression for Student's T statistic. The calculated t is now compared to the appropriate quantile in a Student's T table (Table T-3) for the associated degrees of freedom and the α you have selected for the test. That value is denoted by $t_{1-\alpha}(n-1)$.

Example 9-3:
Fuel pellets and solid lubricant additive

The percent of solid lubricants added to fuel composition before compaction into pellets are recorded for a random sample of size 7 as 1.13, 0.93, 1.06, 1.08, 1.11, 1.19, and 0.97. Does this sample support a claim that the mean percent of solid lubricant does not exceed 1.00?

As the problem is read, this is a one-sided test, where H_0: $\mu = 1.00$, H_1: $\mu > 1.00$. Use $\alpha = .05$, $n = 7$, $df = 6$, and $t_{0.95}(6) = 1.94$.

Calculate: $\bar{y} = 1.067$, $s = 0.091$, $s/\sqrt{n} = 0.034$, and $t = (1.067 - 1.00)/\ 0.0343 = 1.953$.

Since the calculated t statistic is larger than $t_{0.95}(6)$, the null hypothesis that $\mu = 1.00$ is rejected with the claim that μ is larger than 1.00.

For discussion:

- Is a one-tailed alternative hypothesis appropriate for Example 9-3?
- Construct a one-sided 95% confidence interval about μ.

Hypotheses with two-sided alternatives

On certain occasions, you will find that the alternative hypothesis is best written as a two-sided alternative. In that case, you have a critical region composed of disjoint parts. Such a critical region guards against a mean that is too large *or* too small, as compared with the mean specified by H_0. A test that uses a two-sided alternative hypothesis is illustrated by the shipper-receiver example of Chapter 2 (Table 2-4), the data from which are reproduced in Table 9-3. For the purpose of this example, assume that the 10 shipments were collected at random.

Table 9-3:
Shipper-receiver differences

Month	Shipper, S_i	Receiver, R_i	Difference, d_i
1	1471.22	1468.12	3.10
2	1470.98	1469.52	1.46
3	1470.82	1469.22	1.60
4	1470.46	1469.26	1.20
5	1469.42	1465.96	3.46
6	1468.98	1470.80	-1.82
7	1469.10	1467.89	1.21
8	1470.22	1472.28	-2.06
9	1470.86	1469.02	1.84
10	1470.38	1470.16	0.22
Mean			1.021
Std. dev.			1.818

In this example, you need to deal only with shipper-receiver differences, denoted in the last column of Table 9-3 by d_i. To maintain the mnemonic, denote the average difference by $\bar{d}$ and the standard deviation of the differences by s_d. Since s_d is based on $n_d = 10$ differences, it has 9 associated degrees of freedom.

The problem is now stated mathematically:

H_0: $\mu_d = 0$
H_1: $\mu_d \neq 0$
$\alpha = 0.05$,

and the test statistic and the basic calculations are given by

$$t = \frac{\bar{d} - \mu_d}{s_d/\sqrt{n}} = \frac{1.021 - 0.00}{1.818/\sqrt{10}} = 1.76.$$

From Table T-3, you find $t_{0.975}(9) = 2.26$. Because $-2.26 < t < 2.26$, you do not reject H_0.

For discussion:

- What is the physical interpretation of H_0: $\mu_d = 0$?
- Consider again the shipper-receiver example. Since you do not reject H_0, can you say that you have statistical evidence that no real loss occurred? What *can* you say? What would you like to do in future testing?

- If you know that there is a real loss and that μ_d is different from zero, what would you do? (Yes, this is a loaded question.)

- In this problem, you really have two columns of data, one for the shipper and one for the receiver, each with its own mean (as well as its own standard deviation). Why, then, is this topic discussed in this chapter which tests hypothesis about a single mean?

- In the shipper-receiver problem, you found that $-2.26 < t < 2.26$. Which of the following two statements is correct?

  (1) $|t| < 2.26$
  (2) $t < |2.26|$

Getting formal: Testing a hypothesis about a single mean

Here is a set of 11 steps to follow when you test a hypothesis about the *mean of a single group*. The formalism serves as a checklist and drives home the point that certain conditions must be met and dealt with in the process of conducting any test of any hypothesis.

Step 1: **Write the model as $Y = \mu + E$.**
Better yet, write it as $Y_i = \mu + E_i$, $i = 1, 2, ..., n$.

model

The expression of a variable as a function of its components is called a *model*. In this case, each observation, Y or Y_i, is the sum of a constant, μ, and a deviation from that mean, E or E_i. Although we do not dwell on modelling in this chapter, it is, nevertheless, an important ingredient of statistical thinking and development. Indeed, whether we say so or not, carefully formulated models underlie *all* of the estimation and inferential procedures herein.

Step 2: **Write the assumptions:**
(A1') σ^2 is known, or (A1'') σ^2 is unknown.
(A2) μ is a fixed but unknown constant.
(A3) E is distributed as $N(0,\sigma^2)$. (In some cases, you may choose to refine this as $E_i \sim NID(0, \sigma^2)$ where *NID* is read as "normally independently distributed.")

Step 3: **Write the null hypothesis:**
H_0: $\mu = \mu_0$.

Step 4: **Write the alternative hypothesis (i.e., the hypothesis you will adopt if H_0 is rejected):**
H_1: $\mu \neq \mu_0$, **or** H_1: $\mu < \mu_0$, **or** H_1: $\mu > \mu_0$.

Step 5: Select and record the level of significance, α, (i.e., the chance you are willing to take that you will reject H_0 even though H_0 is correct).

Step 6: Write the test statistic. When σ is known, use

$$Z = \frac{\bar{Y} - \mu}{\sigma/\sqrt{n}};$$

when σ is not known, use

$$T = \frac{\bar{Y} - \mu}{S/\sqrt{n}}.$$

Step 7: Determine, from the appropriate table, which values of z or t will lead to the rejection of H_0; (i.e., determine the critical region).

Step 8: Collect the data.

Step 9: If σ is known, calculate $\bar{y}$ and z. Otherwise, calculate $\bar{y}$, s, and t.

Step 10: If z (or t) falls into the critical region, reject the null hypothesis H_0. Otherwise, do not reject the null hypothesis H_0.

What to remember about Chapter 9

Chapter 9 formalized the process of *statistical inference*; that is, it examined the question: "Based upon this set of data, can you draw conclusions on which to base particular actions?" It thus established a foundation for specific material developed in Chapters 10, 11, and 12. Among the special terms employed in the discussion of statistical inference are:

- *null hypothesis*
- *alternative hypothesis*
- *test statistic*
- *critical region*
- *non-critical region*
- *critical value (critical point).*

You encountered two types of errors in hypothesis-testing: *Type I* and *Type II*. The consequences of each type must be considered in any hypothesis-testing. One special concept needed for the analyses of hypothesis-testing procedures, the **power of a test**, was discussed and demonstrated for various scenarios; it provides measures of a statistical test's sensitivity to the truth of alternative hypotheses.

The chapter concluded with a formal protocol for testing a hypothesis about a single mean, whether or not you know the variance of the underlying random variable.

10

Testing statistical hypotheses: variances

What to look for in Chapter 10

Chapter 10 extends the general hypothesis-testing ideas of Chapter 9 to the problem of dispersion as measured by the variance. The discussion begins with an exploration of hypothesis-testing about a single variance, then moves on to the problem of comparing two variances, and concludes with a convenient method for dealing with several variances. Among special terms to look for are:

- *the chi-squared statistic for testing a single variance*
- *the* F *statistic for testing the equality of two variances*
- *the* F_{max} *statistic for testing the equality of several variances.*

Why worry about variances?

Any production line that lays claim to quality output must maintain control over the variability of its product as well as its "average" value. It simply is not enough merely to have a good average. It is equally important for the variability to be consistent; indeed, in some situations consistency is even more important than averages themselves. Variances measure that consistency.

But there is even more to it than that. Many statistical techniques/applications/methodologies make specific assumptions about the variance of a group or about relationships among variances of several groups. These assumptions often are tested by using techniques described in this chapter.

You will become acquainted with three distinct scenarios in which variances are subjected to statistical hypothesis-testing. These tests are used to determine whether:

- a population variance equals a given value
- two groups have equal variances
- the variances of several groups are equal.

Testing the hypothesis that a population variance equals a given value

Suppose that a manufacturing firm claims that the standard deviation of its product is $\sigma = 10$ (or, equivalently, $\sigma^2 = 100$). Note that nothing is said specifically about a claimed or hypothesized value for the mean.

In a statistical framework, the firm's claim leads to a *null hypothesis* that is stated as H_0: $\sigma^2 = 100$.

Formally, the null hypothesis about a specified value of a variance, say σ_0^2, is written as

H_0: $\sigma^2 = \sigma_0^2$.

This null hypothesis is matched against an *alternative hypothesis* which must take one (and only one) of the forms: H_1: $\sigma^2 > 100$ or H_1: $\sigma^2 < 100$ or H_1: $\sigma^2 \neq 100$.

Formally, the *alternative hypothesis* takes one (and only one) of the forms: H_1: $\sigma^2 > \sigma_0^2$ or H_1: $\sigma^2 < \sigma_0^2$ or H_1: $\sigma^2 \neq \sigma_0^2$.

And, as usual, let the level of significance be $\alpha = .05$.

Suppose that the alternative hypothesis is H_1: $\sigma^2 > 100$. You therefore expect to reject H_0 only if the data indicate that the variance is much larger than 100.

As you might expect, the hypothesis is tested by comparing the sample variance, denoted by S^2, to the hypothesized variance σ_0^2. This time the test statistic is a function of the ratio S^2/σ_0^2; it is *not* a function of the difference $(S^2 - \sigma_0^2)$.

Formally, the test statistic is written as

$$X^2(n - 1) = \frac{(n - 1)S^2}{\sigma_0^2}.$$

Recall that the upper-case X in the expression on the left denotes the upper-case Greek letter *chi*; its appearance is identical to the English letter *ex*. The test statistic is distributed as a chi-squared statistic with (n - 1) degrees of freedom. Thus, critical values for this test can be read from Table T-2 as quantiles of the chi-square distribution; e.g., for $\alpha = 0.05$ and $n = 11$, the critical value is $\chi^2_{0.95}(10) = 18.3$. (This specific quantile appears on the opening page of Table T-2, where the use of the chi-squared table is illustrated.)

Example 10-1:
Testing a hypothesis about a single variance

To illustrate the process of testing a single variance, you are interested in testing the manufacturing firm's claim. You set H_0: $\sigma^2 = 100$ against the alternative H_1: $\sigma^2 > 100$. From Table T-2, you find $\chi^2_{0.95}(19) = 30.1$. Suppose your data yield $s^2 = 177.13$ from a sample of size $n = 20$. The degrees of freedom is $df = (20 - 1) = 19$. The test statistic is

$$\chi^2 = \frac{(19)(177.13)}{100} = 33.65.$$

Because $\chi^2 = 33.65$ is larger than $\chi^2_{0.95}(19) = 30.1$, you reject the null hypothesis H_0: $\sigma^2 = 100$, citing statistical evidence that $\sigma^2 > 100$.

For discussion:

- With reference to the conclusion drawn in Example 10-1, could you be wrong? What is the chance of that?
- Suppose that the sample variance had been 96.64. What would your conclusions be?

- Under what circumstances would it be reasonable to take the alternative hypothesis as H_1: $\sigma^2 < 100$? What would your conclusions be if $s^2 = 177.13$? What would they be if $s^2 = 96.6$?

Testing the hypothesis that two groups have equal variances

Let σ_A^2 designate the variance of a product made by Manufacturer A, and let σ_B^2 designate the variance of a "similar" product made by Manufacturer B. The question of the equality of the variances of the two manufacturers is addressed by setting up a null hypothesis as

$$H_0\colon \sigma_A^2 = \sigma_B^2.$$

Because you have no instructions to the contrary, you formulate the alternative hypothesis as

$$H_1\colon \sigma_A^2 \neq \sigma_B^2;$$

that is, you want to protect yourself against either inequality between the variances.

As usual, you set $\alpha = .05$.

From the first population, you draw a sample of n_A observations. You then calculate the sample variance S_A^2 with $df_A = n_A - 1$.

From the second population, you draw a sample of n_B observations. You then calculate the sample variance S_B^2 with $df_B = n_B - 1$.

Let S_{max}^2 and S_{min}^2 denote the larger and smaller of S_A^2 and S_B^2, respectively. Likewise, let df_{max} and df_{min} denote their corresponding degrees of freedom.

Formally, the test statistic is written as

$$F_{max} = \frac{S_{max}^2}{S_{min}^2}.$$

F ratio

This test statistic is called the *F ratio* and is related to Snedecor's *F* distribution whose quantiles appear in Table T-4. In some contexts, the *F ratio* is called the *variance ratio.*

If H_0 is correct, the F ratio should not be much larger than 1 (unity). The question is, as always, how large is large? In statistical language, you are looking for a "critical value" to help you make your decision.

You now need the 0.975^{th} quantile (any idea why?), denoted by $f_{0.975}(df_{max}, df_{min})$. In terms of Table T-4, df_{max} is the "degrees of freedom of the numerator" and df_{min} is the "degrees of freedom of the denominator." Thus, for instance, if $df_{max} = 5$ and $df_{min} = 7$, the critical value is $f_{0.975}(5, 7) = 5.29$.

Example 10-2: Testing the equality of two variances

Suppose you were given these statistics, calculated from two samples, A and B:

Sample	Mean	Variance	n	df
A	18.5	207	16	15
B	21.6	317	25	24

The test statistic is $f = 317/207 = 1.531$. From Table T-4, you find $f_{0.975}(24,15) = 2.70$. Because

1.531 < 2.70, you do not have sufficient evidence to reject the null hypothesis.

For discussion:

- What would you say if the calculated value of the statistic *f* in Example 10-2 had been less than 1?

Testing the hypothesis that the variances of several groups are equal

homo-scedasticity

One of statisticians' all-time favorite words is *homoscedasticity*. No, it's not a disease. It is, rather, a convenient term to say that two or more groups have equal variances. The case for two groups was illustrated in the previous section; the focus of this section is on testing the homoscedasticity of more than two groups.

Statistical literature is replete with procedures that are based on the assumption of homoscedasticity, so it is imperative to have tools for examining data to see if they are consistent with that assumption. One of the more venerable techniques is called *Bartlett's test for homogeneity of variances*; a good reference is Dixon and Massey (1983, pp. 358-360). Many computer packages include Bartlett's test as a standard routine. Although Bartlett's test is a fully rigorous method for examining the homoscedasticity question, it also is a moderately complicated calculational process.

A considerably less-complicated test statistic, called F_{max}, is found in Beyer (1974, pp. 328-329), among others; you apply it in the following fashion:

(1) Start with k groups of observations. (The formal requirement is that the numbers of observations in each of the k groups be equal; i.e., $n_1 = n_2 = \cdots = n_k = n$. For many experimental designs, this requirement is easily met. For others, "approximately equal" sizes are sufficient and do not seriously impair the test.)

(2) State the null hypothesis H_0: $\sigma_1^2 = \sigma_2^2 = \ldots = \sigma_k^2$ and the alternate hypothesis H_1: $\sigma_i^2 \neq \sigma_j^2$ for at least one pair of indices i and j, where $i, j = 1, 2, \ldots, k$.

(3) Set the level of significance at $\alpha = 0.05$.

(4) Calculate the variance for each of the k groups.

(5) Identify the largest and the smallest of the k variances; call them S_{max}^2 and S_{min}^2.

(6) Calculate the test statistic from the ratio of the largest variance to the smallest:

$$F_{max} = \frac{S_{max}^2}{S_{min}^2}.$$

(7) Find the 0.95 quantile, namely the $f_{max,\,0.95}(k, df)$, of the statistic associated with k groups and $df = n - 1$ in Table T-5.

hetero-scedasticity

(8) If F_{max} exceeds $f_{max,\,0.95}(k, df)$, you have statistical evidence of *heteroscedasticity*. Otherwise, you have insufficient evidence against *homoscedasticity*.

Example 10-3:
Testing equality of variances for several groups

A study was conducted to see if materials supplied by six different uranium oxide suppliers were homoscedastic with respect to the percentage of uranium. Eight samples were collected from each of the UO_2 suppliers, giving the following data summary:

Supplier	Mean	Variance	n	df
1	86.2	0.40	8	7
2	88.1	1.49	8	7
3	87.3	3.20	8	7
4	89.9	1.93	8	7
5	86.2	2.35	8	7
6	89.2	1.58	8	7

The critical value for the test is the 0.95 quantile of F_{max} (for $k = 6$ and $df = 7$); i.e., $f_{max,\ (0.95)}(6, 7) = 10.8$. Since $f_{max} = 3.20/0.40 = 8.00$ is smaller than 10.8, you have insufficient statistical evidence to claim heteroscedasticity.

For discussion:

- *Heteroscedasticity*? What's *that*?

- You may run into a problem with F_{max} and other techniques dealing with the testing of equality of variances if one (or more) of the groups has a variance equal to zero. What do you think you should do in this case?

What to remember about Chapter 10

Chapter 10 extended Chapter 9's general hypothesis-testing ideas regarding the mean to the problem of dispersion as measured by the variance. The discussion began with an exploration of hypothesis-testing about a single variance. It then moved on to the problem of comparing two variances and concludes with a convenient method for dealing with several variances. Among the specific techniques introduced in this chapter were:

- *the chi-squared statistic for testing a single variance*
- *the* F *statistic for testing the equality of two variances*
- *the* F_{max} *statistic for testing the equality of several variances.*

You will find reference to these variance-testing techniques in Chapters 11 and 12 which address the testing of hypotheses involving two or more groups.

11

Testing statistical hypotheses: two means

What to look for in Chapter 11

Chapter 11 builds on the foundations laid in Chapters 9 and 10. You will be introduced to some of the hypothesis-stating forms that may be taken when you ask questions about the means of two groups. In the process of building tools for comparing two means you will be introduced to two special concepts:

- *the standard deviation of a difference*
- *pooling the variances of two groups.*

Then you will see four cases in which basic assumptions determine the course of the analysis:

- *paired observations*
- *variances known*
- *variances unknown but assumed equal*
- *variances unknown but* not *assumed equal.*

How might you look at the problem?

Whereas the tests of Chapter 10 for equality of variances ignore group means, the various tests of equality of means—and their subsequent interpretation—are strongly tied to the variances of the associated groups. You will therefore find that the nature of the variances dictates which test statistic is appropriate to your problem.

Often, you will be faced with the task of comparing the means of two or more groups. These comparisons of means are not restricted to the question of whether the means are equal; the comparisons may be made just to see if the means bear some relationship to each other. Hypotheses of interest for such comparisons may take several forms. For two groups,[1] here are three such forms, each given in two equivalent expressions:

Form 1: The means of two populations are equal.

$$H_0\colon \mu_2 = \mu_1$$
$$\text{or } H_0\colon \mu_2 - \mu_1 = 0.$$

Form 2: The mean of one population differs from the mean of a second population by a constant value, C.

$$H_0\colon \mu_2 = \mu_1 + C$$
$$\text{or } H_0\colon \mu_2 - \mu_1 = C.$$

Form 3: The mean of one population is a constant multiple, K, of the mean of a second population.

$$H_0\colon \mu_2/\mu_1 = K$$
$$\text{or } H_0\colon \mu_2 - K\mu_1 = 0.$$

[1] More than two groups are discussed in Chapter 12.

The alternative hypothesis to any of these hypotheses can be written either as a one-sided or as a two-sided hypothesis. For example, the alternative to the null hypothesis in *Form 2* can be written as any of the following:

Form 2a - H_1: $\mu_2 \neq \mu_1 + C$ (a two-sided alternative)

Form 2b - H_1: $\mu_2 > \mu_1 + C$ (a one-sided alternative)

Form 2c - H_1: $\mu_2 < \mu_1 + C$ (another one-sided alternative)

where C is the constant specified in H_0.

Four specific cases regarding two-group means are developed in this chapter. Before describing them, however, two pertinent ideas must be introduced and discussed.

For discussion:

- Examine and verify that the two expressions in each of the null hypotheses *Forms 1*, *2*, and *3* are equivalent.

- Show that *Form 2* is equivalent to *Form 1* when $C = 0$.

- Show that *Form 3* is equivalent to *Form 1* when $K = 1$.

- Decide which of the three null-hypothesis forms you would use to state that:

  - ☐ The average cost of home oil heating did not change between 1991 and 1992.

  - ☐ The time needed to refuel is 20% shorter if refueling begins at sunrise than if refueling begins at noon.

- ☐ The life expectancy of a motor built by Manufacturer A is the same as that built by Manufacturer B.
- ☐ Buffered pain relievers work twice as fast as plain aspirin.
- ☐ Now that a new intrusion-detection system is installed, the response time is two minutes shorter than it was when the old system was in operation.
- ☐ Employee expenses in moving from Region X to Headquarters are 30% more than in moving from Headquarters to Region X.

On the variance of a difference

This chapter focuses on the idea of a difference of two sample means. The null hypothesis is stated conventionally as H_0: $\mu_2 = \mu_1$, although from a mathematical perspective it is more convenient to state it as H_0: $\mu_2 - \mu_1 = 0$. If the difference between two sample means is "small" (i.e., close to zero), there is insufficient evidence to reject the hypothesis. On the other hand, if the difference is sufficiently large, doubt is cast on the correctness of the null hypothesis.

To investigate whether the two populations have the same mean, select a sample from each population in such a fashion that the samples are *independent* of each other. Examine the sample means and compare them with each other. To be able to provide a statistical test of whether the difference between the means is significant, the distribution of the sample mean difference (or some appropriate function of that difference) must be determined.

The range of a difference

Later in this section, you will learn that the *variance of a difference* between two independent random variables equals the *variance of the sum* of those two independent random variables. Because this may be counter-intuitive, we take care examine first the range of the difference of two independent random variables from a discrete uniform distribution associated with dice throwing.

Consider all possible outcomes of the throw of a die: {1, 2, 3, 4, 5, 6}. The *range* of these values is 5 (largest minus the smallest; i.e., 6 - 1).

Consider next a throw of two distinguishable dice and list all possible differences between the outcomes of the first die and the second die, taken in that order. These differences may be as high as +5 and as low as -5. The *range of these differences* is 10, *twice the range of single die*.

The purpose of this dice-throwing example is to show that the variability of the difference between two independent random variables is larger than that of the individual variables. In that example, the observations are discrete and the measure of variability selected is the range.

The standard deviation of a difference in means

In the remaining sections of this chapter, you will deal with continuous data and use the standard deviation (the square root of the variance), rather than range, as a measure of variability.

standardized statistic

In parallel with previous tests of significance of a single mean, we use a *standardized statistic*, which has the following general form:

$$\begin{pmatrix} \textit{standardized} \\ \textit{statistic} \end{pmatrix} = \frac{\begin{pmatrix} \textit{difference} \\ \textit{between} \\ \textit{sample} \\ \textit{means} \end{pmatrix} - \begin{pmatrix} \textit{difference} \\ \textit{between} \\ \textit{hypothesized} \\ \textit{means} \end{pmatrix}}{\begin{pmatrix} \textit{standard deviation} \\ \textit{of difference} \\ \textit{between sample means} \end{pmatrix}}$$

So we need either to know or to estimate the standard deviation (i.e, the square root of the variance) of the difference between two sample means. This is accomplished by calling on the following theorem:

*The **variance of the difference** between two **independent** random variables is the **sum of their variances**.*

The fact that the variance of the difference is larger than the individual components is consistent with the dice-throwing discussion.

Symbolically, the theorem declares that, if $Y_1 \sim ?(\mu_1, \sigma_1^2)$ and $Y_2 \sim ?(\mu_2, \sigma_2^2)$ and Y_1 and Y_2 are independent, then $Z = (Y_1 - Y_2) \sim ?(\mu_1 - \mu_2, \sigma_1^2 + \sigma_2^2)$. (Recall from Chapter 4 that this particular use of the question mark, ?, indicates an unspecified distribution.)

The extension of this theorem to the difference to two independent *means*, $\bar{Y}_1$ and $\bar{Y}_2$, yields:

If $\bar{Y}_1 \sim ?(\mu_1, \sigma_1^2/n_1)$ and $\bar{Y}_2 \sim ?(\mu_2, \sigma_2^2/n_2)$,
then
$\bar{Y}_1 + \bar{Y}_2 \sim ?(\mu_1 - \mu_2, \sigma_1^2/n_1 + \sigma_2^2/n_2)$.

As a direct consequence, the *standard deviation of the difference of two means* is $\sqrt{\sigma_1^2/n_1 + \sigma_2^2/n_2}$. This is known also as the *standard deviation of the mean difference.*

For discussion:

- You must have noticed that, to use the formula for the difference of two means, the means must be independent. This independence is very important. What is meant by the samples being independent? Give an example in which two samples are independent. Give an example in which two samples are not independent.

On pooling the variances of two groups

Suppose you have sample data from each of two groups. If the two groups' variances are not known, you must ask whether the variances are equal because the answer to this question determines which of the several analytical tools should be employed in testing for equality of the groups' means. If you have no theoretical or previous knowledge of the variances' equality, then you may wish to resort to the two-group methodology in Chapter 10. If you decide that the variances are indeed equal, then you go through a step called *pooling the variances.*

pooling variances

Here is the setup: You have two populations with variances σ_1^2 and σ_2^2. Your two samples, based on n_1 and n_2 observations, yield variance estimates S_1^2 and S_1^2, with degrees of freedom (n_1 - 1) and (n_2 - 1), respectively. If the population variances are equal, then it is appropriate to write $\sigma_1^2 = \sigma_2^2 = \sigma^2$ (no need for subscript). Furthermore, you have two independent estimates (S_1^2 and S_2^2) of the common variance σ^2. So you use both pieces of information, S_1^2 and S_1^2, to produce a single estimate of σ^2 that is "better" than either one alone.

weighted average

Pooling variances is akin to averaging the variances. To be more exact, you build a *weighted average* of the two variances, (but *not* of the two standard deviations!) where the weights are the degrees of freedom. This implies that the pooled variance—denoted by S_p^2—will be somewhere between S_1^2 and S_1^2, but closer to the sample variance with the larger number of degrees of freedom. Symbolically, the pooled variance is computed according to the formula

$$S_p^2 = \frac{(n_1 - 1)S_1^2 + (n_2 - 1)S_2^2}{n_1 + n_2 - 2}.$$

The pooled variance (and its square root, the pooled standard deviation, which is denoted by S_p) has $(n_1 + n_2 - 2)$ degrees of freedom.

This variance-pooling procedure can be extended to k groups, as described in Chapter 12 wherein the calculation of a special quantity (called the "Within-groups Sum of squares") is described. The main idea remains the same: You calculate a "weighted average" of the k sample variances using their degrees of freedom for the weights.

The mechanics of pooling two sample variances are illustrated by Example 11-1.

Example 11-1: Pooling variances: An illustration

Consider the fictitious statistics derived from fictitious surveys conducted for two fictitious utilities, as given in Table 11-1. The values are kilowatt-hours used on a specific day in specially metered residences to assess the electricity used for electric dish-washers.

Table 11-1: Pooling variances

Sample	Utility 1	Utility 2
Mean	not required	not required
Variance	106	120
Size	15	18
df	14	17

The sample pooled variance, is calculated as

$$s_p^2 = \frac{(14)(106) + (17)(120)}{(15 + 18 - 2)} = 113.68.$$

As advertised, s_p^2 is between the two sample variances, 106 and 120, and is closer to the variance of the second sample ($S_2^2 = 120$) than the first ($S_1^2 = 106$) because the second sample has more degrees of freedom.

For discussion:

- When you talk about the equality of group variances, do you refer to the population variances or to the sample variances?
- How do you determine if two variances are equal?

- The question about equality of variances may be posed differently: *Is there is any reason for you to believe that the group variances are not equal?* If the answer is "no," you may proceed to pool the sample variances. If the answer is "yes," you may test formally for the equality of the variances.

- If you have two groups with the same sample sizes, do you expect the pooled variance to be the *average* of the individual variances? Can you confirm it in your examples? Can you prove it algebraically?

- If the two sample sizes are the same, would you expect the pooled standard deviation (the square root of the pooled variance) to be the *average* of the individual standard deviations? Can you confirm it in your examples? Can you prove it algebraically?

Case 1: Paired observations

Paired observations are like twins who are expected to grow and behave alike, unless they are treated differently or are exposed to different environments. If one of the twins receives—or is deprived of—a specific food and grows faster than his/her sibling, the diet difference is usually implicated because we are not likely to attribute the growth difference to "stock," or genetic difference. Of course, there may be other factors—genetics or environment—that affect growth; but, in this chapter's narrowly focused presentation, those other factors are dismissed as irrelevant, unlikely, or uncontrolled. The combination of all these other factors is termed "error."

To cite two examples:

> People can be their "own twins" by being treated (at different times) with two medications.
>
> A single bucket of water may be tested for impurities by two different laboratories.

Any data-collection effort should include consideration of whether the observations can be collected meaningfully in pairs. Apart from the greater sensitivity to population mean difference, the statistical methodology is simpler than its counterpart that deals with unmatched observations.

In preparation for the analysis of paired observation studies in which the number of pairs is denoted by n, you write each observation as Y_{ij}, where the first subscript, i, denotes the pair number, $i = 1, 2, \ldots, n$, and the second subscript j, $j = 1, 2$, denotes the member of the pair. Thus, the first reading of the first pair is Y_{11}, the second reading of the 5th pair is Y_{52}, and the second reading of the 222nd pair is Y_{2222}. (When your experimental situation calls for it, you may choose to separate the subscripts and write that observation as $Y_{222,2}$.) To complete our data notation, we use $D_i = Y_{i2} - Y_{i1}$ to denote the difference between the two readings of the same pair and the symbol μ_D to denote the true, but unknown, mean of the population from which the values D_i come. (The decision to subtract the first reading from the second, rather that the second from the first, is arbitrary.)

Our strategy is to construct a mean, $\bar{D}$, of these observed differences and then standardize it. From Chapter 9, we know that

$$T = \frac{\bar{D} - \mu_D}{\frac{S_D}{\sqrt{n}}}$$

is distributed as Student's T statistic with $(n - 1)$ degrees of freedom. Thus, we write our null hypothesis as H_0: $\mu_D = K$ (in which K is an arbitrary constant; often, $K = 0$). Then we pick the level of significance, determine the critical values, and calculate the test statistic. If the test statistic falls into the critical region, the null hypothesis, H_0: $\mu_D = K$, is rejected.

Example 11-2: Paired differences

To illustrate the paired-observations analysis, suppose that your problem is one of verifying the equivalence of two scales used to weigh containers of UO_2. Your plan is to weigh each of n containers once on each of the two scales. You will not be concerned about the scales' equivalence unless the average difference is significantly different from zero. Thus, you establish the null hypothesis H_0: $\mu_D = 0$ and the alternative hypothesis H_1: $\mu_D \neq 0$. As usual, you set $\alpha = 0.05$. You have 10 containers of UO_2 to work with, so you know you have $(n - 1) = 9$ degrees of freedom. From Table T-3, you find the critical value $t_{0.975}(9) = 2.26$; i.e., you will reject the hypothesis if the calculated t is less than -2.26 or larger than +2.26.

Suppose the data you collect are displayed in Table 11-2, where the readings are given in kilograms, along with the relevant statistic.

Table 11-2:
UO_2 container weights, in kg, as reported by two scales

Container i	Scale 1 y_{i1}	Scale 2 y_{i2}	Difference d_i
1	25.6	25.4	0.2
2	21.3	21.1	0.2
3	21.3	21.7	-0.4
4	28.4	28.4	0.0
5	29.9	30.0	-0.1
6	30.0	29.9	0.1
7	23.4	23.1	0.3
8	29.5	29.6	-0.1
9	27.7	27.5	0.2
10	21.3	21.1	0.2
			$n = 10$
			$df = 9$
			$\bar{d} = 0.06$
			$s_d = 0.21$
			$s_d/\sqrt{n} = 0.07$

You calculate the test statistic

$$t = \frac{\bar{d} - 0}{s_d/\sqrt{n}} = \frac{0.06 - 0}{0.07} = 0.86.$$

Since the critical value for $t_{0.975}(9)$ is 2.26, your test statistic does not fall into the critical region. Hence, you have no reason to claim that the two different scales have different means.

For discussion:

- Consider the conclusions of Example 11-2. Can you be wrong? What is the chance of that?
- Could you have used a constant other than zero in the statement of the null hypothesis H_0: $\mu_D = 0$? Would it have made sense?
- Suppose you are now handed a reading on container #11 from scale 1, without a matching reading from scale 2. What would be your most obvious option in analyzing the data? Would it be the right thing to do?

Case 2: Variances known

This may seem to be an optimistic scenario. If you know the variance of each of the independent variables involved, the test statistic is constructed with the help of the standard deviation of a difference in means, discussed earlier in this chapter. The test for comparing the mean of Population A to that of Population B is formally presented by

H_0: $\mu_A - \mu_B = 0$,

H_1: $\mu_A - \mu_B \neq 0$,

$\alpha = \alpha_0$.

The corresponding test statistic is given by

$$Z = \frac{\bar{Y}_A - \bar{Y}_B}{\sqrt{\sigma_A^2/n_A + \sigma_B^2/n_B}}$$

and the critical region by $|Z| > z_{(1 - \alpha/2)}$.

Thus, if $\alpha_0 = 0.05$, then the null hypothesis is rejected if $|Z| > 1.96$. If $|Z|$ does not exceed 1.96, then H_0 cannot be rejected.

Example 11-3:
Mean percent uranium in UO_2 pellets

To illustrate the process of comparing two groups with known variances, consider the manager of a facility producing uranium oxide (UO_2) pellets who wishes to compare the mean percent uranium produced under two different operating processes. Assume that it has been established from experience that the percent uranium of pellets in each process is normally distributed with unknown mean and with variance $\sigma^2 = 0.0055$ regardless of the specific process.

In keeping with the general discussion, you write the hypotheses and the test statistic as:

H_0: $\mu_A - \mu_B = 0$,

H_1: $\mu_A - \mu_B \neq 0$,

$\alpha = 0.05$, and

$$Z = \frac{\bar{Y}_A - \bar{Y}_B}{\sqrt{\dfrac{0.0055}{n_A} + \dfrac{0.0055}{n_B}}}$$

$$= \frac{\bar{Y}_A - \bar{Y}_B}{0.0742\sqrt{\dfrac{1}{n_A} + \dfrac{1}{n_B}}}.$$

Thus, the null hypothesis is rejected if $|Z| > 1.96$; if $|Z|$ does not exceed 1.96, then no conclusion is drawn.

Suppose a random sample of $n_A = 8$ pellets is taken from a batch of pellets made under Process A and a random sample of $n_B = 12$ pellets is taken from the batch made under Process B. Consider the data shown in Table 11-3.

Table 11-3:
Comparing percent of uranium in UO_2 pellets for two processes with known variances

	Process *A*	**Process *B***
	88.056	87.939
	88.088	87.883
	88.044	88.005
	88.015	88.064
	87.897	88.001
	88.039	87.977
	87.950	87.881
	88.113	87.946
		88.107
		87.970
		87.923
		88.119
Sample size	8	12
Sample mean	88.0253	87.9846
Sample variance	0.0050	0.0063
Sample standard deviation	0.0711	0.0790

The calculated statistic is $z = 1.20$. Hence, we do not have sufficient evidence to claim that the two processes have different means. Note that, because $\sigma^2 = 0.0055$ is given, the sample variance is not used in this test.

For discussion:

- This treatment of Case 2 is a direct consequence of the Central Limit Theorem discussed in Chapter 7. What must be demonstrated to connect the theorem to Case 2?
- Show that the test statistic is indeed $z = 1.20$.
- Be honest now. Which of the two processes, A and B, would you prefer? Would you pay a premium to obtain pellets from that process?
- You just realized that you have sample variances available to you. What will you do with those?

Case 3: Variances unknown but assumed equal

Case 3 is perhaps the most common situation involving the test of a hypothesis about two means. The two groups, designated by A and B, are assumed to have equal variances, so that pooling the variances is an important part of the process. The samples sizes from the two groups are n_A and n_B, respectively.

Start by writing the null hypothesis that the group means are equal as

H_0: $\mu_A - \mu_B = 0$.

Next, write the alternative hypothesis as

H_1: $\mu_A - \mu_B \neq 0$, for a two-sided alternative,

or as either

H_1: $\mu_A - \mu_B > 0$ or H_1: $\mu_A - \mu_B < 0$ for the appropriate one-sided alternative.

Finally, set $\alpha = 0.05$, say.

As indicated, the test statistic uses the pooled variance of the two samples, S_p^2, as described earlier in this chapter. The appropriate test is Student's T, given by the formula

$$T = \frac{\bar{Y}_A - \bar{Y}_B}{\sqrt{S_p^2\left(\frac{1}{n_A} + \frac{1}{n_B}\right)}}$$

The calculated T statistic is then compared to appropriate quantiles of Student's T distribution.

For a two-sided alternative hypothesis, reject H_0 if $T > t_{0.975}(n_A + n_B - 2)$. If the alternative hypothesis is H_1: $\mu_A - \mu_B > 0$, reject H_0 if $T > t_{0.95}(n_A + n_B - 2)$. If the alternative hypothesis is H_1: $\mu_A - \mu_B < 0$, reject H_0 if $T < t_{0.95}(n_A + n_B - 2)$.

The procedure is illustrated in Example 11-4.

Example 11-4: *Measuring radiological contamination*

As a follow-up of the decommissioning and cleaning of a nuclear reactor, measurements of radiological contamination were taken by two teams. Suppose you wish to test whether the two teams yield similar averages. To that end, the floor of a specific building was sampled independently by the two teams. The data are summarized in Table 11-4, where radiological contamination is measured in disintegrations per minute (dpm) per

100 cm^2. Assume that the variances of the two teams are equal.

Table 11-4:
Measurements of radiological contamination, in dpm/100cm^2

Sample	Team A	Team B
Mean	1,988	2,008
Variance	2,051	2,447
Size	36	39
df	35	38

Because the population variances are considered equal, you begin by pooling the sample variances. The pooled variance is calculated as

$$s_p^2 = \frac{(36)(2051) + (39)(2447)}{36 + 39 - 2} = 2257.14.$$

The standard error for the mean difference is

$$se = \sqrt{2217.14(\frac{1}{36} + \frac{1}{39})} = 10.98,$$

from which the sampled-based Student's t with $36 + 39 - 2 = 73$ degrees of freedom is calculated as

$$t = 20.0/10.98 = 1.82.$$

Since t falls short of $t_{0.95}(73) = 1.99$, you do not have sufficient evidence to claim that the two teams differ in their means.

For discussion:

- Some authors (e.g., Mendenhall and Ott, 1980, 197-198) observe that when both n_1 and n_2 are at least as large as 30, you may assume that $\sigma_1 = s_1$ and that $\sigma_2 = s_2$ and continue with the analysis given for Case 2. What are the consequences of this procedure?

Case 4: Variances unknown and not equal

Case 4 treats the following situation: You have two independent samples from normal distributions with means μ_A and μ_B and variances σ_A^2 and σ_B^2. Your interest is in testing the null hypothesis H_0: $\mu_A = \mu_B$. Although this sounds like the proverbial piece of cake, this problem has provoked controversy since the early decades of the 20th century. Known as the *Behrens-Fisher* problem, it continues to be an irritation in statistical theory and practice.

Brownlee (1965, pp. 299-303) gives a succinct account of the controversy—and explains and illustrates a pleasing solution in terms of Welch's approximation (1937, 1947).[2] The robustness of the *t* test and the approximation's practicality make it an important tool for dealing with an otherwise impossible situation.

The scenario is simple. As stated, two independent groups are to be tested for equality of means. The variances of the two groups are believed to be different and the sizes of

[2] Another name associated with this process is that of Satterthwaite (1946), who extended Welch's ideas from 2 to *k* groups.

the collected samples are small, say less than 30. This case clearly is not included in Cases 1, 2, or 3.

You write the hypotheses and the test statistic as follows:

H_0: $\mu_A - \mu_B = 0$,

H_1: $\mu_A - \mu_B \neq 0$,

$\alpha = \alpha_0$, and

$$T = \frac{(\bar{Y}_A - \bar{Y}_B) - 0}{\sqrt{\frac{S_A^2}{n_A} + \frac{S_B^2}{n_B}}},$$

where the approximate degrees of freedom, *df*, are given by the expression

$$df = \frac{\left[\frac{S_A^2}{n_A} + \frac{S_B^2}{n_B}\right]^2}{\frac{\left[\frac{S_A^2}{n_A}\right]^2}{n_A - 1} + \frac{\left[\frac{S_B^2}{n_B}\right]^2}{n_B - 1}}.$$

This formula for the degrees of freedom is unusual: it is one of the few instances in hypothesis-testing in which you must determine the degrees of freedom from the sample data themselves.

Example 11-5:
Yield stress of stainless steel pipes

The average yield stress of stainless steel pipes, measured in kilopounds per square inch (ksi), of two manufacturers are to be compared for quality using a statistical hypothesis-testing procedure. You are informed that the variances of the two manufacturers are definitely *not* equal. So you use the procedure outlined in Case 4. Data from an experiment conducted at 100°F are summarized in Table 11-5.

Table 11-5:
Data for comparing yield stress (in ksi) for two manufacturers of steel pipes (test conducted at 100°F)

Sample	**Manufacturer A**	**Manufacturer B**
Sample size	5	8
Degrees of freedom	4	7
Mean	82.3	71.4
Variance	108.16	7.84
Standard deviation	10.4	2.8

As indicated, you write the hypotheses and the test statistic as follows:

H_0: $\mu_A - \mu_B = 0,$

H_1: $\mu_A - \mu_B \neq 0,$

$\alpha = 0.05,$

and evaluate the test statistic

$$t = \frac{(82.3 - 71.4) - 0}{\sqrt{\frac{108.16}{5} + \frac{7.84}{8}}} = 2.29,$$

where the degrees of freedom, *df*, are given by the approximating expression

$$df = \frac{\left[\frac{108.16}{5} + \frac{7.84}{8}\right]^2}{\frac{\left[\frac{108.16}{5}\right]^2}{4} + \frac{\left[\frac{7.84}{8}\right]^2}{7}} = 4.37.$$

Table T-3 provides quantiles for integer degrees of freedom, so some interpolation is needed. You find $t_{0.975}(4) = 2.78$ and $t_{0.975}(5) = 2.57$. The problem becomes one of finding $t_{0.975}(4.37)$. Since 4.37 is 37/100 of the distance between 4 and 5, you find that 37/100 of the distance between $t_{0.975}(4) = 2.78$ and $t_{0.975}(5) = 2.57$ is 37/100(2.78 - 2.57) = 0.065. Thus, the critical value *for this test and these data* is $t_{0.975}(4.37) = 2.78 - 0.065 = 2.715$. Finally, because the test statistic $t = 2.29 < 2.715$, you have insufficient statistical evidence from this experiment that the two manufacturer's stainless steel pipes are different with respect to yield stress.

For discussion:

- Using two different techniques given in Chapter 10. show that the variances in Example 11-5 are indeed different—and hence the variances should *not* be pooled.

- Reconsider Example 11-5, this time with the hypothesis tested as if Case 3 applies. What happens to your decision if you go ahead and pool the variances?

- Three forms of statements about two-mean hypotheses were given at the beginning of this chapter. Work either *Form 2* or *Form 3* all the way through Cases 1, 2, 3, and 4. What are the specific up-side and down-side features of each of the four cases with respect to the type you picked?

- You've been guided through four cases for testing equality of two means. But these four cases are not exhaustive. Can you think of at least one other? (Hint: Examine the assumptions of any one of the four cases and consider what happens when you violate one of them.)

What to remember about Chapter 11

Chapter 11 built on the foundations laid in Chapters 9 and 10. You were introduced to some of the forms that may be taken when you ask questions about the means of two groups. You saw how the processes are linked to understanding the ideas of two special concepts:

- *the standard deviation of a difference*
- *pooling the variances of two groups.*

Then you saw four different cases in which basic assumptions determined the details of the analysis:

- *paired observations*
- *variances known*
- *variances unknown but assumed equal*
- *variances unknown but not assumed equal.*

12

Testing statistical hypotheses: several means

What to look for in Chapter 12

Chapter 12 extends the two-group comparisons of Chapter 11 to several groups. The general technique employed here is called *analysis of variance* (often shortened to the acronym *ANOVA*), a name given to a vast, venerable-and-still-developing body of statistical inferential procedures aimed at comparing the means of several populations. Both the vastness and the venerability of ANOVA preclude a full-blown treatment in this book; however, this chapter does introduce and explore the simplest of all ANOVA applications: the *one-way classification ANOVA*[1]—in which observations are classified according to at least two levels of a *single* criterion. You will be shown the tools and the procedures

[1] We choose to use the simpler term *one-way ANOVA* in the rest of the chapter.

to handle the calculations and to interpret the resultant analysis.[2] Thus, Chapter 12 will show you how to:

- *write a simple data model*
- *recognize the assumptions required for running an analysis of variance*
- *interpret the analysis*
- *identify direct extensions of the simple analysis*
- *appreciate the equivalence of Student's* T *test, as presented in Chapter 11, and the analysis of variance for testing equality of two group means.*

Most important of all, you will learn to recognize when the scope of your problem exceeds the scope of this book.

Setting up a one-way analysis of variance

Because sufficient complexities arise in this chapter regarding analysis-of-variance processes, it seems reasonable to begin with a set of data and to build the ensuing discussion on it. To that end, then, consider the situation posed by Example 12-1 and the related illustrative data in Table 12-1.

Example 12-1: *Comparing mean percent uranium from four production lines*

Suppose your manager asks you to investigate whether four production lines yield the same mean percent uranium. Your manager hands you the numbers displayed in Table 12-1, wishes you good luck, suggests that the weather will be nice at the beach this weekend, and leaves you to your own devices. Oh, yes, one more thing: it's Friday afternoon at 3 o'clock, and the answers are needed

[2] One useful start-up reference to more involved analyses is Neter, *et al.* (1990).

for a 9 o'clock meeting with the manager's managers on the following Monday morning. Oh, yes, there's another one more thing: you are *not* invited to Monday's meeting—your manager will carry the message.

Table 12-1:
A data layout for comparing the mean percent uranium from four production lines

Data Layout	Production Line 1	Production Line 2	Production Line 3	Production Line 4
	87.2	87.7	88.5	87.5
	87.4	87.7	88.9	87.3
	87.5	88.0	88.5	87.2
		87.8	88.6	87.6
		88.2	88.7	87.4
		87.7	88.7	87.9
		87.6	88.9	
		88.3	88.3	
		88.0	88.8	
		87.8	88.4	
		87.6	88.8	
			88.5	
			88.2	
			88.3	
			88.4	

data layout

Table 12-1 is called a *data layout* because it *lays out* the data in a fashion that allows you to look at them in a useful way, irrespective of their original formatting or the sequence in which they were acquired. At least, your manager didn't leave a pile of laboratory notes for you to sift.

A good first thing to do with any new data is to sit down and take a good hard look at it. That means you calculate some basic descriptive statistics and, even more importantly, build a graphical display so you can "see" what the data are trying to tell you. Table 12-1a adds a few rows to the bottom of Table 12-1, each new row containing corresponding descriptive statistics such as the mean and the standard deviation for each of the four production lines. Figure 12-1 shows the values of the individual data with each production line's mean indicated.

Table 12-1a:
The data layout from Table 12-1 with some descriptive statistics added

Data layout with descriptive statistics	Production line 1	Production line 2	Production line 3	Production line 4
	87.2 87.4 87.5	87.7 87.7 88.0 87.8 88.2 87.7 87.6 88.3 88.0 87.8 87.6	88.5 88.9 88.5 88.6 88.7 88.7 88.9 88.3 88.8 88.4 88.8 88.5 88.2 88.3 88.4	87.5 87.3 87.2 87.6 87.4 87.9
Sample size: n_i	3	11	15	6
Sum: $\sum_j y_{ij}$	262.1	966.4	1328.5	524.9
Sum of squares: $\sum_j y_{ij}^2$	22,898.85	84,903.20	117,661.53	46,920.31
Mean: $\bar{y}_i$	87.37	87.85	88.57	87.48
Correction term: $(\sum_j y_{ij})^2/n_i$	22,898.8033	84,902.6327	117,660.8167	45,920.0017
Sum of squared deviations: ssd_i	0.0467	0.5673	0.7133	0.3083
Degrees of freedom: df_i	2	10	14	5
Variance: s_i^2	0.0233	0.0567	0.0510	0.0617
Standard deviation: s_i	0.15	0.24	0.23	0.25

Figure 12-1:
Graphic display of Table 12.1a's four production lines' data values

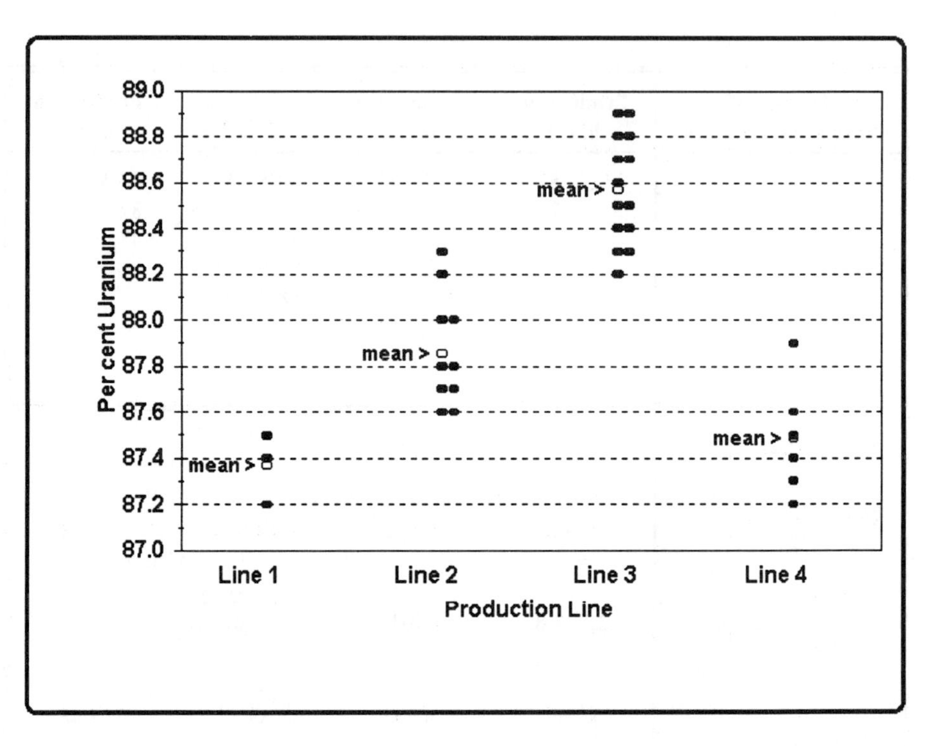

As you can see from Tables 12-1 and 12-1a and Figure 12-1, you are dealing here with data from four samples, with sample sizes $n_1 = 3$, $n_2 = 11$, $n_3 = 15$, and $n_4 = 6$. You might be tempted to apply repeatedly the *T*-test-based methods of Chapter 11. That is, you compare all possible pairs of means, six in all, and try to reach a conclusion. Don't do it. It's just not the right thing.

The underlying principle of the analysis of variance lies in its very name: you will analyze the variance of *all of the data* by dividing it into two or more meaningful parts and determine how those parts relate to each other. This eliminates the probabilistic and interpretive mistakes that would occur if you applied the T test to all possible pairs of means.

Modern communication theory offers a vocabulary that you may find useful. Any set of data may be said to include two components: *signal* and *noise*. ANOVA provide a methodology by which, if the conditions of the data-collection activity are appropriate, you may separate the signal from the noise.

To elaborate on the last point: Suppose you were to administer the same English (or algebra or arithmetic) test to 30 children from the same school, 10 pupils from each of the first, second, and third grades. As you examine the test scores, you observe a pronounced variability among the 30 scores. You ought not to be surprised at this finding. Shouldn't the third-graders perform better than the first- and second-graders? As a matter of fact, you ought not to be surprised that you can tell with relative ease which test score came from which grade; that is, if the scores cluster into three groups, you expect that the 10 lowest scores belong to the first-graders and the 10 highest scores to the third-graders, with the 10 scores for the second-graders falling between those two. You conclude that the signal (the differences *among* the grades) rises above the noise (the differences *within* the grades).

In the ANOVA spirit, the three grades are distinct to the point where you can separate the "total variation" of the scores into two components:

(1) variation reflecting different grades, which might be called "among-grades variation," and

(2) variation reflecting the usual fluctuations you might expect among children in the same class, or "within-grades variation."

As demonstrated in Chapter 5, variation (dispersion) is measured in a number of ways, among them the range, the variance, and the standard deviation. The measure of variation that lends itself most conveniently to ANOVA's quantitative partition is the *sum of squares*, which was introduced in Chapter 5. Reminder: the sum of squares for any univariate dataset is the product of the variance and its degrees of freedom.

The example involving the testing of primary-school children is, perhaps, trivial because you do not expect to need any fancy statistical theory to tell you that there are skill differences among the three grades. So now consider a perhaps less trivial example with essentially the same data structure.

Suppose that you are reviewing the percent-of-uranium readings from 35 randomly selected UO_2 pellets from four production lines, as shown in Table 12-1. Your innate curiosity surely prompts you to ask if the means of the four production lines are alike. What do you do?

You first form some sort of hypothesis. In words, it might read: *The null hypothesis is that the means of the four production lines are all the same.*

total sum of squares (SS_T)

To examine this hypothesis, you employ the ANOVA prescription that partitions the *total sum of squares* (SS_T) —which measures the variation among *all* 35 pellets, regardless of their batch affiliation—into two independent sums of squares:

among-groups sum of squares (SS_G)

The first sum of squares, called the *among-groups sum of squares* (SS_G), measures the variation among the batch means; i.e., among the four values $\bar{Y}_1$, $\bar{Y}_2$, $\bar{Y}_3$, and $\bar{Y}_4$. The among-groups sum of squares is also called the *group sum of squares* or the *among sum of squares*.

within-groups sum of squares (SS_W)

The second sum of squares, called the *within-groups sum of squares* (SS_W), measures the joint variation within the groups. The within-groups sum of squares is also called the *within sum of squares* and, somewhat more commonly, the *error sum of squares*. The within-groups sum of squares is directly associated with the pooled sum of squares obtained from the individual batches, as you will see later in this chapter.

Algebraically, the three sums of squares are related by:

$$SS_T = SS_G + SS_W.$$

This relationship is an identity; i.e., it holds for any set of numbers arrayed in terms of the data layout exemplified by Table 12-1.

For discussion:

- Examine Figure 12-1 carefully. Without using any particular analytical tools, speculate whether:

  - the four production lines are alike?
  - some of the four production lines are alike?
  - none of the four production lines are alike?

  Indeed, what is meant by saying that two or more of the production lines are *alike*?

- If the dataset in Table 12-1 represents a one-way classification, what condition(s) that would render this dataset a two-way classification?

Formulating the generic one-way analysis of variance

one-way classification

Now consider a generic data layout for the simplest ANOVA problem. A data layout for this general case, such as the one given in Table 12-2, is called a *one-way* classification because each observation is classified according to a single criterion—its group affiliation. The data layout shows k groups, each being a sample from one of k populations under study.

Table 12-2:
A data layout for a generic one-way analysis of variance

Data layout	Group 1	Group 2	...	Group i	...	Group k
	Y_{11}	Y_{21}	...	Y_{i1}	...	Y_{k1}
	Y_{12}	Y_{22}	...	Y_{i2}	...	Y_{k2}
	Y_{13}	Y_{23}	...	Y_{i3}	...	Y_{k3}
	$\vdots$	$\vdots$	$\vdots$	$\vdots$	$\vdots$	$\vdots$
	Y_{1j}	Y_{2j}	...	Y_{ij}	...	Y_{kj}
	$\vdots$	$\vdots$	$\vdots$	$\vdots$	$\vdots$	$\vdots$
	Y_{1n_1}	Y_{2n_2}	...	Y_{in_i}	...	Y_{kn_k}

Table 12-2a extends Table 12-2 by adding some specific descriptive statistics to the data layout.

Table 12-2a:
Descriptive statistics associated with the data layout for a generic one-way analysis of variance in Table 12-2

Descriptive statistics	Group 1	Group 2	...	Group i	...	Group k
Sample Size:	n_1	n_2	...	n_i	...	n_k
Sum:	$\sum_j Y_{1j}$	$\sum_j Y_{2j}$	...	$\sum_j Y_{ij}$	...	$\sum_j Y_{kj}$
Sum of squares:	$\sum_j Y_{1j}^2$	$\sum_j Y_{2j}^2$	...	$\sum_j Y_{ij}^2$	...	$\sum_j Y_{kj}^2$
Mean:	$\bar{Y}_1$	$\bar{Y}_2$	...	$\bar{Y}_i$	...	$\bar{Y}_k$
Correction term:	$\frac{(\sum_j Y_{1j})^2}{n_1}$	$\frac{(\sum_j Y_{2j})^2}{n_2}$	...	$\frac{(\sum_j Y_{ij})^2}{n_i}$	...	$\frac{(\sum_j Y_{kj})^2}{n_k}$
Sum of squared deviations:	SSD_1	SSD_2	...	SSD_i	...	SSD_k
Degrees of freedom:	df_1	df_2	...	df_i	...	df_k
Variance:	s_1^2	s_2^2	...	s_i^2	...	s_k^2
Standard deviation:	s_1	s_2	...	s_i	...	s_k

balanced design

If your study is designed to collect the same number of observations from each of the k groups, it is said to have a *balanced design*. If the experiment follows the design (that is, if no unplanned forces intervene), then the data layout also is said to be balanced. This distinction between balanced and unbalanced designs is critical to the data analysis, especially for multi-way classifications; i.e., those designs in which the observations are classified by more than one criterion. Unbalanced designs are inherently more difficult to analyze than balanced designs; worse, even when they yield to the analytical process, they tend to be difficult to interpret and understand.

Examine the symbolic data entries in Tables 12-2 and 12-2a. Each observation has two subscripts: the first denotes group affiliation and the second denotes a sequence number within the group. Should there be some ambiguity (such as the 27th observation in the 54th group versus the 427th observation in the 5th group), the two subscripts may be separated by a comma (e.g., $Y_{54,27}$ versus $Y_{5,427}$). Note also in Tables 12-2 and 12-2a that Group i has n_i observations. Thus, the information in the body of the table does *not* form a rectangle; rather, it forms a set of columns of varying lengths. In some representations of the balanced layout, n needs no subscript because it is constant for each group.

For each of the k groups, you calculate and record several descriptive statistics, including the group's size and its sum, mean, variance, and sum of squares. Let SSD_i denote the sample sum of squared deviations of the ith group for which you have $(n_i - 1)$ degrees of freedom. Formally, SSD_i is given by

$$SSD_i = \sum_{j=1}^{n_i} (Y_{ij} - \bar{Y}_i)^2 = \sum_{j=1}^{n_i} Y_{ij}^2 - (\sum_{j=1}^{n_i} Y_{ij})^2 / n_i .$$

Recall that the summation sign (i.e., the Greek upper-case sigma, Σ) with an underscript j and overscript n_i means to sum the expression immediately to the right of Σ when j assumes the values of 1 through n_i, starting with $j = 1$ and ending with n_i.

Recall also that the variance of the values in the ith group is obtained from

$$S_i^2 = \frac{SSD_i}{(n_i - 1)}.$$

For discussion:

- If $\sum_{j=1}^{n_i} Y_{ij}$ is the sum of all observations in Group i, what meaning do you give to $\sum_{k=1}^{k} \sum_{j=1}^{n_i} Y_{ij}$?

- Describe a situation where observations have a two-way classification. What would you expect the subscript of Y to look like?

- If your intuition suggests that drawing the same number of observations from each group is a good idea, your intuition guides you well. A balanced design is usually desirable, unless your resources dictate otherwise.

- A balanced design, i.e., one in which the intention is to collect equal numbers of observations from each group, does not always meet its intention. Departure from balance may be the result of an *incomplete experiment* or of *missing observations*:

  An *incomplete experiment* occurs when conditions make it impossible to measure everything that was intended in the experimental design. For example, if measurements on some signal-carrying cables are unavailable because those cables are buried inside concrete walls, you have to be sure that the inference of the experiment is not extrapolated to include the buried cables.

Missing observations arise when observations that were properly collected subsequently become lost, folded, spindled, or mutilated. As a user of the data, you will want to ask whether there is any pattern to the missing portions that may impact your conclusions.

- The data analysis for the one-way classification, be it a balanced layout or not, is straightforward. When you graduate from the restrictions of the one-way classification, the analysis of datasets with missing observations becomes complicated. So always keep *balanced design* on your list of data desiderata.

- Still another issue that may come up as you approach your data analysis is that of *outlying observations*. As you review the collected data, you may come across an observation—or *several* observations—that is atypical for the dataset. If the unusual observation is obviously the result of, say, a faulty instrument, then, quite generally, the observation should not be a part of the analysis. An observation which is plausible, yet "unlikely," is called an *outlier* or an *outlying observation*. The question then is how to determine that unlikeliness. The treatment of outliers is beyond the scope of this text. For simple databases, you will find Dixon's Criterion, given by Dixon and Massey (1983, pp. 377-380), useful. *But do watch out!* Sometimes it's the outliers that are the most informative observations in the study. How can that be?

A model for one-way classification

model

The data structure for the layout shown in Tables 12-2 and 12-2a can be described by a *model*. This model shows that every observation Y_{ij} (i.e, the jth observation in the ith group) can be written compactly as:

$$Y_{ij} = \mu_i + E_{ij}; \ i = 1, ..., k; \ j = 1, ..., n_i;$$

where

μ_i = mean of the population from which the ith group was selected,
E_{ij} = error associated with the jth observation of the ith group, and
$E_{ij} \sim N(0, \sigma^2)$.

It is instructive and analytically useful to write the model in an equivalent form:

$$Y_{ij} = \mu + \tau_i + E_{ij}; \quad i = 1, \ldots, k; \quad j = 1, \ldots, n_i;$$

where

μ = the overall mean of all populations considered
$\tau_i = \mu_i - \mu$ = the *contribution* of the ith group that causes a "shift" away from the overall mean, and
$E_{ij} \sim N(0, \sigma^2)$.

If all the means are alike, the contribution τ_i of every group is zero. Thus, in the spirit of statistical hypothesis-testing, asking if all the μ_i are the same is equivalent to asking whether all τ_i are zero.

Assumptions for the one-way ANOVA

Let's make no mistake: we must be crystal clear about the assumptions that underlie the one-way ANOVA. When you perform a one-way ANOVA, you make four important assumptions about the data structure that lead you to the statistical test you perform. If those assumptions are incorrect, so may be your conclusions.

The first assumption is that the data classification scheme, as given by the model, is correct. If, for example, the English-test scores for the 30 children are classified only according to their grade in school, you may very well be ignoring some relevant factors. Is gender a factor? Is the teacher a factor? Is it possible that boys taught by Mrs. Lurie in the second grade do better than girls taught by Mrs. Moore in the third grade?

The second assumption is that the variation within a group is the same as that within any other group. As you've already seen, you have a way of testing the appropriateness of this assumption in Chapter 10, using the F_{max} statistic. Whereas this assumption is often plausible and reasonable, you must be aware that homoscedasticity is a basic assumption in many ANOVA analyses.

The third assumption is that the observations are drawn randomly and independently. It is unfair and statistically unacceptable to demand that the teacher's daughter be a part of the sample, or to insist that, if Jimmy is a part of the sample, Jimmy's brother must be excluded. Selective sampling is always inappropriate to an unbiased statistical analysis.

The fourth assumption is that the data are distributed normally. With experience, you learn to recognize when the normality assumption is not satisfied and then to make fully justified adjustments in your analysis. However, until you gain that experience—or when everything else fails—you can resort to methods reported in Chapter 7, where testing data for normality is discussed.

For discussion:

- The discussion about the four underlying ANOVA assumptions is tied to the data model discussed earlier. Show where and how each of the discussed assumptions can be keyed to the model.

- To get off our soap box for a moment, we note that we have just prescribed a recipe for a futile exercise: We seldom can be sure that *all* of our assumptions are correct and proper. Then we have to wonder about the multitude of statistical analyses conducted daily where many investigators, if not most, live in varying degrees of statistical sin. The important point to remember, however, is that we all must be fully aware of underlying assumptions and be equally aware of the consequences of making an incorrect assumption. Salvation lies, once again, in the *robustness* of the ANOVA.

Stating hypotheses for the one-way ANOVA

Coupled with well-stated assumptions, you must have well-stated hypotheses. If you cannot state your test objectives properly, as reflected in the null and the alternative hypotheses, and/or you cannot follow through on them, you may still get an answer—but will it be to the question you set out to answer?

The null hypothesis for the one-way classification with k groups is stated as

$$H_0:\ \mu_1 = \mu_2 = \mu_3 = \ldots = \mu_k$$

or, equivalently, as

$$H_0:\ \tau_1 = \tau_2 = \tau_3 = \ldots = \tau_k = 0.$$

The alternative hypothesis can be written in several ways, two of which are:

$$H_1:\ \mu_i \neq \mu_j, \quad \text{for some } i, j,$$

and

$$H_1:\ \tau_i \neq 0, \quad \text{for some } i.$$

For discussion:

- Why is it *incorrect* to write the alternative hypothesis as

  H_1: $\mu_1 \neq \mu_2 \neq \mu_3 \neq \ldots \neq \mu_k$?

- Suggest other ways of writing the alternative hypothesis.

- The analysis of more than two groups is always considered a two-sided hypothesis. Why? (*Hint*: Try writing a one-sided alternative hypothesis for, say, three groups).

- Emphasizing the importance of a proper articulation of the hypothesis *and* the selection of appropriate analytical tools, we are reminded of a statement we've heard mentioned many times but whose source we have been unable to trace:

  It is better to have
  an approximate solution to an exact problem
  than
  an exact solution to an approximate problem.

Calculations for the one-way ANOVA

The calculations associated with even the simplest ANOVA can be lengthy, so be sure your calculator batteries are charged before you start. Once you understand the steps involved in the ANOVA construction and have graduated from these studies with *bona fide* certification, your setting up and running an ANOVA computer program would be permissible.

ANOVA table

The results of an analysis of variance are displayed traditionally in an *ANOVA table*, a generic version of which is shown as Table 12-3. When you use this display, you see all the detail needed for drawing your conclusions—and for showing them to others.

Table 12-3:
A generic ANOVA table for a one-way layout

Source of variation	Sum of squares	Degrees of freedom	Mean squares	F
Among groups	SS_G	df_G	$MS_G = SS_G/df_G$	$F_G = MS_G/MS_W$
Within groups	SS_W	df_W	$MS_W = SS_W/df_W$	
Total	SS_T	df_T		

The main body of a one-way ANOVA table shows three rows, each reflecting one of three sources of variation: Among Groups, Within Groups, and Total. But, before you can calculate the entries for the ANOVA table, you need to make three intermediate calculations.

What follows is a side-by-side description of the formal calculating steps shown on the left with the corresponding numerical calculations derived from the problem laid out in Example 12-1 (with its data contained in Table 12-1) shown on the right.

Procedure	**Example 12-1**
Experiment size: $N = \sum_{i=1}^{k} n_i$	Experiment size: $N = 3 + 11 + 15 + 6 = 35$
Grand mean: $\bar{\bar{Y}} = \frac{\sum_{i=1}^{k} \sum_{j=1}^{n_i} Y_{ij}}{N}$	Grand mean: $\bar{\bar{y}} = \frac{(87.2 + 87.4 + \ldots + 87.9)}{35} = \frac{3{,}081.9}{35} = 88.05$
Correction term: $CT = \frac{(\sum_{i=1}^{k} \sum_{j=1}^{n_i} Y_{ij})^2}{N}$ Note that the Correction Term displayed here is identical to the correction term you used in the "working formula" for the variance (Chapter 5), if you regard all of the observations in the experiment as if they belong to a single group.	Correction term: $ct = \frac{(3081.9)^2}{35} = 271{,}374.5031$

Now you are ready to calculate the required entries for Table 12-3. The table is built in pieces, starting with the column labeled **Sum of squares**.

Procedure	Example 12-1
The first calculation is for the bottom row of Table 12-3, in which the **Source of variation** is identified as **Total** to indicate the *un*partitioned source of variation. The **Total Sum of squares** is denoted SS_T. It is calculated by the formula $SS_T = \sum_{i=1}^{k} \sum_{j=1}^{n_i} (Y_{ij} - \bar{Y})^2 = \sum_{i=1}^{k} \sum_{j=1}^{n_i} Y_{ij}^2 - CT$. Note that SS_T is written here in two mathematically equivalent formulas. The first formula is the "definition formula," and the second is the "working formula." For the most part, you use the second expression because it generally is the easier to calculate.	$ss_T = 87.2^2 + 87.4^2 + \ldots + 87.9^2 - 271{,}374.5031 = 271{,}383.8900 - 271{,}374.5031 = 9.3869$
The **Total Sum of squares** has $df_T = N - 1$ associated degrees of freedom. Note that, if you were to divide SS_T by df_T, you would obtain a value that might be called "the variance of the entire data set." But there is no point in doing that: you're dealing with a model and a procedure that are much more complicated than those connected with a single-sample dataset.	$df_T = 35 - 1 = 34$

Procedure	Example 12-1
The second calculation fills in the second row from the top. The **Among-groups Sum of squares** is denoted by SS_G; it is calculated by the formula expressed in two forms: the "definition formula" and the "working formula." $SS_G = \sum_{i=1}^{k} n_i(\bar{Y}_i - \bar{\bar{Y}})^2 = \sum_{i=1}^{k} \frac{(\sum_{j=1}^{n_i} Y_{ij})^2}{n_i} - CT.$ The "working formula" is used in this presentation.	$SS_G = \frac{(262.1)^2}{3} + \frac{(966.4)^2}{11} + \frac{(1,328.5)^2}{15} + \frac{(524.9)^2}{6} - 271,374.5031$ $= 271,382.2544 - 271,374.5031$ $= 7.7513$
The degrees of freedom associated with SS_G is $df_G = (k - 1)$; i.e., df_G equals one less than the number of groups.	$df_G = (4-1) = 3$
The third calculation yields the **Within-groups Sum of squares**, denoted by SS_W, which may be calculated using either of two approaches:	

Procedure	Example 12-1
A direct approach—albeit tedious— pools the sum of squared deviations for each group the same way you pooled the sum of squares for two groups in Chapter 11. Formally, for the ith group, you calculate $SSD_i = \sum_{j=1}^{n_i} (Y_{ij} - \bar{Y}_i)^2 = \sum_{j=1}^{n_i} Y_{ij}^2 - (\sum_{j=1}^{n_i} Y_{ij})^2/n_i$	$ssd_1 = 0.0467$ $sd_2 = 0.5673$ $sd_3 = 0.7133$ $sd_4 = 0.3083$
and then form $SS_W = \sum_{i=1}^{k} SSD_i$ to yield the desired result.	$ss_W = 0.0467 + 0.5673 + 0.7133 + 0.3083 = 1.6356$
The associated degrees of freedom is given by $df_w = \sum_{i=1}^{k} (n_i - 1) = N - k.$	$df_W = (3 - 1) + (11 -1) + (15 - 1) + (6 - 1) = 35 - 4 = 31$
An indirect approach—certainly less tedious—simply computes $SS_W = SS_T - SS_G.$	$ss_W = 9.3869 - 7.7513 = 1.6356$
Like the sums of squares, the degrees of freedom are additive, so that $df_W = df_T - df_G.$	$df_W = 34 - 3 = 31$

Consider now Table 12-3a, which is the generic one-way ANOVA table, Table 12-3, with the numerical results filled in the **Sum of squares** and the **Degrees of freedom** columns for the data in Example 12-1.

Table 12-3a:
The generic ANOVA table for a balanced one-way layout with the second and third columns completed

Source of Variation	Sum of squares	Degrees of freedom	Mean squares	F
Among-groups	$ss_G =$ 7.7513	$df_G = 3$	$ms_G = ss_G/df_G$	$f = ms_G/ms_W$
Within-groups	$ss_W =$ 1.6356	$df_W = 31$	$ms_W = ss_W/df_W$	
Total	$ss_T =$ 9.3869	$df_T = 34$		

The **Total Sum of squares** captures the variation that exists in the entire dataset, no matter what the source (or "cause") of the variation.

The **Among-groups Sum of squares** captures the variation that is caused by differences among the means of the groups. If these means were identical, then this sum of squares also would be zero. If the means were not identical, then this sum of squares increases directly as the means become more unalike.

The **Within-groups Sum of squares** captures the variation inherent within the groups, irrespective of where the groups are located with respect to each other.

Thus, by comparing the sum of squares for **Among-groups** with the sum of squares for **Within-groups**, you expect to have a test statistic for addressing the basic null hypothesis, namely,

H_0: $\mu_1 = \mu_2 = \mu_3 = \ldots = \mu_k$.

But these two sums of squares are not directly comparable. In order to attain comparability, you must form the penultimate column in Table 12-3a, labeled **Mean squares**.

Entries under **Mean squares** are formed by dividing the **Among-groups Sum of squares** and the **Within-groups Sum of squares** by their respective **Degrees of freedom**. As mentioned earlier in this chapter, there is no need to form an entry corresponding to the **Total Sum of squares**.

Once you have calculated the Mean squares, the final step requires formation of the test statistic $F = MS_G/MS_W$. The decision occurs when you compare the calculated F with a critical value obtained from Table T-4 for your choice of the level of significance and the appropriate degrees of freedom.

You are now ready to complete the ANOVA table, here displayed in Table 12-3b.

Table 12-3b:
The generic ANOVA table for a balanced one-way layout with all columns completed using data from Example 12-1

Source of Variation	Sum of squares	Degrees of freedom	Mean squares	F
Among-groups	7.7513	3	7.7513/3 = 2.5838	2.5838/0.0528 = 48.94
Within-groups	1.6356	31	1.6356/31 = 0.0528	
Total	9.3869	34		

The critical value used as a rejection criterion for this test is $f_{0.95}(df_G, df_W)$, found in Table T-4, assuming that $\alpha = 0.05$. Entries in that table are functions of α, df_G (referred to as *degrees of freedom in the numerator* and denoted in Table T-4 as df_1), and df_W (referred to as *degrees of freedom in the denominator* and denoted in Table T-4 as df_2). If the calculated ratio, f, exceeds the critical value, the means of the groups are declared significantly different.

Consistent with earlier stated philosophy/policy/practice, if $f < f_{0.95}(df_G, df_W)$, you state that you have insufficient evidence to make the claim that the population means are different.

Interpolating in Table T-4, you find that the critical value for $\alpha = 0.05$, $df_G = 3$, and $df_W = 31$ is, approximately, $f_{0.95}(3, 31) = 2.91$. Because the calculated $f = 48.97$ is larger than 2.91, you reject H_0 and conclude that the four group averages are *not* equal.

For discussion:

- The number of significant figures in the final statistics (such as mean and standard deviation), is usually one more than that reported in the original data. Intermediate calculations should use more significant figures than that. There is no harm done, nor additional effort involved, in letting the calculator/computer carry as many significant figures it can carry.

- It is important to recognize the additive nature of the sum of squares and of the degrees of freedom in the ANOVA table. Note the parallelism between the model and the ANOVA. Just like the model, where the response Y_{ij} is regarded as the sum of a group effect (τ_i) and a random error (E_i), so is the **Total Sum of squares** the sum of the **Among-groups Sum of squares** and **Within-groups Sum of squares**. This additivity is precisely what renders ANOVA a powerful systematic methodology for comparing group performance.

- Verify that $f_{0.95}(3, 31) = 2.91$.

- The Within-groups Mean square can be calculated by summing the SSD_i ($i = 1, 2, ..., k$) and dividing the total by ($N - kI$). Show how this "direct approach" is a direct extension of the variance-pooling procedure described in Chapter 11.

- Consider the general one-way ANOVA and its hypothesis-testing process. If you cannot state that the population means are not equal, can you at least claim that the sample means are different? What would you accomplish by such a claim?

Multiple-range tests

If the value of the test statistic *f*, calculated from the ANOVA, falls short of the critical value found in the *F* table, you generally say that you have insufficient evidence to claim that the population means are not equal; essentially, you are no longer concerned with the hypothesis that the means are equal.

Suppose you have statistical evidence to reject the hypothesis of equal means. If you have more than two means, you may (and, indeed, you should) be curious as to which of those means are different and to what degree they are different. Is just one of the means very much different from the others? Do they form several clusters of means? Are they such that each mean is different from each of the other means?

Thus, when you study the alternative hypotheses in a one-way ANOVA applied to three groups, you soon recognize that the alternative hypothesis (H_1: $\mu_i \neq \mu_j$, for some i, j) may not be very informative. For example, letting the three groups be denoted by A, B, and C, you may not know what triggered the statistical significance: It may be because $\mu_A \neq \mu_B$ or because $\mu_A \neq \mu_C$ or because $\mu_B \neq \mu_C$, or, perhaps, because no two of the three means are alike.

You may be tempted to test the groups pair-wise separately; that is, Group A vs. Group B, Group A vs. Group C, and Group B vs. Group C. Unfortunately, for k groups with $k > 2$, this direct approach is incorrect because the probability of finding at least one significant difference among the $k(k - 1)/2$ possible differences[3] increases as k increases. In other words, the probability of Type I error, that you carefully set at 0.05, will be much larger than 0.05 as you increase the number of comparisons. For example, if four groups are tested in a pair-wise fashion, the probability that at least one pair of the six possible means will be identified as different can

[3] Although no proof of this statement is given here, you may satisfy your curiosity by trying the formula on, say, $k = 2, 3, 4, 5$.

be as high as 0.26, an error rate which may be too high for your taste.

multiple-range tests

A widely used solution to this problem employs one of the many techniques that are classified as *multiple-range tests*. Also known as *multiple comparisons*, these techniques are designed to help you identify which means are different, while still essentially preserving the probability of Type I error. The numerous multiple-range tests may leave you apprehensive because, if you apply two or more of them to the same set of multi-group data, you will not necessarily obtain identical results. Each multiple-range test is constructed to meet somewhat different assumptions and offer differing degrees of conservatism. A full menu showing the variety and techniques of multiple-range testing is beyond the scope of this text; see Toothaker (1991), for example, for an extended discussion of this knotty and often perplexing and sometimes controversial problem.

Duncan's multiple-range test

However, despite these drawbacks, multiple-range tests provide an important decision-making tool. One of the most venerable—and, at the same time, most useful and intuitive—is *Duncan's multiple-range test* (Duncan, 1955). See Steel and Torrie (1980, Chapter 8) for a thorough discussion. Another exposition of Duncan's multiple-range test is offered by Bowen and Bennett (1988, pp. 256-261); the essence of their approach is laid out in Example 12-1a.

Example 12-1a: Duncan's multiple-range test applied to data in Example 12-1

Recall from the ANOVA table, Table 12-3b, that the size of the calculated statistic, f, led to the rejection of the hypothesis of equal means. To apply Duncan's multiple-range test, you assemble and record several "building-block" values from the preceding analyses.

The experiment size:

$N = 35$;

the $k = 4$ sample sizes:

$n_1 = 3, n_2 = 11, n_3 = 15, n_4 = 6$;

the $k = 4$ means:

$\bar{y}_1 = 87.37, \bar{y}_2 = 87.85, \bar{y}_3 = 88.57, \bar{y}_4 = 87.48$;

and the Within-groups Mean square (from Table 12-3b):

$MS_W = 0.0528$.

You proceed with these calculations and comparisons:

harmonic mean

- Calculate the *harmonic mean* of the k sample sizes:

$$\begin{aligned}\bar{n}_h &= k/(1/n_1 + 1/n_2 + \ldots 1/n_k) \\ &= 4/(1/3 + 1/11 + 1/15 + 1/6) \\ &= 6.0829.\end{aligned}$$

  [Note that, if the sample sizes are all the same (i.e., $n_1 = n_2 = \ldots = n_k = n$), then $\bar{n}_h = n$, say, and no special calculation is required.]

- Calculate a *pseudo* standard deviation for each sample mean:

$$S^*_{\bar{y}_i} = \sqrt{\frac{MS_W}{\bar{n}_h}} = \sqrt{\frac{0.0528}{6.0829}} = 0.0932.$$

From Table T-7 with $\alpha = 0.05$ and $N - k = 35 - 4 = 31$, record the values of $q_{0.95}(p, N - k)$ for each $p = 2, 3, \ldots, k$; for this example, after some interpolating, these are:

$q_{0.95}(2, 31) = 2.89$
$q_{0.95}(3, 31) = 3.04$
$q_{0.95}(4, 31) = 3.12$.

- Calculate $R_p^* = q_{0.95}(p, N - k)S_{\bar{y}_i}^*$ for each $p = 2, 3, \ldots, k$. These are the critical values against which differences in means are compared; for this example, these are:

$R_2^* = 2.89(0.0932) = 0.2693$
$R_3^* = 3.04(0.0932) = 0.2832$
$R_4^* = 3.12(0.0932) = 0.2907$.

- Arrange the sample means in ascending order; for this example:

$\bar{y}_1 = 87.37$
$\bar{y}_4 = 87.48$
$\bar{y}_2 = 87.85$
$\bar{y}_3 = 88.57$.

- Denote the *rank* of the smallest mean by 1, the next smallest by 2, $\cdots$, and the largest by k. Next, for any two means, let p denote the difference between the corresponding ranks plus 1. Thus, for comparing the largest and the smallest means among k means, $p = (k - 1) + 1 = k$.

- Calculate the *differences between the largest mean and each of the other means*, compare them to the values of the corresponding R_p^*, beginning with the *smallest*, and determine which of the differences are statistically significant; for this example:

$$\bar{y}_3 - \bar{y}_1 = 1.20 > R_4^* = 0.2907$$
$$\bar{y}_3 - \bar{y}_4 = 1.09 > R_3^* = 0.2832$$
$$\bar{y}_3 - \bar{y}_2 = 0.72 > R_2^* = 0.2693.$$

- Here, all three differences are significantly different. You conclude that **Line 3**'s mean is larger than any of the means of **Line 1**, **Line 2**, and **Line 4**.

- Next, compare the *second largest* mean with each of the *smaller* means, beginning with the smallest, and determine which of the differences are statistically significant; for this example:

$$\bar{y}_2 - \bar{y}_1 = 0.48 > R_3^* = 0.2832$$
$$\bar{y}_2 - \bar{y}_4 = 0.37 > R_2^* = 0.2693.$$

- Here, both differences are significantly different. You conclude that **Line 2**'s mean is larger than either **Line 1**'s or **Line 4**'s.

- Continue in this fashion until the two smallest means are compared; for this example, compare the *third largest* mean with the *smallest* mean and determine if the difference is statistically significant:

$$\bar{y}_1 - \bar{y}_4 = 0.11 < R_2^* = 0.2693.$$

- Here, the difference is *not* significantly different. You conclude that **Line 1**'s mean is *not* larger than that of **Line 4**.

A graphical display, described by Duncan (1986, p. 757-760), among others, is useful for describing multiple-range test results. Write the means in an increasing sequence, left-to-right, and underline those collections of means that are *not declared to be significant* by the multiple-range test; for this example, the display

Group:	**Line 1**	**Line 4**	**Line 2**	**Line 3**
Mean:	**87.37**	**87.48**	**87.85**	**88.57**

indicates that only the means of **Line 4** and **Line 1** are *not statistically significantly different*; i.e., they form a group, as do each of **Line 2** and **Line 3** individually.

For discussion:

- Suggest other ways to present the results of Duncan's multiple-range test.

- Compare the results of Duncan's multiple-range test to the plotted data in Figure 12-6. Any surprises?

Toward a more generalized ANOVA

As mentioned at the beginning of this chapter, the vastness of the analysis of variance literature and the scope of this text preclude a detailed treatment of the subject. Still, it is important to indicate some of the types of generalizations you may encounter as you explore ANOVA further.

two-way classification

The first generalization may seem like only a small step toward a more complex design and analysis, but it leads to the important *two-way classification* layout. Here, you have two factors. Either, both, or none of them may affect the experimental results, and you need ways to ask questions about those factors.

To illustrate, consider an example in which the subject is the performance of thermoluminescent dosimeters (TLDs). Suppose each of three TLD producers makes two types: Type 1 is worn on the shirt pocket and Type 2 is worn at the waist on a belt. With a proper ANOVA design and corresponding experimental data, you can answer two basic questions with a single experiment:

(1) Are the three producers' products equally sensitive; that is, if their TLDs are exposed to the same sources of radioactivity, will they yield the same readings?

(2) Are the two types of TLDs equally sensitive; that is, does it make any difference if you wear the shirt-pocket or the wrist-band device?

Suppose that five TLDs are now available for each of the two types from each of the three producers and that they are simultaneously exposed to the same high-energy gamma field for a fixed amount of time. At the end of the period of exposure, all 30 TLDs are read by the same technician using the same technique. A model describing these 30 readings can be written as

$$Y_{ijk} = \mu + \phi_i + \tau_j + E_{ijk}; \quad i = 1, 2, 3; \; j = 1, 2; \; k = 1, 2, 3, 4, 5;$$

where

Y_{ijk} = the reading of the kth TLD of the jth type from the ith producer,

μ = the *overall* mean,

ϕ_i = the *effect* of the ith producer,

τ_j = the *effect* of the jth type of TLD, and

E_{ijk} = the *error* associated with Y_{ijk}, such that

$$E_{ijk} \sim N(0, \sigma^2).$$

Since the same number of TLDs is used for each combination of producer and type, the layout is "balanced." The model assumes that the *effects* of each producer and each type are strictly additive. In the ANOVA table, the total sum of squares is broken down into three separate sums of squares: one corresponding to producer, one to type, and one to error.

The second generalization builds on the first, this time with a model that extends the first by inserting an odd-looking term:

$$Y_{ijk} = \mu + \phi_i + \tau_j + (\phi\tau)_{ij} + E_{ijk}; \quad i = 1, 2, 3; \; j = 1, 2; \; k = 1, 2, 3, 4, 5.$$

interaction

This odd-looking term, $(\phi\tau)_{ij}$, is called *interaction*. It is designed to detect situations in which, for example, the difference between Type 1 and Type 2 TLDs are not the same for all three producers; here, a more specific label is *producer-type interaction*.

The third generalization also builds on the first, this time with a subtle change in the model:

$$Y_{ijk} = \mu + P_i + \tau_j + E_{ijk}; \quad i = 1, 2, 3; \quad j = 1, 2; \quad k = 1, 2, 3, 4, 5.$$

Instead of the term ϕ_i, originally designed to capture the effects of the three producers, the model now contains P_i. If it suggests a random variable to you, then you've got it, by George! The idea arises if the three producers are selected as a *random sample* of all possible producers. For this situation, estimation and/or hypothesis-testing on the P_i is not feasible—but what *is* feasible is the estimation of the variance of the population of producers from which these three were selected.

On the equivalence of the *T* test and the ANOVA for two groups

Example 12-2 shows the equivalence of the *T* test and the ANOVA when the means of two groups are compared against a two-sided alternative.

Example 12-2: Number of trials-to-failure of relays—the one-way ANOVA applied to two groups

The number of times a relay was exercised before failure was measured on 10 relays, four of Type 1 and six of Type 2, with results shown in Table 12-4. Is the average number of trials-to-failure the same for both types?

Table 12-4:
Trials-to-failure for two types of relays

	Type 1	Type 2
	974	778
	1,031	993
	1,249	1,011
	1,332	805
		966
		633
Sample size:	4	6
Sum:	4,586	5,186
Mean:	1146.5	864.3
Variance:	29,337.67	22,578.27
Standard deviation:	171.3	150.26
Sum of squares of deviations:	88,013.00	112,891.35

Try working through the ANOVA calculations. Note that if you use the working formulas, the grand mean is not needed.

$$ct = (974 + 1031 + \ldots + 633)^2/10 = 9549198.40$$

$$ss_T = 974^2 + 1031^2 + \ldots + 633^2 - ct = 391987.60$$

$$ss_G = 4586^2/4 + 5186^2/6 - ct = 191083.27$$

The remaining calculations for the ANOVA are given in Table 12-5.

Table 12-5:
Analysis of relay trials-before-failure for two types

Source of variation	Sum of squares	Degrees of freedom	Mean squares	*f*
Between relay types	191,083.27	1	191,083.27	7.61
Within relay type	200,904.33	8	25,113.04	
Total	391,987.60	9		

The critical value for the *F* statistic is obtained from Table T-4; it is $f_{0.95}(1, 8) = 5.32$. Because the calculated $f = 7.61 > f_{0.95}(1, 8) = 5.32$, you have evidence to claim that the mean times-to-failure for the two types of relays are different.

Since two groups are involved in this problem, you could have used the methodology introduced in Chapter 11 to test whether $\mu_1 = \mu_2$. The pooled variance is calculated as (88013.00 + 112891.35)/(3 + 5) = 25113.04, which is the same as the Within-groups Mean square obtained from the ANOVA table. Student's *t* statistic is calculated as

$$t = \frac{1146.50 - 864.33}{\sqrt{25113.04(1/4+1/6)}} = 2.76.$$

From Table T-3, the critical value for Student's *t* is $t_{0.975}(8) = 2.306$. The calculated t statistic exceeds this value, and the hypothesis of equality of means is rejected.

As you might expect, the two tests are indeed equivalent. Note that a unique relationship exists between a one-way classification ANOVA for two groups and a two-sided Student's *t* test with assumed homoscedasticity; i.e., $t^2 = f$, as you can verify from the sample calculation. The critical values are similarly related; i.e., $f_{0.95}(1, 8) = 5.32 = [t_{0.975}(8)]^2 = [2.31]^2$.

For discussion

- Why do you consult Student's *T* table at the 0.975 quantile whereas you consult the *F* table at the 0.950 quantile?

- For a number of degrees of freedom of your choice—call it df_{yc}—verify that $[t_{0.975}(df_{yc})]^2 = f_{0.95}(1, df_{yc})$. If df_{yc} is not included in the table, select a df_{yc} that is accommodated there.

- Choose an α level (for which you have table values) and verify that $[t_{(1-\alpha)}(df)]^2 = f_{(1-\alpha)}(1, df)$.

- Ponder the vicissitudes of technical language in this chapter: Although you analyze the data for equality of ***means***, you do it through an analysis of ***variance***.

What to remember about Chapter 12

Chapter 12 introduced the analysis of variance (ANOVA), one of the major analysis tools available for the purpose of comparing multi-group data. You participated in the construction of an ANOVA for the one-way classification (where observations are classified by group using a single classification criterion) and were pointed toward even more complex data structures. You also encountered the following concepts:

- *the virtue of a balanced design and layout*
- *total sum of squares*
- *among-groups sum of squares and mean square*
- *within-groups sum of squares and mean square*
- F *statistics for ANOVA*
- *the model for one-way classification ANOVA*
- *assumptions in applying ANOVA*
- *multiple-range tests*
- *modelling the two-way layout*
- *equivalence of Student's t and ANOVA's F for two groups.*

When next faced with a problem involving multi-group comparisons, you should be able to:

- *review an ANOVA table for a simple design and interpret the results*
- *recognize when the assumptions required for data analysis using ANOVA methods are reasonable*
- *construct an ANOVA table for a simple investigation*
- *recognize when the scope of your problem exceeds the scope of this book.*

13

An overview of regression

What to look for in Chapter 13

Chapter 13 eases you into the topic of *regression*,[1] a set of statistical techniques designed to support the investigation of relationships among two or more variables. The discussion begins with a reminder of some algebraic and geometric ideas you used when you plotted given functions by locating points on a coordinate system under the watchful eye of your algebra teacher.

This time, however, your task is reversed—you are given a set of points and are told to "fit a curve" to these points; that is, you are to determine a function that can be associated with that set of points "in a reasonable manner." This new task opens the path to an exploration of the processes called regression.

[1] According to Kendall and Buckland (1971, p. 127), the term *regression* "... was originally used by Galton [Sir Francis Galton, 1822-1911, an English biostatistician and cousin of Charles Darwin] to indicate certain relationships in the theory of heredity but it has come to mean the statistical method developed to investigate those relationships." We opt for this second meaning in this book.

As the exploration unfolds, you will be reacquainted with terms you learned in algebra and geometry:

- *independent variable*
- *dependent variable*
- *function*
- *ordered pairs*
- *slope*
- *intercept*
- *parameter*,

you will be reminded of terms from the analysis of variance (Chapter 12):

- *model*
- *error*,

and you will encounter the ideas of:

- *curve-fitting*
- *simple linear regression*.

Recalling some algebra and geometry

independent, dependent variables

function, ordered pairs

At some time in your mathematics training, you learned to plot functions, both linear and non-linear. You learned that you usually start with an *independent variable*, almost always called x. You learned that x was linked to a *dependent variable*, almost always called y. The values of y depended on x in some straightforward fashion; i.e., the dependent variable y was a *function* of the independent variable x. You learned to build a table of ordered pairs of values, designated by (x, y), and then to plot those pairs of values on a coordinate system and connect them.

To illustrate, suppose you are given the function $y = 3x + 2$ and are asked to plot it. You pick some convenient values for the independent variable x and calculate the corresponding values of the dependent variable y. Some selected (x, y) pairs are displayed in Table 13-1.

Table 13-1:
Selected values of *x* and corresponding calculated values of *y* for the function $y = 3x + 2$

x	-2	-1	0	1	2	3	4	5
y	-4	-1	2	5	8	11	14	17

The (x, y) pairs then are used to produce a two-dimensional graph of the given function, as shown in Figure 13-1.

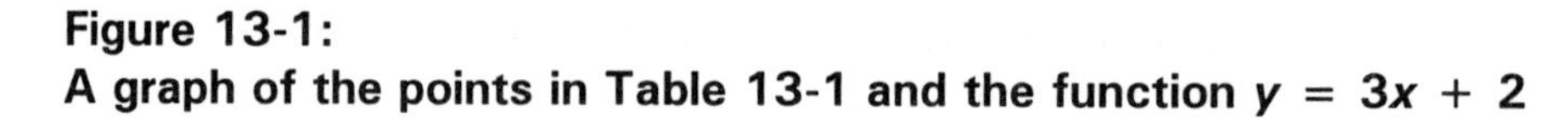

Figure 13-1:
A graph of the points in Table 13-1 and the function $y = 3x + 2$

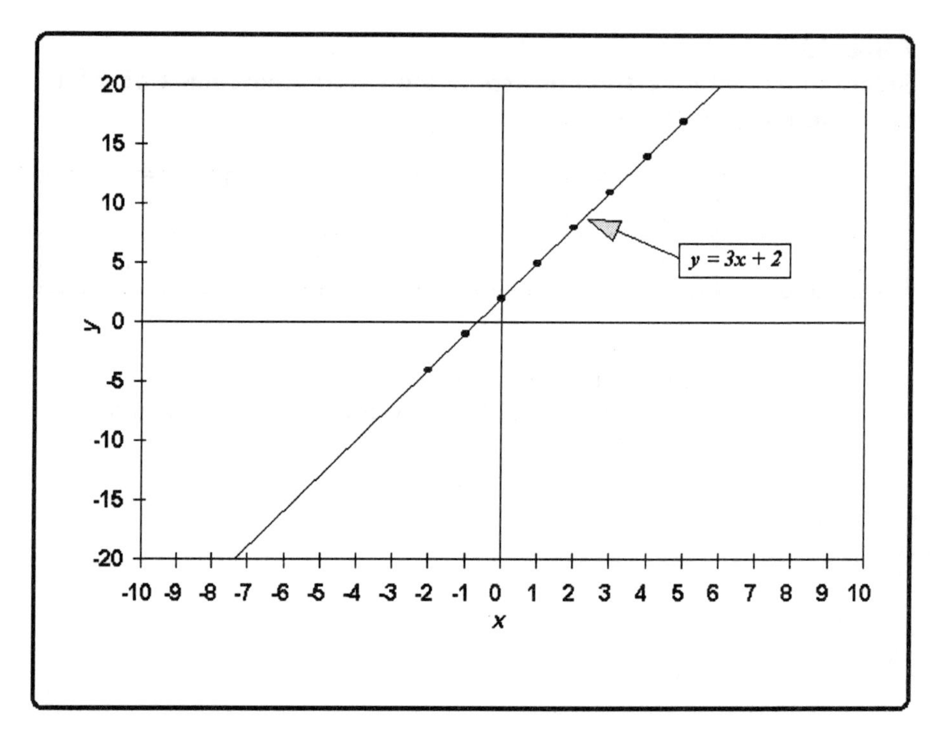

slope, intercept

Figure 13-1 shows a straight line. It has a *slope* of 3 and and an *intercept* of 2. The slope is the coefficient of the independent variable x; it shows that, for every unit of increase (decrease) in x, you have 3 units of increase (decrease) in y. The intercept is the value of y for which $x = 0$. Notice especially that the line "goes through" each of the points corresponding to the ordered pairs.

You may recall that "two points determine a straight line" and may wonder why eight points appear in Table 13-1 and Figure 13-1. Operationally, it's simply good practice to plot a few extra points to be certain that you set up the graph correctly. As it turns out, of course, any two of the points in Table 13-1 are indeed sufficient to draw the line that links them in Figure 13-1. Moreover, given any pair of points, you can determine the slope and the intercept of the straight-line function that links them.

The line in Figure 13-1 is a specific straight line; it is *the* line with slope 3 and intercept 2. You may recall a widely used notation for a *generic straight line*: $y = mx + b$. *In this form, the slope is m and the intercept is b.* The symbols m and b are the *parameters* of the general straight line.

parameters

Once you prove something is true for the parametric form of the line, $y = mx + b$, you can apply your findings to *any* straight line, both algebraically and graphically. For example, if a general ordered pair (x_0, y_0) is known to lie on the generic line $y = mx + b$, then you can conclude immediately that the slope is $m = (y_0 - b)/x_0$.

For discussion:

- Use the pairs of points (-2, -4) and (3, 11) in Table 13-1 to confirm that the line's slope is 3 and its intercept is 2.

- For the generic straight line $y = mx + b$ and the two symbolic points (x_1, y_1) and (x_2, y_2), find the expressions for the slope m and the intercept b that are functions of the two points.

- In the expression $y = mx + b$, m represents the slope and b represents the intercept of the line. In statistics and in other disciplines, you will encounter other symbols used to express a straight line. Examine the four expressions below; in each case, indicate which symbol denotes the slope and which denotes the intercept:

  (1) $y = bx + a$
  (2) $y = \beta_0 + \beta_1 x$
  (3) $y = \alpha + \beta x$
  (4) $y_i = \alpha + \beta x_i$.

This list is far from exhaustive. But you need not worry about the variety of expressions because you now know how to identify the line, and its intercept and slope, by just looking at the expression for that line.

- Suppose you come across an expression like $y = p + qx + rz$. What do you suppose this expression describes?
- Guess what $y = a + bx + cx^2$ describes.
- What does the expression $y = cx$ describe?
- If $y = 5 - 2x$, what does the negative sign tell you?
- Consider the expression $y = 4.5$. Is this a straight line? Does it satisfy the conditions of the generic line given by $y = mx + b$? If it is a line, what are its slope and its intercept?
- What would you make of Table 13-1 if the pair (3, 12) had been included? How would that affect the line in Figure 13.1?

On the role of regression

Now consider Table 13-2. It has much the same appearance as Table 13-1. It holds eight pairs of what can be taken as (x, y) values. But what is the function that links the y values to the x values?

Table 13-2:
Eight pairs of values of *x* and *y*

x	-2	-1	0	1	2	3	4	5
y	-3	-2	4	1	11	12	12	18

Try plotting the pairs shown in Table 13-2. You should get something like Figure 13-2.

Figure 13-2:
A graph of the points in Table 13-2

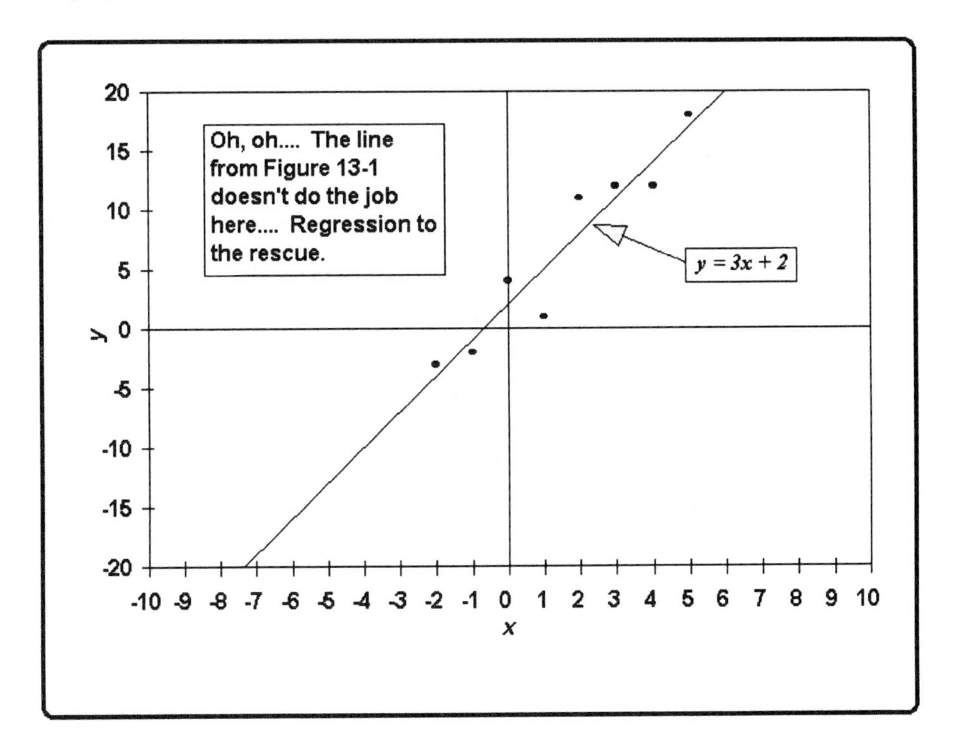

The points in Figure 13-2 tend to "move" from the lower left to the upper right. But they certainly do not lie on a straight line. What might you wish to do in this case?

The differences between Figures 13-1 and 13-2 contain the essence of the problem that *regression* is designed to tackle. In a fully mathematical setting, you know the function that links the values of y to the values of x, and your analysis is predicated on that knowledge. By contrast, in a statistical setting, you are given the (x, y) pairs, and your task is to find a link among the pairs.

If you have exactly two (x, y) pairs, the task in either setting is easy. Two points determine a straight line, and little more remains to be determined. However, when you have more than two (x, y) pairs *and* an inadequate scenario for linking them, then regression comes into play and shows its worth.

At the risk of stating the obvious: *statistical concepts and tools cannot and do not create miracles*. You cannot produce a straight line that goes through every one of an arbitrary set of points if the points are not already on a straight line. Rather, statistical methodology produces a compromise line which meets some desirable criteria. The nature of that line and of those criteria are explored in Chapter 14.

curve-fitting

The process of fitting a model (an equation, a curve) to a set of points often is called *curve-fitting*. This presentation focuses on the case in which there is one independent variable, say x, associated with the dependent variable, say y.

simple, linear

A model with a single independent variable is said to be *simple*. If, in addition, the model is in the form $y = a + bx$, then the model is also said to be *linear*. The technique for constructing and analyzing such a model is called *simple linear regression*. If a model has more than one independent variable (e.g., $y = a + bx + cz$), the procedure for constructing the associated curve is called *multiple linear regression*.

Modeling an imperfect line: Dealing with error

A straight line like $y = \alpha + \beta x$ is at best an imperfect representation of data such as those in Table 13-2. No matter what values you pick for the parameters α and β, the resulting line will "miss" one or more of the points. Thus, for this problem, the model

$$Y = \alpha + \beta x$$

is at best incomplete for this situation. One way of completing the model is to add an "*error*" term, denoted by E_i, yielding

$$y_i = \alpha + \beta x_i + E_i,\ i = 1, 2, \ldots, n.$$

This revised model forms the basis for Chapter 14's extended discussion of simple linear regression.

For discussion

- How does the model with an error term accommodate data like those in Table 13-2? Does it work equally well for Table 13-1?

- Do you see a similarity between the error term for simple linear regression and the error term in the model for one-way analysis of variance (Chapter 12)? Do they serve the same purpose?

- The basic simple linear regression model developed in this chapter has this appearance:

  $$y_i = \alpha + \beta x_i + E_i,\ i = 1, 2, \ldots, n.$$

  Suppose you encounter a model that looks like this:

  $$y_i = \gamma - \delta x_i + E_i,\ i = 1, 2, \ldots, n.$$

  What do you suppose is intended in the second model? Do you think that the appearance of the term $-\delta x$ makes any difference in your analysis?

- As you may have guessed by now, there are non-linear (simple and multiple) regression processes as well as those designed for linear processes. Although they are not discussed in this book, these non-linear models have been studied extensively and used extensively in scientific and technical work. These models are *non-linear in their parameters*. An example is

$$y_i = \alpha e^{\beta x_i} + E_i.$$

You will find introductory discussions of non-linear models in Draper and Smith (1981) and Weisberg (1985). A fuller discussion, including multiresponse parameter estimation and models defined by systems of differential equations appears in Bates and Watts (1988).

What to remember about Chapter 13

Chapter 13 introduced the concepts of simple linear regression. You observed the similarities and the differences among data points and lines constructed from and equation and an equation constructed from data points. You reviewed the fundamental straight-line ideas of:

- *independent variable*
- *dependent variable*
- *function*
- *ordered pairs*
- *slope*
- *intercept*
- *parameter*,

you were reminded of terms from the analysis of variance (Chapter 12):

- *model*
- *error*,

and you encountered the ideas of:

- *curve-fitting*
- *simple linear regression*.

Simple linear regression and correlation

What to look for in Chapter 14

Chapter 14 provides tools to construct a regression line to fit a set of data points. Following a short discourse on regression principles, you will be given explicit formulas and directions to manipulate a set of data and to calculate such statistics as:

- *the estimate of the slope*
- *the estimate of the intercept*
- *the estimates of several regression-related variances.*

You will learn how to test the hypotheses that:

- *the slope of the regression line is zero*
- *the slope equals a prescribed constant*
- *the intercept equals a prescribed constant.*

You will be shown how to construct:

- *a regression analysis table*
- *a confidence interval for the slope*
- *a confidence interval for the intercept.*

You will encounter a distinction between two important statistical ideas: *regression* and *correlation*. Attention will be given to:

- *estimating the correlation coefficient*
- *testing the hypothesis that the correlation coefficient equals a prescribed constant (including zero)*
- *constructing a confidence interval for the correlation coefficient.*

A model for simple linear regression

Effective discussion of simple linear regression begins with a *model*. Recall from Chapter 12 that a model is a symbolic expression for the data under study; in this case, the model is a mathematical expression that relates a dependent variable, designated by an upper-case Y, to an independent variable, designated by a lower-case x.

Continuing from the material in Chapter 13, consider this model for a set of data consisting of n ordered pairs (x_i, Y_i):

$$Y_i = \alpha + \beta x_i + E_i, \quad i = 1, 2, \ldots, n,$$

where

n = number of (x_i, Y_i) ordered pairs (points),
x_i = the value of the independent variable for the ith point,
Y_i = the value of the dependent variable for the ith point,
α = an unknown constant (the intercept),

β = an unknown constant (the slope), and
E_i = the "error" associated with the ith point.

For discussion:

- In a perfect world, all *n* points would lie on a straight line, say $Y = \alpha + \beta x$, as suggested by the model; in that perfect world, you would also know the intercept and the slope, α and β, exactly. But there's little doubt that we live in an imperfect world: data points that should lie on a straight line simply refuse to toe the line. The failure of a point to coincide with the value described by the model is classified in statistical parlance as an *error*, written in the model with an upper-case *E*. This error may be attributed to experimental error, to experimenter error, to modeling error, or to a combination of these and other errors.

- We are especially careful with our choices of the symbols employed in describing and working with the data and the model. A lower-case letter *x* indicates a fixed value that is measured without error. This *x* contrasts with the associated upper-case *Y* which indicates a random variable. Thus, supposing that your process is the measurement of the cumulative drop in air pressure inside a containment vessel as a function of time, you can fix *x* at will (say, every half-hour from the sealing of the vessel) and measure or calculate the associated pressure at each of these times. Whereas *x* is treated as if it is known exactly, *Y* is treated as if it is subject to random fluctuations and sensibly modeled as a random variable. (This very setup appears and is explored in more detail in Example 14-1.)

- The intercept and the slope are constants; they are written with Greek letters, α and β, to denote parameters that are to be estimated from the data because, as a rule, they are unknown.[1]

---

[1] They *do* look Greek to you, don't they?

"Looking" at the data

Suppose you are given the set of $n = 5$ data points displayed in Table 14-1.

Table 14-1:
Some simple data for starting the study of simple linear regression

x	1	2	3	5	7
y	4	3	4	8	9

scatter diagram

You are trying to identify a line that relates the dependent variable y to the independent variable x. Before you do anything else, it's simply good practice to take a "look" at the data. You draw an x-y grid and place the points on the grid, thus creating a *scatter diagram* for the data, as shown in Figure 14-1.

Figure 14-1:
Scatter diagram for the data in Table 14-1

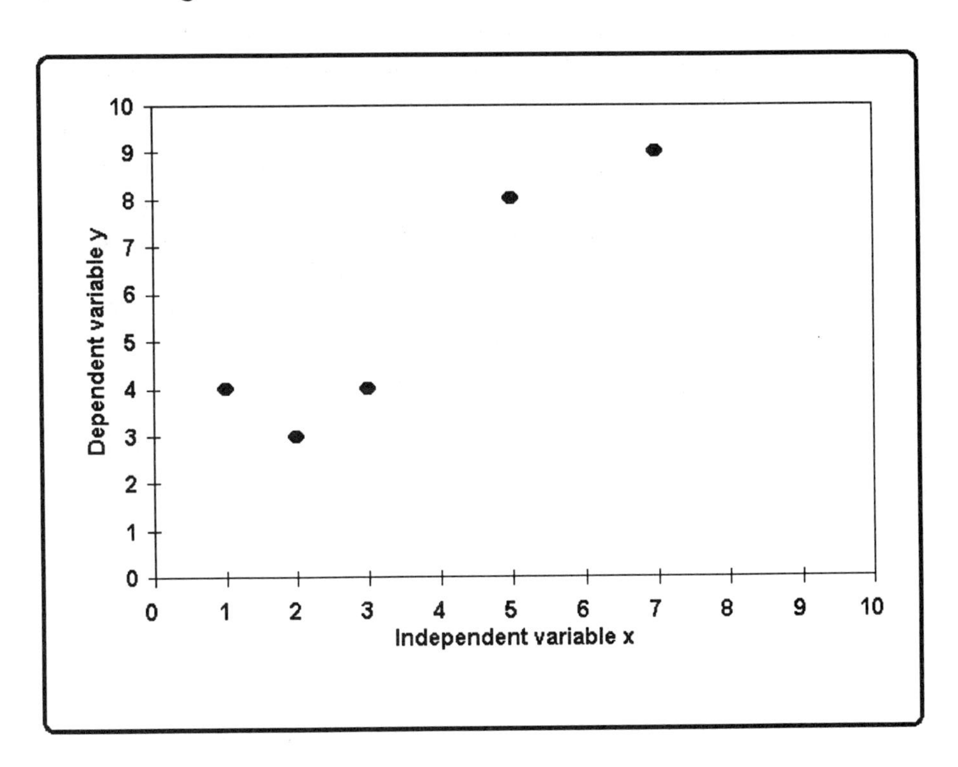

But just how *do* you go about selecting a line that will summarize these data? As with most analyses, it pays to start with some simple ideas and then build upon their results. Why not call on the average value of the observed y values? The mean is a well-established measure of location. The mean value of the five values of y in Table 14-1 is $\bar{y} = 5.6$. The corresponding line is written $y = 5.6$; it is displayed with the label **Line 1** in Figure 14-2, along with two other candidate lines, **Line 2** and **Line 3**, more about which is developed later in this chapter.

On representing (x, y) data

empirical line Your task is to find an *empirical line* that captures the nature of the (x, y) pairs of data *and* provides an intercept and a slope that estimate the simple linear regression model's parameters. Although you might scan ("eyeball") the data and arrive at a decent representation of the data, such a line seldom is acceptable. For one thing, your repeated scans could yield different lines, as might single or repeated scans made by a trusted and capable colleague. Only in the very most time-constrained circumstance should you settle for this scanning procedure and, even then, for only the crudest purposes.

Any empirical line, no matter how it is derived, can be expressed in a form that provides estimates of each of the n values Y_i:

$$\hat{Y}_i = A + Bx_i \, , \quad i = 1, \ldots , n,$$

where

A is an estimator of the model's intercept α,
B is an estimator of the model's slope β, and
$\hat{Y}_i$ is an estimator of the model's Y_i for the fixed value x_i.

No matter how you arrive at the values of A and B, you are performing an act of estimation in a statistical sense; that is, you are using the data at hand to determine something about the parameters of an underlying model.

The term $\hat{Y}_i$, pronounced *Y-hat-sub-i*, is an estimator of the true-but-unobserved random variable Y_i and thus carries a "hat" (caret) to distinguish it from the observed y_i. This notation follows the well-established and widely used practice of "hatting" a symbol for a parameter to distinguish an estimate from the parameter itself. For example, an estimator of a mean with the symbol μ is indicated by writing $\hat{\mu}$. Just remember that $\hat{Y}_i$ and $\hat{\mu}$ are both impostors—they wear hats and masquerade as the real things.

The estimators of the intercept and the slope, A and B, are functions of the sample you *happen* to collect—and of your method of estimation. Because the Y_i are random variables, it follows that A and B are to be treated as random variables. Thus, they are written using upper-case English letters.

You will find that the simple linear regression procedure helps you with a number of different objectives in the analysis of datasets containing two variables. But, first, you must decide if your process and the data it yields meet the assumptions required for simple linear regression.

Assumptions for simple linear regression

Any statistical treatment of a set of data is founded on a collection of assumptions about the dataset and its structure. Regression analysis is no exception, so here are the assumptions critical to simple linear regression. These assumptions are of two classes:

- basic assumptions which must be met for the construction of the regression coefficients

- extended assumptions which must be met for the construction of statistical tests of significance and confidence intervals.

The basic assumptions required for the construction of simple linear regression coefficients are:

- The model is appropriate; it is both simple *and* linear.
- The intercept and the slope, α and β, are unknown constants to be estimated from the data.

- Each x_i is measured without error.
- The variance of Y is constant for all x values. That means that, if you were to imagine all possible Y observations for a selected value of x, the associated population variance, σ^2, would be the same, regardless of which value of x is selected.[2]

The one extended assumption used in this book to perform hypothesis-testing and confidence-interval construction in a simple linear regression analysis is:

- The error E_i associated with Y_i is distributed normally with mean 0 and variance σ^2; symbolically, $E_i \sim N(0, \sigma^2)$ which is consistent with Chapter 7.[3]

These assumptions are displayed graphically in Figure 14-1a which is oriented specifically to reflect the five points of Table 14-1 and Figure 14-1. A wider-angle view of the same data and the assumptions appears in Figure 14-1b.

[2] If this assumption cannot be met, then construction of the regression equation requires "weighted regression" techniques that are beyond the scope of this book. See Draper and Smith (1981) or Neter, *et al.* (1990) for details.

[3] Regression analysis with *non-normal errors* is beyond the scope of this book.

Figure 14-1a:
Illustrating the assumptions of simple linear regression with the data in Table 14-1

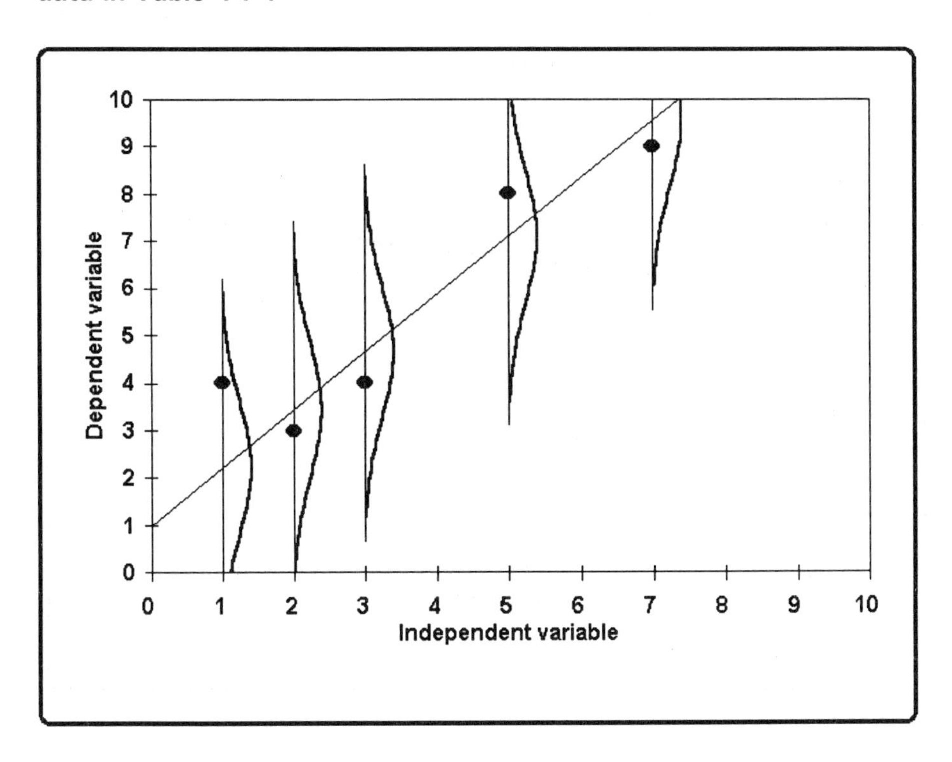

Figure 14-1b:
A wider-angle view of Figure 14-1b

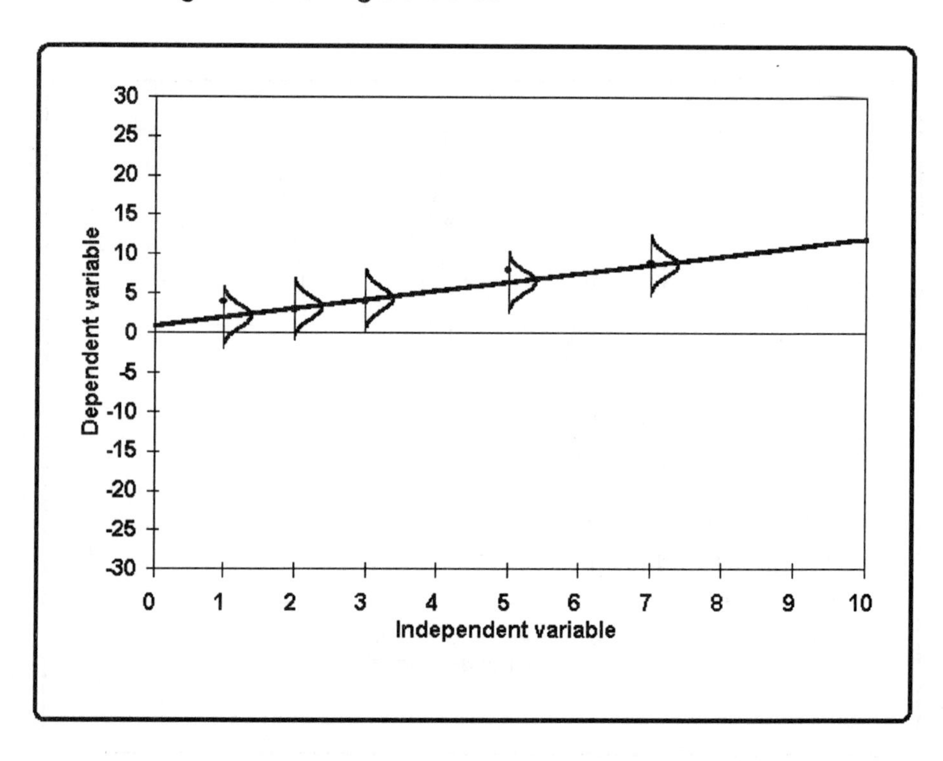

Estimating the intercept and the slope

Given a set of n (x_i, y_i) points, you have many choices to designate a line that summarizes these points. For instance, you could "estimate a line" merely by inspecting the data and guessing at numeric values for A and B to substitute for α and β in the model. But you wouldn't do *that*, or would you?

If you are unaware of any particular methodology for estimating the required line, or if you are seeking a somewhat more objective approach than guessing, you may even want to consider an earlier suggestion of first "fitting" the data with a horizontal line that goes through

$\bar{y}$, the mean of the sample observations. In particular, if your observed x and y are unrelated, this line makes sense. For example, if the model's Y represents the yearly energy consumption by a rental household and x represents the landlord's body weight, you might as well ignore the landlord's weight and estimate the energy consumption of the ith household by $\hat{Y}_i = \bar{Y}$. However, even when x and Y *are* declared to be related for theoretical or practical reasons, it often is useful to employ that horizontal line as a *baseline* or *standard* against which other lines are compared.

Let's apply this principle to the data in Table 14-1 and displayed in Figure 14-1. Recall the mean of the y_i is $\bar{y} = (4 + 3 + 4 + 8 + 9)/5 = 28/5 = 5.6$. This provides the line $y = 5.6$ as our first "candidate" to summarize the data in Table 14-1. It is shown in Figure 14-2 with the label **Line 1**.

Figure 14-2 contains two other lines for consideration. **Line 2** has the equation $y = 2x - 2$. **Line 3** has the equation $y = (8/9)x + 2$.

Figure 14-2:
Three candidate lines for the data in Table 14-1

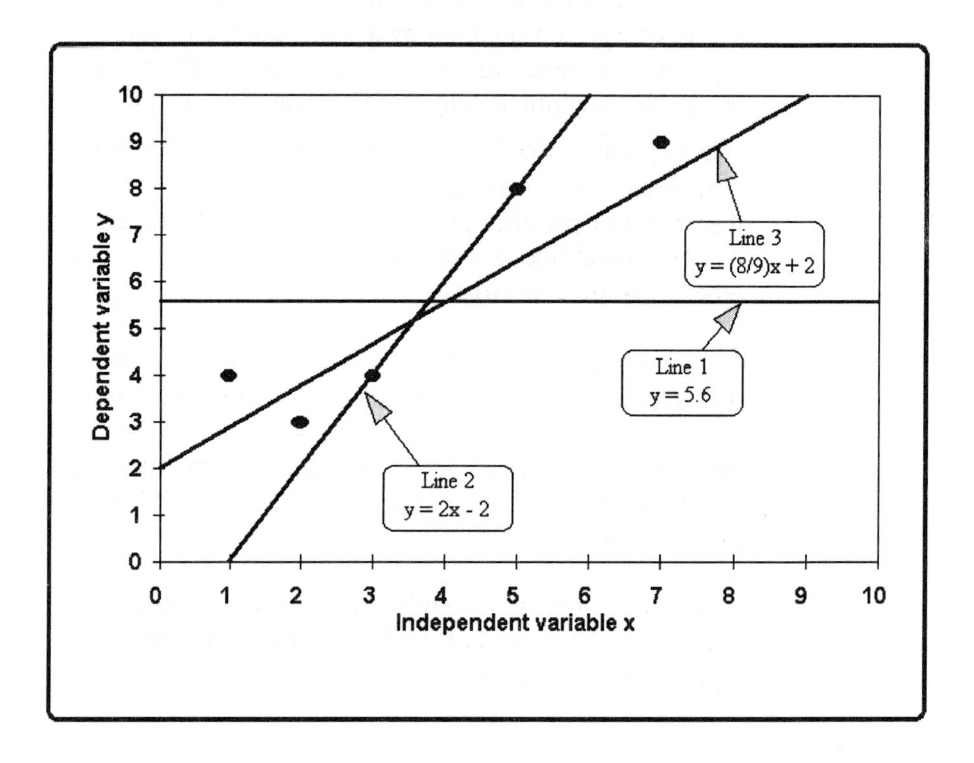

But, with the horizontal line in hand, what will you do next? The problem is that there are many ways to guess at a line, so you have many "candidate" lines to fit the data. Some candidates are "better" than others and some are "worse"; some are clearly different and some are nearly indistinguishable from each other. As a means of illustrating the problem, recall Figure 14-2. It shows the data points given in Table 14-1 and Figure 14-1 along with three candidate, albeit arbitrary—and, yes, even exaggerated—lines for your consideration.

For discussion:

- Which of the three candidate lines in Figure 14-2 appear acceptable? Why? Which lines *don't* you like? Why? Which line do you like the "best"? Why? Could there be still another line you would like even better? How do you know?

Regression and the least squares criterion

Consider **Line 3** in Figure 14-2. Pick any of the five points in the figure and draw a vertical line connecting that point to **Line 3**; the length of this line represents the vertical distance between the point and **Line 3**. Repeat this process for all the other points as shown in Figure 1-3.

Figure 14-3:
Displaying the distances between the points in Table 14-1 and Line 3

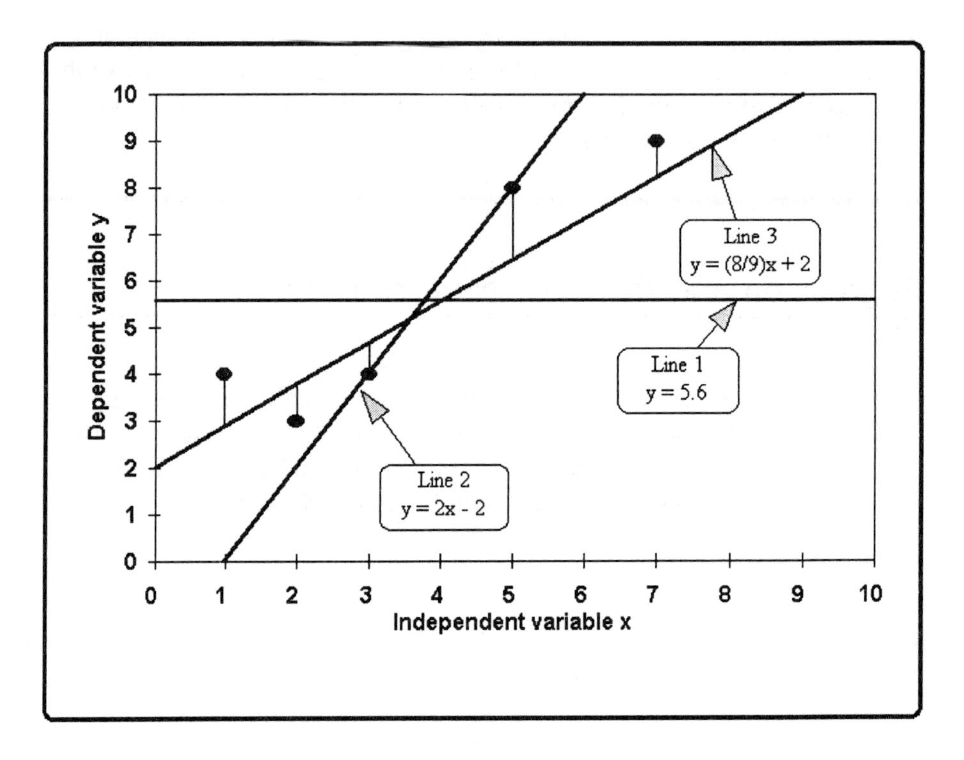

Next, square each distance and add those squares to obtain a "sum of squares of vertical deviations of the points from **Line 3**" and call it SS_3. Repeat the process for **Line 1** and **Line 2** to form SS_1 and SS_2, respectively. Finally, compare the three sums of squares and select the line with the smallest sum of squares. Table 14-2 gives the results of these calculations.

Table 14-2:
Calculations used to select from among the three lines in Figure 14-3

x	1	2	3	5	7	Sum of Squares
y	4	3	4	8	9	
Line 1						
$\hat{y}_i$	5.6	5.6	5.6	5.6	5.6	SS_1 = 35.40
$y_i - \hat{y}_i$	-1.6	-3.6	-1.6	2.4	3.4	
$(y_i - \hat{y}_i)^2$	2.56	12.96	2.56	5.76	11.56	
Line 2						
$\hat{y}_i$	0	2	4	8	12	SS_2 = 26
$y_i - \hat{y}_i$	4	1	0	0	-3	
$(y_i - \hat{y}_i)^2$	16	1	0	0	9	
Line 3						
$\hat{y}_i$	2.89	3.78	4.67	6.44	8.22	SS_3 = 5.3384
$y_i - \hat{y}_i$	1.11	-0.78	-0.67	1.56	0.78	
$(y_i - \hat{y}_i)^2$	1.2321	0.6084	0.4489	2.4336	0.6084	

From this set of three lines, you select **Line 3** because it has the smallest sum of squares. You should wonder, however, if there is a "better" line which has a still smaller sum of squares. Indeed, is there a "best" line, such that no other line can have a smaller sum of squares for the given set of data? Using regression methodology, you can readily construct such a line; i.e., you can find and express the *unique* line with the smallest sum of squares of deviations.

method of least squares

The theoretical development of this approach, often called the *method of least squares* will be found in Neter, *et al.* (1990, Chapter 2), Draper and Smith (1981, Chapter 2), and many others. (The term *least squares* summarizes a "mouthful" of methodology; it is a shorthand way of saying "the smallest sum of squares of deviations between a set of points and a fitted function when those deviations are measured along the vertical axis.")

The essence of the method of least squares can be stated as follows:

For a set of n *ordered pairs,* (x_i, Y_i), $i = 1, 2, \ldots, n$, *as long as at least two of the* x *values are different, the least squares line is unique and the estimators of its slope,* B, *and its intercept,* A, *are given by two equations:*

$$B = \frac{\sum_{i=1}^{n} (x_i - \bar{x})(Y_i - \bar{Y})}{\sum_{i=1}^{n} (x_i - \bar{x})^2}$$

$$= \frac{\sum_{i=1}^{n} x_i Y_i - (\sum_{i=1}^{n} x_i)(\sum_{i=1}^{n} Y_i)/n}{\sum_{i=1}^{n} x_i^2 - (\sum_{i=1}^{n} x_i)^2/n}$$

$$A = \bar{Y} - B\bar{x}.$$

The second expression for B is called the "working formula."

The calculations can be simplified by using notation similar—*but not identical*—to that used in Chapter 12, where the summation sign (Σ) indicates summing over the index i from 1 to n. Let

$$S_{xx} = \Sigma x^2 - (\Sigma x)^2/n,$$

$$S_{YY} = \Sigma Y^2 - (\Sigma Y)^2/n, \text{ and}$$

$$S_{xY} = \Sigma xY - (\Sigma x)(\Sigma Y)/n,$$

from which the formula for the slope becomes

$$B = S_{xY}/S_{xx}$$

and the formula for the intercept remains unchanged as

$$A = \bar{Y} - B\bar{x}.$$

As these expressions are applied to sets of observations, the upper-case-indicated random variables are replaced by lower-case-indicated numerical values.

For discussion:

- Examine the expression for S_{YY} given above. Does it remind you of anything you encountered earlier in this book?

- Note that, although the term S_{YY} is not used in the construction of A or B, it comes into play at later stages, when the quality of the regression line is examined and when correlation methods are used.

- The terms *least squares* and *regression* are unfortunately often used interchangeably. *The method of least squares* is simply one of many methods for determining the parameters of the model of interest; another is the *method of least absolute values*. Nothing inherent in the method of least squares carries such requirements as equal variance and normally distributed errors; those are specific requirements of *regression*.

Example 14-1: Pressure stabilization

Table 14-3 contains a fully worked example in which a regression line is fitted to 10 data points. Here, the independent variable x represents the time, in hours, since the beginning of an experiment designed to measure the change in pressure in a pressurized system. The dependent variable Y measures the cumulative drop in pressure since the beginning of the test (i.e., $x = 0$), in hundredths of pounds-per-square-inch (psi/100) in the system. The analysis assumes that, at least for the time under study, the cumulative drop in containment pressure is linear.

Table 14-3:
Calculations for simple linear regression applied to Example 14-1

Data setup, initial calculations, estimated line, and deviations:

Index	Time (hrs)	Pressure drop, (psi/100)	Initial calculations			Predicted pressure drop	Deviation (residual)
i	x_i	y_i	x_i^2	y_i^2	$x_i y_i$	$\hat{y}_i$	$y_i - \hat{y}_i$
1	0.5	1.00	0.25	1.00	0.50	1.19	-0.19
2	1.0	2.00	1.00	4.00	2.00	2.03	-0.03
3	1.5	4.00	2.25	16.00	6.00	2.86	+1.14
4	2.6	3.00	6.76	9.00	7.80	4.69	-1.69
5	3.4	6.00	11.56	36.00	20.40	6.02	-0.02
6	4.6	8.00	21.16	64.00	36.80	8.01	-0.01
7	5.5	11.00	30.25	121.00	60.50	9.51	+1.49
8	6.6	11.00	43.56	121.00	72.60	11.34	-0.34
9	6.7	12.00	44.89	144.00	80.40	11.51	+0.49
10	7.5	12.00	56.25	144.00	90.00	12.84	-0.84
Sums	$\Sigma x =$ 39.9	$\Sigma y =$ 70.00	$\Sigma x^2 =$ 217.93	Σy^2 660.00	$\Sigma xy =$ 377.00	$\Sigma \hat{y} =$ 70.00	$\Sigma(y-\hat{y}) =$ 0.00

Intermediate calculations:

$S_{xx} = \Sigma x^2 - (\Sigma x)^2/n = 217.93 - (39.9)^2/10 = 58.7290$

$S_{yy} = \Sigma y^2 - (\Sigma y)^2/n = 660.00 - (70.0)^2/10 = 170.0000$

$S_{xy} = \Sigma xy - (\Sigma x)(\Sigma y)/n = 377.00 - (39.9)(70.0)/10 = 97.7000$

$\bar{x} = (\Sigma x)/n = 39.9/10 = 3.99$

$\bar{y} = (\Sigma y)/n = 70.0/10 = 7.00$

Regression equation calculations:

$b = S_{xy}/S_{xx} = 97.7000/58.7290 = 1.6636$

$a = \bar{y} - b\bar{x} = 7.00 - (1.6636)(3.99) = 0.3622$

The estimated equation is $\hat{y} = 0.362 + 1.664x$. For each x, the corresponding $\hat{y}_i$ is shown above, as is the difference $(y_i - \hat{y}_i)$.

The precept that one good look is worth a thousand guesses is as valid in statistics as it is in poker and bridge. Accordingly, Figure 14-4 displays the 10 data points from Example 14-1 (Table 14-3).

Figure 14-4:
Data from Example 14-1 and the regression line calculated in Table 14-3

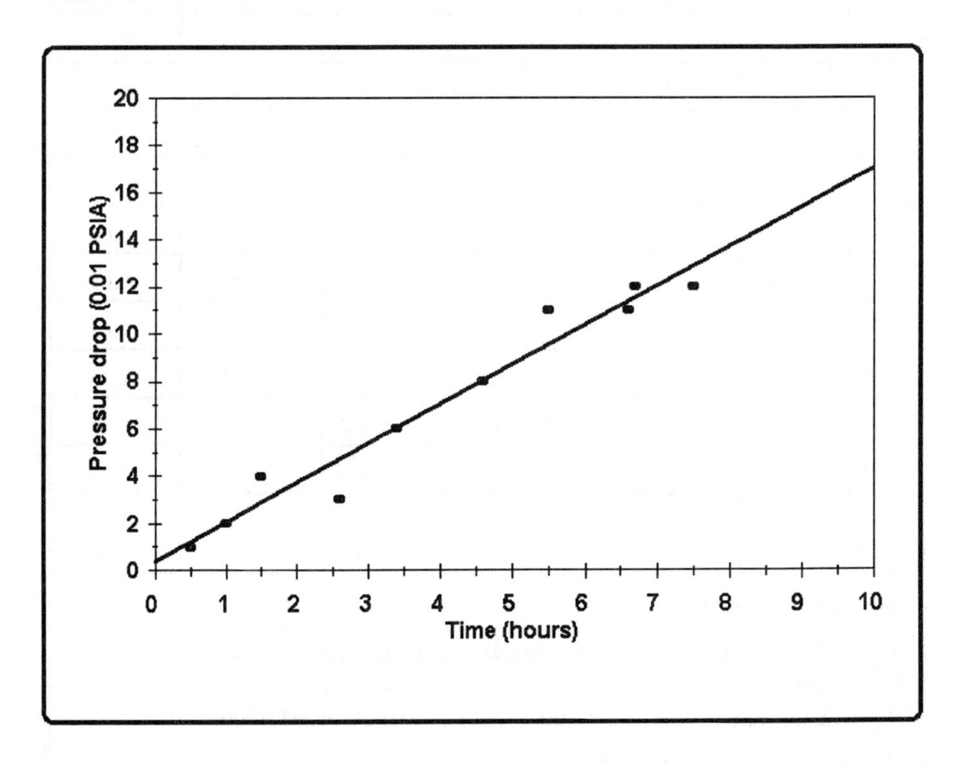

Examination of Figure 14-4 reveals a general lower-left to upper-right trend in the plotted points. No particularly striking curving trend is apparent. Thus, the linear model appears to be a good start in characterizing Table 14-2's data.

predicted value

residuals

One feature of Table 14-3 is worth pointing out: Following the calculations of b and a, you may wish to calculate the value of y as determined by the regression line. This means that, for each x value of interest, you calculate $a + bx$ to yield an associated $\hat{y}$, a *predicted value*. The $\hat{y}$ values for Example 14-1 are given in the next to the last column of Table 14-3. The deviation (or difference) between the observed y and the predicted $\hat{y}$ values are given in the very last column of Table 14-3. Clearly, if those deviations—called *residuals*—are small, the regression equation fits the data well.

For discussion:

- The formal definition for the regression line requires that at least two of the x values be different. Why is this necessary?

- Show that the regression line goes through the point $(\bar{x}, \bar{y})$. This bit of information is particularly valuable when you draw the regression line through a set of points. To that end, calculate the point $(\bar{x}, \bar{y})$ and consider only lines that pass through that point to expedite your selection.

- Examine the data in Figure 14-4 and try to guess the pressure drop at 5.0 hours from the graph and from the regression equation. Are the two results in reasonable agreement? What about 10 hours? What about 20 hours?

- The **prediction equation** is a special application of the regression equation. It is called the *prediction equation* because it is used to predict a "future" y value for *any* fixed value of x, not necessarily limited to values in your dataset. For example, you may wish to use the prediction equation to estimate the cumulative drop in pressure at 8 hours after the beginning of the experiment. Of course, you must be very careful in extrapolating beyond the values collected during the experiment. To emphasize the point, consider this: What would you be willing to guess is the drop in pressure 2,000 hours after the beginning of the experiment?

Sources of variation in simple linear regression

The model for a simple linear regression and the prediction equation are given by two expressions:

Model: $Y_i = \alpha + \beta x_i + E_i, i = 1, \ldots, n$

Regression equation: $\hat{Y}_i = A + Bx_i$.

Consider now a mathematical identity for any point (x_i, Y_i) in the dataset:

$$(Y_i - \bar{Y}) = (Y_i - \hat{Y}_i) + (\hat{Y}_i - \bar{Y}).$$

The three components described in the mathematical identity have both geometric (as shown in Figure 14-5) and statistical interpretations (as explained in the following text).

Figure 14-5:
Graphical interpretation of error components

Using Line 3 from Figure 14-3 to interpret the three components of the identity $(Y_i - \bar{Y}) = (Y_i - \hat{Y}_i) + (\hat{Y}_i - \bar{Y})$ for a selected point $(x_s, y_s) = (5, 8)$

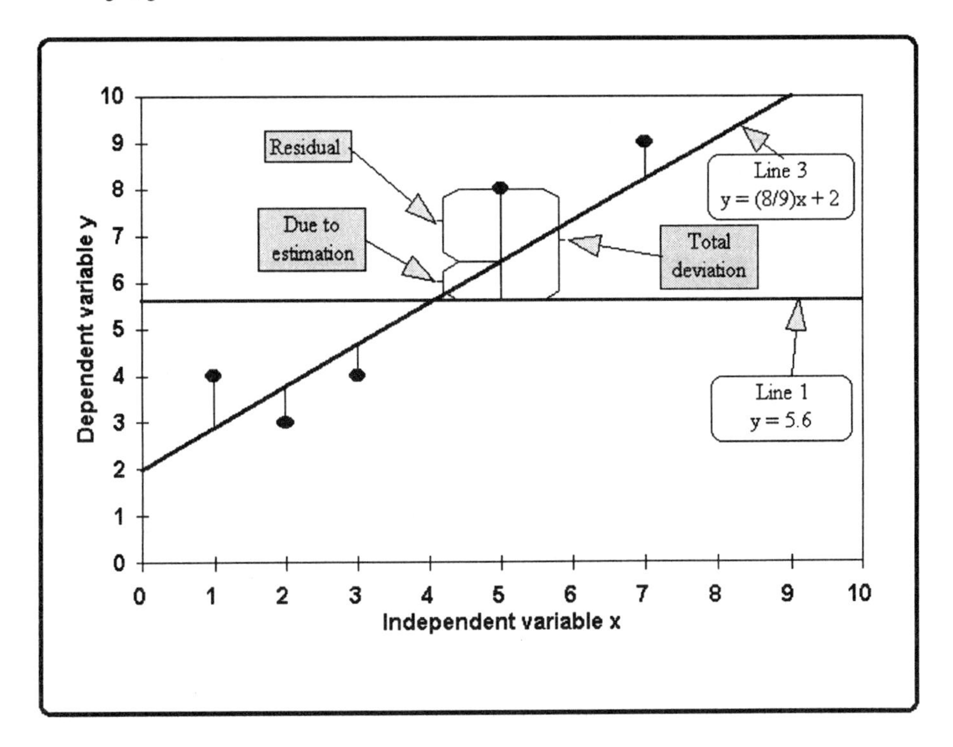

The term $(Y_i - \bar{Y})$ is represented in Figure 14-5 by the distance between the 4th observed point, i.e., $(x_s, y_s) = (5, 8)$, and a point on the horizonal line, $y = 5.6$, that represents $\bar{y}$, the mean of all five observations, so that the term has the value $(8 - 5.6) = 2.4$. This particular distance represents the basic element used in the calculation of the sample variance $S^2 = \Sigma(Y_i - \bar{Y})^2/(n - 1)$. The term $(Y_i - \bar{Y})$ is called the *total deviation*.

total deviation

The term $(Y_i - \hat{Y}_i)$ is represented in Figure 14-5 by the distance between the observed point (x_i, Y_i) and the "fitted point" $(x_i, \hat{Y}_i)$. For the data in Figure 14-5, the fitted point is (5, 6.44), this term has the value (8 - 6.44) = 1.56. A small value for this term indicates that the estimated line provides a "reasonable" fit through, or near to, the point (x_i, Y_i). A large value for this term indicates that the estimated line does not fit well, at least in the neighborhood of the point (x_i, Y_i). The term $(Y_i - \hat{Y}_i)$ usually is called the ith *residual*. (Two other commonly used expressions for this term are *departure from the fitted line* and *error*). Indeed, this term estimates the error E_i used in the model to accommodate departures from strict linearity.

residual

The term $(\hat{Y}_i - \bar{Y})$ is represented in Figure 14-5 by the distance between the predicted value of Y at the ith point and the mean $\bar{Y}$. For the data in Figure 14-5, this term has the value (6.4 - 5.6) = 0.8. If $\hat{Y}_i$ and $\bar{Y}$ are near each other, this means that the estimated line does little to improve the fit beyond that offered by horizontal line $\hat{Y}_i = \bar{Y}$, at least for the ith data point. If $\hat{Y}_i$ and $\bar{Y}$ are far apart, this indicates that the estimated line assigns a predicated value for the ith point that appears to be an improvement over the sample mean. As suggested in this context, the term $(\hat{Y}_i - \bar{Y})$ is said to be *due to regression* because it assesses the "improvement" made by the estimation procedure over the simple use of the mean value.

due to regression

The residuals from Example 14-1 are calculated and listed in the right-most column of Table 14-3.

For discussion:

- The relation $(Y_i - \bar{Y}) = (Y_i - \hat{Y}_i) + (\hat{Y}_i - \bar{Y})$ is said to be a mathematical identity. What is a mathematical identity? How is a mathematical identity different from a mathematical equation?

- Consider the residuals from Table 14-3. Note that, apart from rounding error, they sum to zero. This property suggests that, although sometimes your predicted value exceeds the observed Y and at other times falls short of the observed Y, the residuals of a regression-derived line sum to zero for the values of x used in the regression.

- In this regression equation, Y is written as a linear function of x. At times, you may be tempted to use the regression equation and predict x from Y by solving for x as a function of Y. For example, if $Y = 3 + 2x$, a simple algebraic manipulation may entice you to use an inverse-prediction equation of $x = (Y - 3)/2$. In the strongest terms, our advice is to resist such temptation. If you realize later in the game that x should be treated as the dependent variable (and the variability of y is sufficiently smaller than that of x), your best bet is to start the experiment and the analysis over and construct the appropriate regression with the roles of the dependent and independent variables properly defined. Failing that, spend some time with the literature of inverse-prediction equations and in the presence of people who have ventured therein. While you are in the waiting room, for example, you will benefit from a visit with Brownlee (1965, pp. 346-349).

Simple linear regression analysis

A sensible question you may ask at this point is whether the effort involved in fitting a line to your data is warranted. Put another way: Is the dependent variable really a linear function of the independent variable, or can you do just as well by using the sample mean $\bar{Y}$ as a predictor regardless of the value of x? To answer this and other questions, you must delve deeper into regression analysis. From this point on, the assumption of normal distribution of error should be deemed valid, because it forms the basis for all of the hypothesis-testing procedures discussed here.

trend

One natural question: "Is there evidence of a *trend* in the data?" Thus, in Example 14-1, the question is whether there is evidence for a constant drop in pressure over the time interval for which measurements are available. For a straight line, a trend is captured by the slope. So you set out to test the hypothesis H_0: $\beta = 0$. The alternative hypothesis is usually in the form of H_1: $\beta \neq 0$. Of course, you *could* write the alternative hypothesis as a one-sided hypothesis, either as H_1: $\beta > 0$ or as H_1: $\beta < 0$.

The regression analysis is based on some mathematical and statistical facts, the most important being the ability to partition the variability of observations around the sample mean into meaningful components. To gain appreciation for the partition, recall the mathematical identity discussed earlier in this chapter:

$$(Y_i - \bar{Y}) = (Y_i - \hat{Y}_i) + (\hat{Y}_i - \bar{Y}).$$

Using this identity, the following expression also can be shown (*cf.* Neter, *et al.*, 1990, p. 90) to be an identity:

$$\Sigma(Y_i - \bar{Y})^2 = \Sigma(Y_i - \hat{Y}_i)^2 + \Sigma(\hat{Y}_i - \bar{Y})^2.$$

The first identity offers a geometric interpretation of the variation associated with an individual observation. The second identity extends the first by expressing the joint variation of all the points in the sample. This identity partitions the variation of *all* of the sample observations about the sample mean into meaningful components as discussed below. As in Chapter 12, the measure of variation that lends itself to meaningful partition is the sum of squares.

Total Sum of Squares, SS_{Tot}

The term $\Sigma(Y_i - \bar{Y})^2$ is called the *Total Sum of Squares.* Denote it as SS_{Tot}. For the *n* points (x_i, Y_i), SS_{Tot} has $(n - 1)$ associated degrees of freedom. The formula for SS_{Tot} is denoted by S_{YY}. It is applied to the data in Table 14-3 where it is written as SS_{yy}.

Residual Sum of Squares, SS_{Res}

The term $\Sigma(Y_i - \hat{Y}_i)^2$ is called the *Residual Sum of Squares*. This sum of squares, designated by SS_{Res}, has $(n - 2)$ associated degrees of freedom. Although you can calculate the Residual Sum of Squares by creating all the residuals, squaring them, and summing the squares, a simpler procedure is given in the second paragraph below.

Regression Sum of Squares, SS_{Reg}

The term $\Sigma(\hat{Y}_i - \bar{Y})^2$ is called the *Regression Sum of Squares* and is denoted by SS_{Reg}. For a simple linear regression, the associated number of degrees of freedom is always 1. You calculate the Regression Sum of Squares by this formula:

$$SS_{Reg} = S_{xY}^2/S_{xx}.$$

With SS_{Tot} and SS_{Reg} in hand, you now can calculate the Residual Sum of Squares by this formula:

$$SS_{Res} = SS_{Tot} - SS_{Reg}.$$

To summarize: The total variation among the observations in the sample is separated into two independent components: SS_{Reg} attempts to explain the data behavior through a regression model, while SS_{Res} measures the "unexplained" variation.

These three different sources of variation are now entered into a *regression analysis table*, as shown generically in Table 14-4.

Table 14-4:
A generic regression analysis table

Source of variation	Sum of squares	Degrees of freedom	Mean square	F
Regression	SS_{Reg}	$df_{Reg} = 1$	$MS_{Reg} = SS_{Reg}/df_{Reg}$	$F = MS_{Reg}/MS_{Res}$
Residual	SS_{Res}	$df_{Res} = n - 2$	$MS_{Res} = SS_{Res}/df_{Res}$	
Total	SS_{Tot}	$df_{Tot} = n - 1$		

Return now to Example 14-1 and retrieve the necessary calculations as needed from Table 14-2. They are:

$$SS_{Tot} = 170.00$$
$$SS_{Reg} = (97.70)^2/\ 58.729 = 162.5311$$
$$SS_{Res} = SS_{Tot} - SS_{Reg} = 170.00 - 162.5311 = 7.4689.$$

The rest is easy. Just as in the analysis of variance tables of Chapter 12, the Regression and the Residual Mean squares are obtained by dividing their Sums of squares by their corresponding Degrees of freedom. Thus, the Regression Mean square is calculated as $MS_{Reg} = SS_{Reg}/df_{Reg}$ and the Residual Mean square as $MS_{Res} = SS_{Res}/df_{Res}$. The appropriate statistic for testing the significance of the slope is $F = MS_{Reg}/MS_{Res}$. Construction of these Mean squares and the test statistic F should be familiar; you went through the same steps in the construction of the analysis of variance tables in Chapter 12.

The regression analysis table for Example 14-1 is displayed in Table 14-5, where the calculated value of the test statistic f is 174.07.

Table 14-5:
The regression analysis table for Example 14-1

Source of variation	Sum of squares	Degrees of freedom	Mean square	f
Regression	162.53	1	162.53	174.07
Residuals	7.47	8	0.93	
Total	170.00	9		

Recall that this study was conducted to ask and answer a question like this: Is the pressure drop constant over the time of the experiment? If the regression is found to be "significant," then you reject the hypothesis of constant pressure drop; if it is found "not significant," then you have insufficient evidence against the hypothesis of constant pressure drop.

To determine the significance, you compare the statistic $f = 174.07$ to the critical value of $f_{0.95}(1, 8) = 5.32$. The calculated statistic exceeds the 95th percentile of the F distribution with corresponding degrees of freedom, so that the null hypothesis of $\beta = 0$ is rejected—a conclusion you might have reached by merely looking at the data.

Note that the analysis of regression table and the associated F-distributed test statistic are constructed to test only a two-sided hypothesis. If you need a one-sided alternative hypothesis, you compare your statistic to a tables value of $f_{0.90}(1,8)$ and examine the direction of the slope. You may find it useful to sketch the F distribution in order to visualize this.

One more thing: Earlier in the chapter we stated that a requirement for the construction and the analysis of the regression line is that the variance, σ^2, of the observed value Y is constant for all values of x. Now that the regression analysis is complete, the estimator of σ^2 is obtained from the Residual Mean square of the regression analysis table. Consistent with previous notation, you denote this estimator by S^2.

For discussion:

- Using the analysis of Table 14-5, you reject the null hypothesis, claiming significant time trend. Could you have been wrong? What is the chance of that? What would have been your conclusions if the calculated f had been 1.86? Could you be wrong? What would you have concluded if the calculated f had been -1.86?

- Examine the Table 12-2, the generic ANOVA from Chapter 12 and compare it to the generic regression analysis table, Table 14-3. What are the similarities between the two tables? Where do they differ?

- The analysis in Tables 14-2 and 14-3 tests an hypothesis about the slope of the regression line. Note, however, that no test for the intercept is given in the table we constructed.

- Note the strong resemblance between the Mean square for Error (MS_E) of the ANOVA in Table 12-2 and the Mean square for Residuals (MS_{Res}) in the regression analysis in Table 14-2. Could they both be estimating the same thing? If so, what might it be?

- The similarity between the ANOVA table and the regression analysis table of this chapter is no coincidence. Both tables are based on the same types of assumptions and tests of hypotheses that can be written in mathematically equivalent forms. For this reason, many authors call the regression analysis table an analysis of variance table.

Student's *T* statistic in regression applications

A statistical analysis based solely on a regression analysis table is limited for three reasons:

(1) The regression analysis table tests only whether the regression line is horizontal. Thus, a test investigating whether the regression slope equals a constant other than zero cannot be executed directly through that table.

(2) The test based upon the regression analysis table is formally a two-sided test (H_1: $\beta \neq 0$). If a one-sided test is required, additional, although minor, steps need be taken, as suggested earlier in this chapter.

(3) The regression analysis table does not provide a tool for constructing confidence intervals for the slope, for the intercept, or for a predicted value of Y for a given value of x, say x_0.

What the regression analysis table *does* give you, however, are systematic steps to calculate S^2—the estimate of the common variance σ^2—a quantity pivotal to many statistical tests associated with regression.

Before discussing these statistical tests, let's gather our thoughts and recall several relevant properties about regression and introduce a few new ones, basing these remarks on Neter, *et al.* (1990, pp. 69-71).

- B, the estimator of the slope β, is a random variable.
- B is an unbiased estimator of β (if the model is correct).
- If the observations Y_i are distributed normally, then the estimator B is also distributed normally.

- The variance of B is σ^2/S_{xx}.

- The estimator of the variance of B is S^2/S_{xx}.

- The estimator of the standard deviation of B is $\sqrt{S^2/S_{xx}}$. You will often find this estimator called "the standard error of B."

- Similar properties apply to A, the estimator of the intercept α.

Using these properties, you can readily construct a number of useful statistics. For example, consider the statistic in the formulation:

$$T = \frac{B - \beta}{\sqrt{S^2/S_{xx}}}$$

which is distributed as Student's T with $(n - 2)$ degrees of freedom.

You can use this statistic to test the hypothesis that β equals a given constant, say β_0 (i.e., H_0: $\beta = \beta_0$). All you have to do is to substitute β_0 for β in the formulation given above. For example, if the alternative hypothesis is set as H_1: $\beta \neq \beta_0$, with level of significance $\alpha = 0.05$, you reject H_0 if the absolute value of T exceeds $t_{0.975}(n - 2)$. For H_1: $\beta > \beta_0$, you reject H_0 if T exceeds $t_{0.95}(n - 2)$. Similarly, for H_1: $\beta < \beta_0$, you reject H_0 if T is negative and $T < t_{0.05}(n - 2)$ or, equivalently, $|T| > t_{0.95}(n - 2)$.

For discussion:

- In testing for significant trend, regardless of the direction of the trend (i.e. H_1: $\beta \neq \beta_0$), you may run both the F test (in the regression analysis table) *and* Student's T test. If you compare the calculated statistics t and f, you will discover that t^2 and f are numerically identical.

- In a similar vein, compare Tables T-3 and T-4, to see that for any number of degrees of freedom, df, $t^2_{0.975}(df)$ equals $f_{0.95}(1, df)$. This equivalence holds for all levels of significance, α.

Example 14-1a:
Pressure stabilization (continued)

Suppose that the experiment whose data are given in Table 14-1 was designed to test the hypothesis that the containment pressure drops at a rate not larger than 0.01 psi/hr. In terms of the units in which these data are recorded, this hypothesis is equivalent to H_0: $\beta = 1$. Thus, the null and alternative hypotheses are written as follows:

H_0: $\beta = 1$, H_1: $\beta > 1$, and $\alpha = 0.05$.

The test statistic is:

$$T = \frac{B - 1}{\sqrt{S^2/S_{xx}}}$$

which for these data yields the value:

$$t = \frac{1.664 - 1}{\sqrt{0.934/58.729}} = 5.27.$$

Thus, $t = 5.27 > t_{.950}(8) = 1.86$, and so the null hypothesis is rejected with the claim that the rate of drop in the system's pressure with respect to time is higher than 0.01 psi/hr.

Example 14-2:
Coefficient of expansion of a manufacturer's metal rods

Suppose you wish to test a manufacturer's claim that the expansion coefficient of a metal rod is at most 0.0012. Suppose, further, that from data from a sample of size $n = 20$ readings, you calculate $b = 0.0014$, $ms_{Res} = 0.00033$, and $S_{xx} = 312.25$.

You start with the null and alternative hypotheses:

H_0: $\beta = 0.0012$, H_1: $\beta > 0.0012$, and $\alpha = 0.05$.

Continue with the test statistic and its calculated value,

$$T = \frac{B - 0.0012}{\sqrt{S^2/S_{xx}}}$$

and

$$t = \frac{0.0014 - 0.0012}{\sqrt{0.00033/312.25}} = 0.1945,$$

so that $t = 0.1945 < t_{0.95}(18) = 1.734$. You do not reject H_0, stating that there is insufficient evidence against the manufacturer's claim that the coefficient of expansion is at most 0.0012.

For discussion:

- In Example 14-2, regarding the expansion coefficient, you obtained a sample of 20 readings. Does it make a difference if the 20 readings were made on one rod, on a single reading on each of 20 rods, or on some combination of the two?

- Are you comfortable seeing that the sample expansion coefficient is higher than claimed, and yet H_0 is not rejected? What can you do about it?

- If you wish a test with greater sensitivity to departure from the manufacturer's claim, you may elect to use a larger sample size. What would you expect a larger sample to do to S_{xx} and to T? Are these consequences of a larger sample desirable?

Constructing a confidence interval for the slope

A point estimate for the slope, given by B, is often unsatisfying; it leaves you wondering about how far off you might be in your estimation of β. Your level of anxiety may be reduced by the use of a confidence interval. The advantage of the confidence interval over a point estimate is that it provides a "margin of play," giving you measurable assurance that the interval includes the true, yet unknown, slope.

The confidence interval for the slope is written in one of the following forms:

- For the endpoints of a (1 - α)100% two-sided confidence interval, you calculate

$$B \pm t_{(1 - \alpha/2)}\sqrt{S^2/S_{xx}}\,.$$

- For the endpoint of an upper (1-α)100% confidence interval, you calculate

$$B + t_{(1 - \alpha)}\sqrt{S^2/S_{xx}}\,.$$

- For the endpoint of an lower (1-α)100% confidence interval, you calculate

$$B - t_\alpha\sqrt{S^2/S_{xx}}\,.$$

Example 14-1b:
Pressure stabilization (continued again)

Construct a one-sided 95% confidence interval for the rate of drop in the system's pressure with respect to time. Undoubtedly, you are interested in an upper confidence interval to see whether the pressure drop is excessive. The required endpoint of the interval is given by calculating:

$$1.664 + 1.86\sqrt{0.934/58.729} = 1.899.$$

For discussion:

- What is meant by "a 95% confidence interval for the rate of drop; in the system's pressure with respect to time"? Specifically, it's 95% of what?

- Calculate a 95% confidence interval for the coefficient of expansion given by Example 14 -2.

Hypothesis-testing and confidence-interval construction for the intercept

Hypothesis-testing and confidence-interval construction for the intercept are built on the following properties (based on Neter, *et al.*, 1990, pp. 71-73):

- A, the estimator of the intercept α, is a random variable.
- A is an unbiased estimator of α.
- If the observations Y_i are distributed normally, so is the estimator A.
- The variance of A is $\sigma^2(1/n + \bar{x}^2/S_{xx})$.
- The estimate of the variance of A is $S^2(1/n + \bar{x}^2/S_{xx})$.
- The estimator of the standard deviation of A (often called the standard error of the intercept) is

$$S_A = S\sqrt{\frac{1}{n} + \frac{\bar{x}^2}{S_{xx}}}\ .$$

From these properties, you construct the standardized statistic of the estimator A:

$$T = \frac{A - \alpha}{S\sqrt{1/n + \bar{x}^2/S_{xx}}}.$$

This statistic is distributed as Student's T with $(n - 2)$ degrees of freedom.

Now you may test whether α equals a given constant, say H_0: $\alpha = \alpha_0$, by substituting α_0 for α in the expression.

If the alternative hypothesis is set as H_1: $\alpha \neq \alpha_0$, you reject H_0 at the 0.05 level of significance if the absolute value of the calculated t exceeds $t_{0.975}(n - 2)$. For H_1: $\alpha > \alpha_0$, you reject H_0 if t exceeds $t_{0.95}(n - 2)$. For H_1: $\alpha < \alpha_0$, you reject H_0 if t is less than $t_{0.05}(n - 2)$, or equivalently, if $-t$ is larger than $t_{0.95}(n - 2)$.

Endpoints for a $(1 - \alpha)100\%$ two-sided confidence interval for the line's intercept are given by

$$A \pm t_{(1 - \alpha/2)}(n - 2) \times S_A$$

where $t_{(1 - \alpha)}(n - 2)$ is from Students' T table value for $\alpha = 0.05$ and $df = n - 2$.

One-sided tests and confidence intervals for the intercept are similarly constructed.

For discussion

- Construct a 95% confidence interval for the intercept of the pressure stabilization problem, Example 14-1.
- What interpretation will you give to that intercept?

Prediction with simple linear regression

Consider the use of the estimated equation, $Y = 0.3623 + 1.6636x$ as a "prediction equation" associated with the data given in Table 14-1. For each value of x that you place in that equation, you get a corresponding value of Y. To keep things straight, let $Y|x$ (read: "Y given x") designate the "predicted value" for x. Thus, for x = 5, $Y|5 = 8.68$ is the value that's "predicted." Similarly, $Y|5.5 = 9.51$ and $Y|20 = 33.63$ are "predicted" for $x = 5.5$ and $x = 20$, respectively.

Be very careful here. Since the x values in the original dataset are between 1 and 10, the prediction of y for $x = 20$ is considered an "extrapolation." Extrapolations must be accompanied by caution. If you are not cautious, you risk violating at least one of the fundamental assumptions of linear regression methodology: that the model is linear over the entire set of independent values for which it is applied.

The prediction equation offers a single predicted value from a single predictor x. But you should not be willing to bet heavily on this predicted value being correct, even if is it the best you have. Only a few, if any, of the original data rest on the prediction line. At best, you would expect the value you wish to predict to be near the prediction line.

Since the prediction may be in error, you hedge it. You do that by obtaining bounds which you are pretty sure—say, with probability $(1 - \alpha)$—to contain the true-but-unknown value $Y|x$. Endpoints for a two-sided $(1 - \alpha)$ confidence interval for $Y|x$ is given (see, for example, Neter, *et al.*, 1980, pp. 79-85) by:

$$A + Bx \pm t_{(1-\alpha/2)}(n - 2) \times S \times \sqrt{1 + \frac{1}{n} + \frac{(x - \bar{x})^2}{S_{xx}}}$$

where S is the square root of the Residual Mean Square error in the regression analysis table.

If you plot these endpoints for a series of values of x, you will see two curves, one lying entirely above and the other lying entirely below the regression line. Together, these curves form a *confidence belt* which is narrowest when $x = \bar{x}$, as seen in Figure 14-6.

Example 14-1c:
Pressure stabilization (continued one more time)

Your boss wants to know what you think will be the cumulative pressure drop when you are 8 hours into the test. What do you give as an answer?

Of course, you use the regression equation

$$Y_8 = 0.3623 + 1.6636(8) = 13.67$$

and report accordingly.

But then your boss says, "I may not know statistics, but I know what I like. And I like confidence in my answers. Just how confident *are* you that the pressure drop will be 13.67?"

Your personal confidence exceeds all bounds as you evaluate these endpoints for a 95% confidence interval:

$$0.3623 + 1.6636 \times 8 \pm 2.31 \times 0.966 \times \sqrt{1 + \frac{1}{10} + \frac{(8 - 3.99)^2}{58.7290}}$$

which yields 11.06 and 16.29 as the lower and upper endpoints of the 95% confidence interval.[4]

The locus of the endpoints of two-sided 95% confidence intervals on predicted values of x are shown in Figure 14-6, which is based on the data in Example 14-2.

[4]Although the dialog between you and your boss could continue indefinitely—or at least until $x = 8.01$—and it could be both entertaining and enlightening to continue to observe the process, we think it discreet to draw the curtain on that little drama right here.

Figure 14-6:
Endpoints of two-sided 95% prediction intervals for Example 14-1

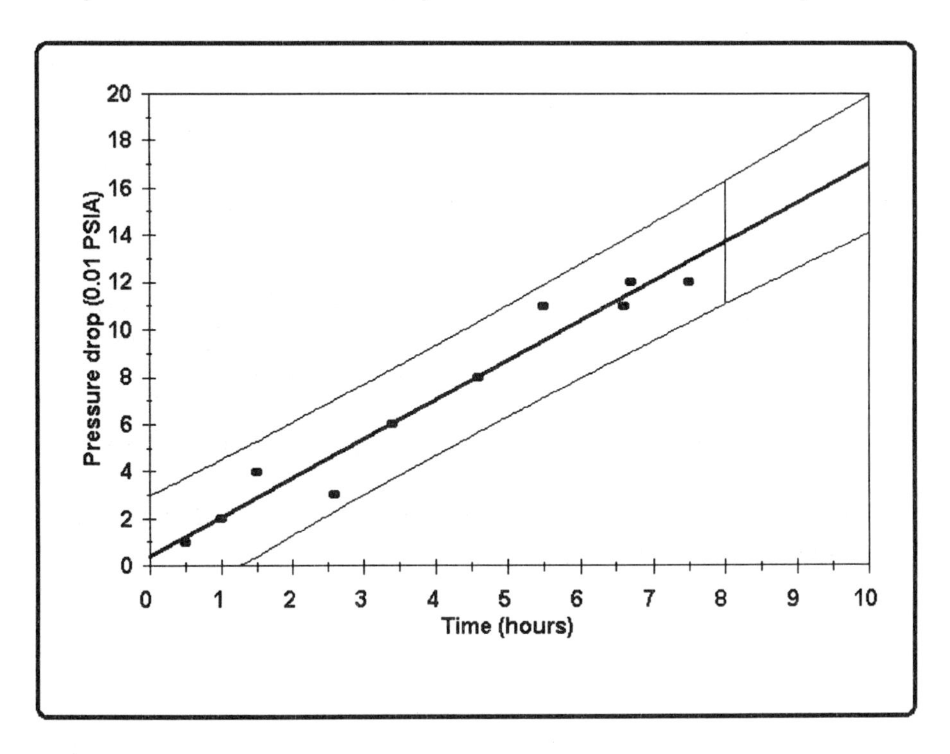

For discussion:

- Verify that the confidence belt for predicted values is narrowest for $x = \bar{x}$ and that the belt "flares out" away from $\bar{x}$. What does it tell you about the nature of prediction?

- A one-sided confidence interval may be more meaningful to your application. What two changes must you make to make the confidence belt one-sided?

The correlation coefficient

The previous sections of this chapter concentrated on how to "best" fit a line to a set of points and culminated in the procedures grouped under the topic of *regression*. Now it's time to add another device to your toolkit: the *correlation coefficient*, a device designed to assist your investigation of the strength of a linear relation between two random variables.

correlation coefficient

Consider a sample of size n of a two-dimensional random variables, say (X_i, Y_i), $i = 1, 2, \ldots, n$. This simply means that, in contrast to the one-dimensional random variables that appear in the bulk of this book, you observe two values at a time—and you do so a total of n times.

sample linear correlation coefficient

The *sample linear correlation coefficient* for the two variables X and Y, is given by

$$R_{XY} = \frac{S_{XY}}{\sqrt{S_{XX}\, S_{YY}}}$$

where S_{XX}, S_{YY}, and S_{XY} are the indicated sums of squares and cross-products, the computing formulas for which are

$$S_{XX} = \Sigma X^2 - (\Sigma X)^2/n,$$

$$S_{YY} = \Sigma Y^2 - (\Sigma Y)^2/n, \text{ and}$$

$$S_{XY} = \Sigma XY - (\Sigma X)(\Sigma Y)/n,$$

as given earlier in this chapter.

In the remainder of this chapter, we omit the adjective *linear* although, as you will see, linearity will always be there implicitly. Also, when there is no ambiguity, we omit the descriptive word *sample*.

Before illustrating the use of the correlation coefficient, it's good practice to make sure that you do not perform the right operation on the wrong patient. To that end, here are three underlying assumptions and requirements that validate the use of the correlation coefficient:

- The two random variables X and Y are linearly related. In the next *For discussion* section, you will find an example in which the correlation coefficient is calculated for a set of points that are *not* linearly related.

- There is no designation of independent and dependent roles to be played by X and Y. This is in marked contrast to regression methods, where x and Y are classified—even defined—as independent and dependent variables, respectively. Correlation treats each variable with equal respect, just as you and I and Rodney Dangerfield desire. Indeed, notice especially that the lower-case x, used in regression to indicate a fixed value, appears as an upper-case X to indicate a random variable when correlation is the topic.

- Statistical tests about the correlation coefficient require that X and Y have a bivariate normal distribution, but they do not necessarily have equal means or standard deviations.

The sample correlation coefficient R_{XY} has several properties, five of which are listed below:

- R_{XY} is symmetric in X and Y; that is, if you reverse the roles of X and Y, you find that $R_{XY} = R_{YX}$.

- R_{XY} is a random variable; it is expected to be different from one sample to the next. This is the reason that the sample correlation coefficient is designated with an upper-case symbol.

- A mathematical truth is that R_{XY} cannot be larger than +1 nor smaller than -1; i.e., you *always* have $-1 \leq R_{XY} \leq +1$, no matter what the individual values of the (X, Y) pairs.

- Consistent with your intuition, there also is a population correlation coefficient. This parameter, denoted by the lower-case Greek letter ρ_{XY} (read: "rho-sub-*XY*"), also is symmetric in X and Y; i.e., $\rho_{XY} = \rho_{YX}$. It also is bounded between -1 and +1; i.e., you *always* have $-1 \leq R_{XY} \leq +1$, no matter what the values of the individual (X, Y) pairs.

- **Heads up!** Be careful not to interpret the correlation coefficient as an indication of cause-and-effect. To reinforce the point, take an extreme case: Suppose that a study shows a strong correlation (i.e., "close to +1") between the annual number of children born in a certain country and the annual number of storks reported migrating into that country's bird sanctuaries. Does that suggest that storks deliver babies or that babies deliver storks or neither or both?

Example 14-3:
Correlating the cost of crude oil and the cost of premium gasoline

Table 14-6 gives the average prices of a barrel of crude oil and the retail price of a gallon of gasoline in a fictitious location over eight consecutive years. Calculate the correlation coefficient for the two prices.

Table 14-6:
Calculations for Example 14-3 of the correlation coefficient

Data setup and calculations

Index	Crude oil, $/barrel	Premium gasoline, $/gallon	Initial calculations		
i	x	y	x^2	y^2	xy
1	19.00	0.82	361.00	0.6724	15.58
2	24.00	0.97	576.00	0.9409	23.28
3	31.00	1.15	961.00	1.3225	35.65
4	32.00	1.33	1024.00	1.7689	42.56
5	33.00	1.30	1089.00	1.6900	42.90
6	28.00	1.15	784.00	1.3225	32.20
7	30.00	1.35	900.00	1.8225	40.50
8	23.00	1.29	529.00	1.6641	29.67
Sums	$\Sigma x =$ 220.00	$\Sigma y =$ 9.36	$\Sigma x^2 =$ 6224.00	$\Sigma y^2 =$ 11.2038	$\Sigma xy =$ 262.34

Intermediate calculations:

$S_{xx} = \Sigma x^2 - (\Sigma x)^2/n = 6224.00 - (220.0)^2/8 = 174.0000$
$S_{yy} = \Sigma y^2 - (\Sigma y)^2/n = 11.2038 - (9.36)^2/8 = 0.2526$
$S_{xy} = \Sigma xy - (\Sigma x)(\Sigma y)/n = 262.3400 - (220.00)(9.36)/8 = 4.9400$

Correlation coefficient calculation:

$$r_{xy} = \frac{S_{xx}}{\sqrt{(S_{xx})(S_{yy})}} = \frac{4.9400}{\sqrt{(174.0000)(0.2526)}} = 0.7451$$

How do you interpret the calculated correlation coefficient r_{xy} = 0.7451? Notice that r_{xy} is positive, indicating that the costs of crude oil and premium gasoline generally move in the same direction. If you had, rightly or wrongly, modeled these data as a regression problem, the slope of the regression equation also would be positive, thus telling you the same general story.

When you report a study involving correlation, the magnitude of the correlation coefficient may or may not be meaningful to your readers. In a discipline such as physics or chemistry, a correlation coefficient lower than, say, 0.95, may suggest sloppy experimentation and/or irreproducible results. In other disciplines, such as psychology or medical clinical trials, a correlation coefficient larger than, say, 0.5 may be a real finding and cause for celebration.

As always, it's a good idea to plot your data to examine the nature of the relationship between X and Y. In Example 14-3, the price of one commodity rises with the price of another, although there are some "glitches" that may or may not be explainable. With such a picture at hand, a value of R = 0.75 appears to be telling.

Example 14-3 encourages a repeat of our earlier warning against using a "high" correlation coefficient to suggest causality. Here, for instance, it may appear that the cost of crude oil drives the cost of gasoline. It is conceivable, however, that the cost of (and the demand for) gasoline drives up the cost of the oil. Indeed, it's even conceivable that both of these costs are driven by still a third factor.

And still another warning: The numerical interpretation of a correlation coefficient appears easy. But it's an inviting trap, so you must be on the lookout. For example, you cannot conclude that a correlation coefficient of 0.80, say, is twice as good as that of 0.40.

On the calculational similarities between regression and correlation

Although regression and correlation methods are based on considerably different sets of assumptions, the two methods have much in common in terms of the calculations involved. If you ignore for the moment the independent-dependent relation between the two variables *X* and *Y*, the correlation coefficient can, and often does, seem to apply to regression-type data. Clearly, if it appears that a linear trend relates *Y* to *X*, you would expect that the regression slope and the correlation coefficient would both be statistically significant.

coefficient of determination

The mathematical connections between correlation and regression are straightforward. If the Regression Sum of squares, SS_R, is divided by the Total Sum of squares, SS_T (c.f., Table 14-2), then the resultant ratio is the *square* of the correlation coefficient. Formally, $R^2 = SS_R/SS_T$, is called the *coefficient of determination*. For example, the coefficient of determination in Example 14-1 (Table 14-3) is 162.53/170.00 = 0.956, from which $r = 0.98$. Notice that R^2 gives the fraction of the data variability, as measured by the Total Sum of squares, that is "explained" by the Regression Sum of squares. For this reason, many writers prefer the use of the coefficient of determination, R^2, to the correlation coefficient as a measure of the quality of the fit.

For discussion:

- Give examples where a very large correlation coefficient, say $R = 1$ for a population or $r = 1$ for a sample, is not meaningful.

- Some writers denote the sample correlation coefficient by $\hat{\rho}$ (read: "rho-hat"). Is this symbology consistent with your statistical upbringing?

- Consider the five points (x_i, y_i) given in this table:

x_i	1	2	3	4	5
y_i	1	4	9	4	1

If you plot these points, you obtain a fully predictable pattern, at least within the range of the x_i. Proceed to calculate the correlation coefficient, and you will obtain $r_{xy} = 0$. How would you explain this correlation? Does the term *linear* seem appropriate here?

- What is the coefficient of determination in Example 14-2.

Testing the correlation coefficient

If you wish to test whether X and Y are uncorrelated (that is, you wish to test the null hypothesis H_0: $\rho = 0$), you use the following test statistic which is distributed as Student's T statistic with $(n - 2)$ degrees of freedom:

$$T = R \sqrt{\frac{n - 2}{1 - R^2}}.$$

Formally, you write H_0: $\rho = 0$, H_1: $\rho \neq 0$, $\alpha = 0.05$ (two-sided). If the calculated t value exceeds $t_{0.975}(n - 2)$ or falls short of $t_{0.025}(n - 2) = -t_{0.975}(n - 2)$, you reject the null hypothesis H_0.

If the alternative hypothesis is one-sided, you select the critical value from Table T-3 accordingly.

In Example 14-1, $r = 0.9778$, so that

$$t = .9778 \sqrt{\frac{10 - 2}{1 - (0.9778)^2}} = 13.199.$$

The numerical result of this test is in exact agreement with that of the regression analysis (or its equivalent two-sided T test). In the regression analysis of Example 14-1 (displayed in Table 14-5), $f = 174.04$, which—apart from rounding error —is identical to $t^2 = (13.199)^2$. Because t^2 with $(n - 2)$ degrees of freedom equals f with 1 and $(n - 2)$ degrees of freedom, the equivalence of the two statistics is verified for this example.

If you wish to test whether the correlation coefficient differs from a non-zero constant, you use another technique, called Fisher's Z-transformation. The reason for the transformation is that the sample correlation coefficient, R, is not distributed normally, unless $\rho = 0$. Fisher's Z statistic, given by the expression

$$Z = \frac{1}{2} \ln \frac{1 + R}{1 - R},$$

however, does follow the normal distribution. The mean and the variance of the random variable Z are given respectively by

$$mean[Z] = \frac{1}{2} \ln \frac{1 + \rho}{1 - \rho}$$

and

$$var[Z] = \frac{1}{n - 3}.$$

To use Fisher's Z statistic effectively, you need to bring it into a standardized form. This is achieved by dividing the difference (Z - *mean*[Z]) by the square root of *var*[Z]. Consistent with conventions established earlier in this book, the standardized normal should be called Z—but this designation is pre-empted by Fisher's use of Z. So, for this section alone, we designate the standard normal by $\boldsymbol{J}$ (***Just** for this section*). The constructed standard normal statistic is

$$\boldsymbol{J} = \frac{\frac{1}{2}\ln\frac{1+R}{1-R} - \frac{1}{2}\ln\frac{1+\rho}{1-\rho}}{\sqrt{\frac{1}{n-3}}}.$$

To illustrate the use of the statistic $\boldsymbol{J}$, consider again Example 14 -2, where you wish to test H_0: $\rho = 0.80$ against H_1: $\rho \neq 0.80$, and the sample correlation coefficient, based on 8 observations, yields $r = 0.75$. You complete the calculations as follows:

Step 1: Under H_0: $\rho = 0.80$, *mean*[Z] is calculated as

$$mean[Z|\rho = 0.80] = \frac{1}{2}\ln\frac{1+0.80}{1-0.80} = 1.0986.$$

Step 2: For a sample of size $n = 8$, the standard deviation associated with Fisher's Z is

$$S[Z] = \frac{1}{\sqrt{8-3}} = 0.4472.$$

Step 3: The sample z transformation of $r = 0.75$ is calculated as

$$z(r = 0.75) = \frac{1}{2} \ln \frac{1 + 0.979}{1 - 0.979} = 0.9618.$$

Step 4: Putting the pieces together, you calculate the observed standardized normal variable (once again, note that the random variable J changes to an observed value j when it is replenished with data)

$$j = \frac{0.9618 - 1.0986}{0.4472} = -0.3059.$$

Step 5: Since $-1.96 < j < 1.96$, you do not reject H_0. You state that you have insufficient statistical evidence against the claim that $\rho = 0.80$.

Confidence interval for the correlation coefficient

Using procedures learned earlier, you can employ Fisher's Z transformation to construct one- or two-sided confidence intervals for the correlation coefficient ρ. This procedure is conducted in two stages:

First stage: Construct confidence limit(s) for the mean of Fisher's Z transformation; i.e., *mean*[Z].

Second stage: Invert the confidence limit(s) to capture the correlation coefficient; i.e., ρ.

Assume that a $100(1 - \alpha)\%$ two-sided interval is desired. Then the details of the two stages are these:

First stage: To construct a confidence interval on *mean*[Z] when the variance is known to be $1/(n - 3)$, you construct the interval's endpoints by calculating

$$\frac{1}{2} \ln \frac{1 + R}{1 - R} \pm z_{(1-\alpha/2)} \sqrt{\frac{1}{n - 3}}.$$

Designating these 95% confidence limits by L_1 and L_2 leads to the statement that you are 100(1 - α)% confident that

$$L_1 < \frac{1}{2} \ln \frac{1 + \rho}{1 - \rho} < L_2.$$

Second stage: You manipulate this double inequality to capture ρ by treating each side as an equality and solving for the unknown parameter ρ in terms of the endpoints L_1 and L_2. Solving for ρ in terms of L_i ($L_i = L_1$ or $L_i = L_2$,) you obtain the respective lower and upper confidence limits on ρ as

$$R_1 = \frac{e^{2L_1} - 1}{e^{2L_1} - 1}$$

and

$$R_2 = \frac{e^{2L_2} - 1}{e^{2L_2} - 1}.$$

For Example 14-2, you calculate a 95% two-sided confidence interval for the correlation coefficient between the costs of crude oil and premium gasoline as follows:

You have these statistics with which to work:

$r = 0.7451$, $n = 8$, and Fisher's $z = 0.9618$.

First stage: To construct a 95% two-sided confidence interval on the mean, *mean*[Z], when the variance is known to be $1/(n - 3) = 1/(8 - 3) = 1/5 = 0.2$, you construct the interval's endpoints by calculating

$$0.9618 \pm 1.960 \sqrt{\frac{1}{5}} = 0.9618 \pm 0.8765,$$

from which you obtain the limits for the mean of Fisher's Z as $L_1 = 0.0853$ and $L_2 = 1.8383$. Thus, you are $100(1 - \alpha)\%$ confident that

$$0.0853 < \frac{1}{2} \ln \frac{1 + \rho}{1 - \rho} < 1.8383.$$

Second stage: You manipulate this double inequality to capture ρ by treating each side as an equality and solving for the unknown parameter ρ in terms of the endpoints $L_1 = 0.0853$ and $L_2 = 1.8383$. Solving for ρ, you obtain the respective lower and upper confidence limits on ρ as

$$R_1 = \frac{e^{2(0.0853)} - 1}{e^{2(0.0853)} - 1} = 0.0851$$

and

$$R_2 = \frac{e^{2(1.8383)} - 1}{e^{2(1.8383)} + 1} = 0.9506.$$

The considerable length of this 95% confidence interval (0.0851, 0.9506) doesn't seem very informative. Notice, however, that it is derived from the endpoints of the interval on *mean*(Z), itself of considerable length. This condition, therefore, is directly attributable to the small sample size, despite the seemingly "strong" value of the sample correlation coefficient itself.

For discussion:

- What conclusions from Example 14-3 do you draw about the correlation between the price of crude oil and the cost of premium gasoline, as analyzed and displayed in Table 14-6?

- Examine the data in Example 14-3 (and Table 14-6) regarding the cost of gasoline. Are you satisfied with the data? How would *you* have collected and recorded the data? Is something missing? Is something misleading?

To wrap things up, this one's just for fun ...

... unless you really are in the movie-making business. *A scenario:* You are the data-whiz for a movie mogul. He swings into your office, hands you this *Time* (August 23, 1993, p. 61) table, and tells you to plan next year's production for the studio. You sit back and consider: (1) The mogul owns the studio. (2) You have a mortgage. (3) You have a nice car. (4) You have these data.

1993 Summer Movies

Title	Rating	Earnings in millions as of 8/13/93	Projected earnings*
Jurassic Park	PG-13	$297	$320
The Firm	R	$137	$160
Sleepless in Seattle	PG	$100	$120
In the Line of Fire	R	$80	$125
Cliffhanger	R	$80	$85
Free Willy	PG	$51	$80
Last Action Hero	PG-13	$48	$50
Dennis the Menace	PG	$47	$50
The Fugitive	PG-13	$45	$150
Rookie of the Year	PG	$44	$50

* From industry analysts

How do these data stand up to Huff's criteria (Chapter 1)? You also know **OSDAR** (also Chapter 1); does this database hold up under **OSDAR**'s spotlight? *[The car.]* Can you use these data to win an Oscar for the mogul's company? Or, at least, to make a pile of money? *[The house.]* Or, are they even worth spending any time on?

Have you ever encountered problems parallel to these in your daily work? How do you characterize those problems? What did/can you do to deal with them? Do you think other people might encounter similar problems. How would you counsel them?

What to remember about Chapter 14

Chapter 14 laid out and sharpened the tools needed to construct a regression line to fit a set of data points. You were given explicit formulas and directions to manipulate a set of data and to calculate such statistics as:

- *the estimate of the slope*
- *the estimate of the intercept*
- *the estimate of the variance.*

You were shown how to test the hypotheses that:

- *the slope of the regression line is zero*
- *the slope equals a prescribed constant*
- *the intercept equals a prescribed constant.*

You were shown how to construct:

- *a regression analysis table*
- *a confidence interval for the slope*
- *a confidence interval for the intercept.*

You compared and contrasted two important statistical ideas: regression and correlation. Special attention was given to:

- *estimating the correlation coefficient*
- *testing the hypothesis that the correlation coefficient equals a prescribed constant (including zero)*
- *constructing a confidence interval for the correlation coefficient.*

15

More on probability

What to look for in Chapter 15

Chapter 15 reviews the elements of probability, some of which were introduced in Chapter 3 where probability and statistics were compared and contrasted. It also introduces a few new elements that are essential in the development of Chapters 16 through 21. Here you will meet, again or for the very first time, the concepts of:

- *experiment, outcome, and event*
- *probability*
- *marginal and joint probabilities*
- *complementary events*
- *mutually exclusive events*
- *cumulative probabilities*
- *conditional probability*
- *independent events*.

You will also find rules for carrying out probability calculations under a variety of conditions, such as:

- *when events are independent*
- *when events are mutually exclusive*
- *when events are not mutually exclusive*
- *when events are cumulative.*

Down to basics

Even though the topic called *probability* is generally considered to have originated in gambling games in 16th- and 17th-century Europe, it remains a continuously expanding complex human endeavor. Kendall and Buckland (1971, p. 118) have this to say:

> **Probability** A basic concept which may be taken either as undefinable, expressing in some way a 'degree of belief', [sic] or as the limiting frequency in an infinite random series. Both approaches have their difficulties and the most convenient axiomization of probability theory is a matter of personal taste. Fortunately both lead to much the same calculus of probabilities.

With *that* thought thoroughly digested, let's get started.

experiments

In our pursuit of knowledge, we often conduct *experiments*. From these experiments, we obtain data that may help us confirm or refute some stated hypothesis or hypotheses. Indeed, many experiments are designed and conducted with the direct intention of examining a hypothesis.

In Chapter 3, we offered this definition:

> **An experiment** is a planned inquiry to obtain new facts or to confirm or deny the results of previous experiments, where such inquiry will aid in decision-making.

observations, univariate, multivariate

The experimental facts and results are called *observations*, which are categorized further as being either *univariate* or *multivariate*. Examples of univariate observations include a head *or* a tail in the toss of a single coin, the number of pumps (out of 20) that do not start on demand, the number of automobile accidents at the corner of Fifth and Main in Yourtown, USA, during the year 1966, and the average weight of 100 fuel rods selected from a specified manufacturing process. Examples of multivariate observations are the height, age, and weight of first-grade students at the end of their first semester; the make, model, color, age, and tail-pipe emissions of automobiles passing a certain point on a certain highway on a certain day; and the type of business, its floor area, and its use of energy for heating and air-conditioning in a specific building in a certain city in a certain year.

outcome

An observation derived from an experiment is called an *outcome*. An outcome of an experiment involving a quantity on an interval scale can be any one of a nondenumerable (i.e., uncountable) infinity of values. An outcome involving a discrete variable is a denumerable (i.e., countable) infinity of values.

sample space, sample points

The set of all possible outcomes of any experiment is called the *sample space of the experiment*. The elements of a sample space are called *sample points*. Here are some examples:

- When you toss a coin one time, the sample space has two sample points: $\{H, T\}$.

- When you attempt to start 20 pumps and count the number that do not start, the sample space has 21 sample points: $\{0, 1, 2, \ldots, 20\}$.

- When you count the accidents at Fifth and Main, the sample space has an unknown number of sample points, all of which are integers: $\{0, 1, 2, 3, \ldots\}$.

- When you weigh and then average the weights of 100 fuel rods, the sample space for that average weight has a nondenumerable infinity of sample points: it is the set of all values, say x, that lie above some lower bound, say L, and below some upper bound, say H. In set theory notation, you might find this idea expressed as $\{x \mid L < x < H\}$ where L and H are determined by the specified manufacturing process.

event

An *event* is a collection of outcomes; i.e., an event is a subset of the sample space.

For example, for the single throw of a six-sided die, the sample space can be indicated by {1, 2, 3, 4, 5, 6}. An event called "Even" occurs when the outcome is 2 or 4 or 6. An event called "Odd" occurs when the outcome is 1 or 3 or 5.

As indicated by the Kendall-Buckland quotation, probability appears for most of us to be a matter of personal choice. For example, in everyday discourse, we use such phrases as "It'll prob'ly rain tomorrow" or "The [favorite team] probably will win." It's when we try to perform quantitative analyses in support of decision-making that these "degrees of belief" let us down.

Instead, let us proceed to a few basic ideas derived from Brownlee (1965, pp. 6-24). The approach taken by Brownlee *and* the one adopted in this text is *axiomatic*. We start with the axiom (i.e., a basic, self-evident truth; a basic proposition stipulated to be true) that it is possible to associate with every point A_i in the sample space a special kind of non-negative number, called the probability of A_i and denoted by the symbol $Pr\{A_i\}$. These probabilities satisfy the condition that they sum to 1.

These conditions are summarized in two important expressions:

(1) $0 \leq Pr\{A_i\} \leq 1$

(2) $Pr\{A_1\} + Pr\{A_2\} + \ldots = 1.$

For discussion:

- What do the terms "denumerable" and "nondenumerable" mean to you?

- Drawing a single card from a well-shuffled standard deck of 52 cards (composed of four suits of 13 cards each) provides homely but easily described illustrations of these ideas. Because there are exactly 52 cards which might be drawn, we conclude that there are 52 points in the sample space. Moreover, they are easily enumerated: A♣, K♣, Q♣, J♣, 10♣, . . ., 2♠; they are fully displayed in an array like this:

  A♣ K♣ Q♣ J♣ 10♣ 9♣ 8♣ 7♣ 6♣ 5♣ 4♣ 3♣ 2♣

  A♦ K♦ Q♦ J♦ 10♦ 9♦ 8♦ 7♦ 6♦ 5♦ 4♦ 3♦ 2♦

  A♥ K♥ Q♥ J♥ 10♥ 9♥ 8♥ 7♥ 6♥ 5♥ 4♥ 3♥ 2♥

  A♠ K♠ Q♠ J♠ 10♠ 9♠ 8♠ 7♠ 6♠ 5♠ 4♠ 3♠ 2♠

  Because it is difficult to argue *a priori* that any one of the 52 cards is favored over the others, we conclude that the probability of being drawn is the same for each card; i.e.,

  $Pr\{A♣\} = Pr\{K♣\} = \ldots = Pr\{2♠\}.$

  These probabilities must sum to one. It follows that each probability is exactly 1/52.

- Does this analysis of the 52-card deck probabilities depend upon the way you display the cards; that is, would you get a different answer (i.e., different probabilities for the cards) if, say, you displayed the cards in the sequence {2♣ 3♣ ... *K*♠ *A*♠}?

Probabilities associated with equally likely outcomes

equally likely

Consider a sample space containing exactly N outcomes, each of which has the same probability of occurring; i.e., the outcomes are *equally likely*. Now define an event E, and let N_E be the number of outcomes in the event E. Then the *probability of the event E* is expressed as

$$Pr\{E\} = \underbrace{(1/N) + (1/N) + \ldots + (1/N)}_{N_E \text{ terms}} = N_E/N.$$

For example, consider a group of 1,000 cables, with 900 functional ones and 100 defective ones. Let E be the event of finding a defective cable when a single cable is selected. The probability of inspecting a single cable and finding it defective is

$$\begin{aligned} Pr\{E\} &= Pr\{\text{cable is defective}\} \\ &= \underbrace{1/1{,}000 + 1/1{,}000 + \ldots + 1/1{,}000}_{100 \text{ terms}} \\ &= 100/1{,}000 = 0.10. \end{aligned}$$

Starting with $0 \leq N_E$ and $N_E \leq N$, you have confirmed the non-negativity of the probabilities and can write, for any event E,

$$0 \leq Pr\{E\} \leq 1.$$

For discussion:

- How would *you* define and/or interpret *probability*?

- The expression of $Pr\{E\} = N_E/N$ for equally likely events is sometimes referred to as the *classical definition of probability* and sometimes called the frequentist's definition. Some authors, e.g., Ott and Mendenhall (1985, Chapter 11), call it the classical *interpretation* of probability to avoid the word "definition." Note that we did not call this expression a definition, either, principally to avoid a malpractice action brought by picky-picky probabilists and/or superscrupulous statisticians. In fact, this expression is a theorem which follows from the axiomatic approach adopted in this book.

- How would you define equally likely *outcomes* now?

- What would you expect the term *equally likely events* to mean? Are the events "odd" and "even" equally likely?

- In the audit process of travel vouchers, you may select 10% of the vouchers submitted during the month for extensive review. Suggest a way for random sampling of 10% of the vouchers. Are all the vouchers equally likely to be selected? Do you really want all vouchers to be selected with equal probability? Why, do you suppose, this wide-spread, nearly ubiquitous, "10% sample" was called for?

- How would you select an agency's employees for random drug-testing?

- Every evening, a restaurant polls its customers and offers a free dinner to customers celebrating their birthday on that day. Describe this promotion in terms of an experiment, equally likely outcomes, and the probability that a customer randomly coming in will be treated to a free dinner.

- What does $Pr\{E\} = 0$ mean to you?

- What does $Pr\{E\} = 1$ mean to you?

- Your good friend states that the probability of a tornado hitting your community in the next five years is $Pr\{tornado\} = -1$. What do you think your friend meant to say? How would you have stated it?

More terminology and more rules of probability

marginal probability

The probability of a single event E, $Pr\{E\}$, is often called the *marginal probability* of E. The usefulness of this special terminology becomes apparent when you need to contrast marginal probability with the probability of several events occurring simultaneously.

complementary event

If an event E does not occur, then the *complementary event*, denoted by E^c, *must* occur.[1] Some authors use the $\bar{E}$ (read E-bar) notation to denote the complement of E.

In light of this definition and because E and E^c account for all of the points in the sample space, it follows that

$$Pr\{E\} + Pr\{E^c\} = 1.$$

This simple-appearing statement is extremely powerful when it comes to solving certain types of probability problems. For numerous situations, finding $Pr\{E\}$ is too complex a task—but finding $Pr\{E^c\}$ may be the proverbial "piece of cake." When that happens, you take advantage of the fact that $Pr\{E\} = 1 - Pr\{E^c\}$, and you're done.

joint probability

The probability that two events, call them E_1 and E_2, happen simultaneously is called the *joint probability* of E_1 and E_2. This joint probability is denoted by $Pr\{E_1 and E_2\}$ and is read as "the probability of E_1 *and* E_2."

[1] This complementary event is sometimes called the *E-complement*.

mutually exclusive events

Two events, E_1 and E_2, are said to be *mutually exclusive* if the occurrence of either one of the two events precludes the occurrence of the other event. That is, no sample point lies in E_1 and E_2 simultaneously, so that the joint probability is zero. This is expressed as

$$Pr\{E_1 and E_2\} = 0.$$

To illustrate the property of mutual exclusivity, if a switch can be set in only one of two positions (i.e., open or closed), then the two possible outcomes of the switch's inspection are mutually exclusive.

If two events are mutually exclusive, then the probability of *either* one occurring is the sum of the probabilities of each one occurring. This yields

$$Pr\{E_1 or E_2\} = Pr\{E_1\} + Pr\{E_2\}.$$

If the k events E_1, E_2, ..., E_k are mutually exclusive, there are *no* sample points that belong simultaneously to any pair, E_i and E_j for any combination of i and j, of these events. Then the probability of any one of the events occurring is the sum of the probabilities associated with the individual events; i.e.,

$$Pr\{E_1 or E_2 or \ldots or E_k\}$$
$$= Pr\{E_1\} + Pr\{E_2\} + \ldots + Pr\{E_k\}$$
$$= \Sigma\ Pr\{E_i\}.$$

Suppose that your organization receives a shipment of batteries. You test the quality of the entire shipment by examining 100 batteries from the lot. Suppose further that the entire lot is acceptable if no more than two of the 100 examined batteries fail to hold their charges. Obviously, you accept the lot if 0 *or* 1 *or* 2 batteries in the sample fail. These outcomes are mutually exclusive, so that

$$Pr\{\text{shipment is accepted}\} = Pr\{0\} + Pr\{1\} + Pr\{2\}.$$

cumulative probability

Compared with probabilities associated with individual events, the battery-inspection example illustrates the concept of *cumulative probability*, an idea applied primarily to mutually exclusive events. As the name suggests, a cumulative probability gives you the probability of more than one event occurring, beginning with a reasonable starting event and proceeding through a reasonable sequence and ending with an equally reasonable event, as shown in the example above.

If two events are *not* mutually exclusive, then the probability of either one occurring is the sum of the probabilities of each one occurring *minus* the probability of both occurring. This is written as

$$Pr\{E_1 or E_2\} = Pr\{E_1\} + Pr\{E_2\} - Pr\{E_1 and E_2\}.$$

conditional probability

If it is known that an event E_2 has occurred, then the probability that a different event E_1 has occurred is called the *conditional probability* of E_1, given E_2. This probability is denoted by $Pr\{E_1 | E_2\}$.

To illustrate, if a component fails a test the first time, the conditional probability that the same component will pass the test at the next inspection is $Pr\{P | F\}$. If one defective cable is found in the sampling of a bundle of cables, the conditional probability of finding no defective cables in the next inspected bundle is written as $Pr\{0 | 1\}$.

independent events

Two events E_1 and E_2 are said to be *independent* if and only if $Pr\{E_1 | E_2\} = Pr\{E_1\}$. As you see, if E_1 and E_2 are independent, then knowledge of the probability of E_2 is irrelevant in the calculation of the probability of E_1.

For example, in many human genetic studies, the assumption is made that the sex of a forthcoming child is independent of the sexes of the children already in the family. Thus, in a family with five children, all of them girls, the probability that the next baby will be a girl may be written as $Pr\{G \mid GGGGG\} = Pr\{G\}$, where $GGGGG$

represents the event that there are already five girls in the family.

As another example, let the V denote the event that Valve 1 is open and V^c that it is closed. Similarly, let W denote the event that Valve 2 is open and W^c that it is closed. If the status of Valve 1 and the status of Valve 2 are independent, then all of the following probability statements are correct:

$$Pr\{W|V\} = Pr\{W\}$$
$$Pr\{W|V^c\} = Pr\{W\}$$
$$Pr\{V|W\} = Pr\{V\}$$
$$Pr\{V|W^c\} = Pr\{V\}.$$

An important probability law for the *joint probability* of the two events E_1 and E_2 is given by the expression

$$Pr\{E_1 and E_2\} = Pr\{E_1|E_2\}Pr\{E_2\} = Pr\{E_2|E_1\}Pr\{E_1\}.$$

If the two events E_1 and E_2 are independent, this equation simplifies to

$$Pr\{E_1 and E_2\} = Pr\{E_1\}Pr\{E_2\}.$$

Conversely, if the probability expression

$$Pr\{E_1 and E_2\} = Pr\{E_1\}Pr\{E_2\}$$

holds for the two events E_1 and E_2, then the events are independent.

To illustrate, consider a commuting problem involving two bridges. Suppose that the probability of a traffic delay on Bridge 1 is *not* independent of a traffic delay on a parallel Bridge 2; i.e., the overflow from either bridge burdens the other. In this case,

$$Pr\{\text{delay on both}\}$$
$$= Pr\{\text{delay on 1}|\text{delay on 2}\}Pr\{\text{delay on 2}\}.$$

If the bridges' traffic delays are independent of each other, then

$$Pr\{\text{delay on both}\} = Pr\{\text{delay on 1}\}Pr\{\text{delay on 2}\}.$$

Summarizing some useful rules of probability

Here, in one place, is a restatement of each of the probabilistic rules encountered so far. Although a number of other rules could be added, this list contains the ideas that will be called upon in Chapters 16 through 21.

- The probability of an event E cannot be less than zero or greater than 1; that is,

  $$0 \leq Pr\{E\} \leq 1.$$

- An event E either occurs or it does not occur; that is,

  $$Pr\{E\} + Pr\{E^c\} = 1.$$

- The probability of two events, E_1 and E_2, both occurring is

  $$\begin{aligned} Pr\{E_1 and E_2\} &= Pr\{E_1 | E_2\}Pr\{E_2\} \\ &= Pr\{E_2 | E_1\}Pr\{E_1\}. \end{aligned}$$

- If the two events E_1 and E_2 are *mutually exclusive*, then the probability of either event occurring is the sum of their marginal probabilities; that is,

  $$Pr\{E_1 or E_2\} = Pr\{E_1\} + Pr\{E_2\}.$$

- If the two events E_1 and E_2 are *not mutually exclusive*, then probability of either event occurring is the sum of their marginal probabilities less their joint probability; that is,

$$Pr\{E_1 or E_2\} = Pr\{E_1\} + Pr\{E_2\} - Pr\{E_1 E_2\}.$$

- If the two events E_1 and E_2 are independent, their joint probability equals the product of their marginal probabilities; that is,

$$Pr\{E_1 E_2\} = Pr\{E_1\} Pr\{E_2\}.$$

- If the k events E_1, E_2, ... , E_k, are mutually exclusive, then the probability of *at least* one of them occurring is the sum of their marginal probabilities; that is,

$$Pr\{E_1 or E_2 or \ldots or\ E_k,\} = \Sigma Pr\{E_i\}.$$

Example 15-1: Gender and hiring practice

Table 15-1 contains the fictional probabilities of selecting male and female applicants for a position in a certain agency. The three selection options are: (1) do not hire, (2) select on a trial basis, and (3) select for a permanent position.

Table 15-1:
Example showing independent events

Selection option	Gender		Marginal probability of hiring
	Male	Female	
Do not hire	0.24	0.36	0.60
Trial basis	0.12	0.18	0.30
Permanent	0.04	0.06	0.10
Marginal probability of gender	0.56	0.44	1.00

Examine Table 15-1 to recognize that the probability of each joint event shown in the shaded area equals the product of the corresponding marginal probabilities. This assures probabilistic independence of gender and hiring.

There are a few other features to this table you will recognize:

The events "Do not hire," "Trial basis," and "Permanent" are mutually exclusive. An applicant *must* fall into one of the three categories, and into *only* one. Thus, the probabilities of these events are additive. Since these events exhaust all possibilities, their probabilities sum to 1.00. Similarly, the "male" and "female" events in the gender category also are additive. Note that the probabilities within a row or a column also sum to their marginal probabilities. For example, for the line labeled "Trial basis," you have Pr(*Male and Trial basis*} = 0.12 and Pr(*Female and Trial basis*} = 0.18; these two probabilities sum to Pr(*Trial basis*} = 0.30.

Studying the layout of Table 15-1 will reinforce the idea behind the term *marginal probability*.

You may wonder how such a table might look when the probabilities are *not* independent. Table 15-2 is a modification of Table 15-1, designed to illustrate such dependence.

Table 15-2:
Example showing dependent events

Selection option	Gender		Marginal probability of hiring
	Male	Female	
Do not hire	0.40	0.20	0.60
Trial basis	0.12	0.18	0.30
Permanent	0.04	0.06	0.10
Marginal probability of gender	0.40	0.60	1.00

That **Selection option** and **Gender** are *not* independent is shown in at least one cell in which the product of the marginal probabilities does not equal the joint probability. Can you find such a cell?

Example 15-2:
Earthquakes and tornados

The probability of a severe earthquake in a specific location in the year 2010, call it event E, is calculated as 0.03. The probability of a severe tornado during the same year at the same location, call it event T, is 0.06. The two events are considered independent. What is the probability of both events occurring in the year 2010?

$$Pr\{QandT\} = Pr\{Q\}Pr\{T\}$$
$$= (0.03)(0.06) = 0.0018.$$

Are these two events (the earthquake and the tornado) mutually exclusive? Are they independent? Are they equally likely?

Example 15-3: Elevator failure

Suppose the probabilities of failure-to-operate for elevators A, B, and C in a given moment of time are $Pr\{A\} = 0.01$, $Pr\{B\} = 0.02$, and $Pr\{C\} = 0.10$; suppose also that these three events are independent. What is the probability that both elevators A and B will be out of service tomorrow at 10:00 a.m.? What is the probability that all three will be out of service tomorrow morning at 10 a.m.?

$$Pr\{AandB\} = Pr\{A\}Pr\{B\} = (0.01)(0.02) = 0.0002.$$

$$Pr\{AandBandC\} = Pr\{A\}Pr\{B\}Pr\{C\}$$
$$= (0.01)(0.02)(0.10) = 0.00002.$$

Example 15-4: Telephone call routing

Assume that a telephone call from Washington to Seattle is routed through Chicago or Denver with 0.4 and 0.3 probability, respectively. What is the probability that the next call will be routed through *neither* of these cities?

The routings through Chicago or Denver are mutually exclusive. Therefore, $Pr\{\text{Chicago}or\text{Denver}\}$ $= Pr(\text{Chicago}\} + Pr\{\text{Denver}\} = 0.4 + 0.3 = 0.7$. Thus, $Pr\{\text{neither Chicago nor Denver}\} = 1.0 - 0.7$ $= 0.3$.

For discussion:

- Describe events associated with your work which are (i) clearly independent, (ii) clearly not independent, (iii) clearly mutually exclusive, (iv) clearly not mutually exclusive, and (v) not clearly any of the above.

- Reconsider Tables 15-1 and 15-2. What similarities do you note between these tables and the contingency tables in Chapter 4?

- Referring to Example 15-3, is it ever reasonable to assume that the failure events of the three elevators are mutually exclusive?

A group exercise

This exercise begins with a partial set of probabilities that are intended to help you review some of the concepts learned in this chapter. Table 15-3 gives the probabilities of failure of new motors (no repairs allowed) during the first, second, third, fourth, and fifth years after installation. Complete the table and then answer a few questions. (The empty cells contain lower-case letter indicators in parenthesis to simplify identification of the table's cells.)

Table 15-3:
Table for group exercise

Failure time	Marginal probability of failure	Cumulative probability of failure
First year	0.25	(b)
Second year	0.20	(c)
Third year	0.10	(d)
Fourth year	0.15	(e)
Fifth year	0.20	(f)
Beyond fifth year	(a)	(g)

(1) The six rows in the body of Table 15-3 describe different events. Are these events mutually exclusive? Why?

(2) Calculate the probability that a motor will fail after five years of service. Enter your answer in (a).

(3) Calculate the probability that a motor will fail before the beginning of the second year. Enter your answer in (b).

(4) Calculate the probability that a motor will fail before the end of the second, third, fourth, fifth, and beyond the fifth years. Enter your answers in (c), (d), (e), and (f), respectively.

(5) Now go back and calculate item (g). What does this probability represent?

(6) Calculate the probability that a motor will not fail in the first year? In the second year? Before the end of the second year?

(7) Using the constructed table, show two ways to calculate the probability that the motor will be in service for at least three years.

(8) Calculate the probability that a motor will fail during the third year *or* the fourth year *or* the fifth year after installation.

What to remember about Chapter 15

Chapter 15 identifies and explains several concepts, specifically:

- *probability*
- *marginal and joint probabilities*
- *mutually exclusive events*
- *cumulative probabilities*
- *conditional probability*
- *independent events*
- *complementary events.*

You examined some examples and participated in some exercises to learn to calculate these probabilities in real-life situations. The understanding of these concepts and the ability to calculate selected probabilities are essential to grasping and interacting with the rest of this book.

16

Hypergeometric experiments

What to look for in Chapter 16

Chapter 16 is the first of a series of three chapters, each dealing with a specific discrete distribution. Of these three chapters, only this one is dedicated to sampling from a strictly finite population.

Among the special terms in this chapter are:

- *attributes*
- *sampling without replacement*
- *factorial function*
- *binomial coefficient*
- *hypergeometric distribution function*.

In your encounter with Chapter 16, you will gain tolerance for—if not achieve comfort with—the hypergeometric distribution and learn to:

- *recognize experiments related to the hypergeometric distribution*

- *calculate probabilities associated with hypergeometric distribution*
- *calculate the mean and standard deviation for hypergeometric variables*
- *recognize the difficulties in hypergeometric probability calculations and how to avoid some of them*
- *test hypotheses about the hypergeometric distribution.*

Sampling for attributes in finite populations

attribute

Inferences derived from many types of data, primarily measured on interval or ratio scales, typically concentrate on such measures as the mean and the variance. On the other hand, inferences derived from discrete data typically focus on the composition of a population, e.g., the proportion of the population having a specific *attribute* (characteristic). In a deck of bridge cards, the attribute may be "heart" for a given card. In an urn containing marbles of different colors, the attribute may be "size" (or "color" or "condition") for a given marble. In a jar of cookies, an attribute of considerable interest may be "broken" for a given cookie.

dichotomous (binary)

Sampling for attributes typically requires that every element in the population can be classified unambiguously into one of two groups. Such classification labels the outcome as being a *dichotomous*[1] (i.e., a *binary*) response, and the associated event is, similarly, dichotomous (binary).

Examples of dichotomous outcomes (events) are {yes, no}, {on, off}, {defective, not-defective}, {male, female}, and {employed, unemployed}. If an outcome of an experiment lands in a grey area (e.g., maybe, or almost, or undetermined), then you must look

[1] From the Greek word **dichotomy**: division into two parts.

for another approach (perhaps using the *multinomial distribution*[2]) to analyze your data. Alternatively, you may agree to classify a grey area response like "slightly pregnant" consistently and unambiguously as one event or the other. However, it is imperative that the disposition of the "close-call" outcomes be determined *before* data collection; certainly, you must make such decisions before you have an opportunity to interject either your own or your management's bias into the experiment's analysis.

A reminder: The test of hypothesis about a population usually is made by first examining a sample from that population and determining a summary (i.e., a statistic like the mean) of the sample. Then you ask whether the sample's summary contradicts (in a statistical sense) the hypothesis you have formulated according to a set of rules called the test statistic. You answer this question by calculating the probability of obtaining the specific sample composition given the hypothesis, rejecting that hypothesis if the sample results are too "unlikely."

Sampling without replacement: An introduction to the hypergeometric distribution

Consider a population of N items, M of which have an attribute of interest. If you draw a sample of n items from that population, you observe H items with that attribute. The number H is a random variable because you don't know its value before you draw the sample. However, you *do know* that H is a non-negative integer and that it can neither be larger than n (the sample size) nor larger than M (the number of items with the attribute in the entire population). Mathematically, you write $H = 0, 1, \ldots, \min(M, n)$.

[2] See, for example, the discussion in Brownlee (1965, pp. 206-210).

Your problem is to determine whether the number of items in the sample with the attribute of interest (the value of H) is reasonable relative to a given hypothesis. To that end, you need to calculate the probability of finding H items in a sample of size n.

In performing these calculations, you recognize that the probability of obtaining an item with the attribute affects the probability that the next item in the sample will have that attribute. This is illustrated in a simple example. Suppose an urn holds exactly two white and two red marbles. The probability of drawing a white marble from the urn in the first draw is clearly 1/2. The probability of drawing a white marble in the second draw depends on what is drawn (and not replaced) in the first draw: That probability is 1/3 if a white marble is drawn in the first try, but it is 2/3 if the first marble drawn is red.

The sampling scenario alluded to in the preceding paragraph is called *sampling without replacement from a finite population*. Such a sampling plan's characteristics are fully described by the *hypergeometric distribution*. On the other hand, if your plan insists that drawn marbles be returned to the urn immediately after each is inspected, then your scheme is called *sampling with replacement from a finite population*. The binomial distribution (discussed in detail in Chapter 17) is then the appropriate distribution to consider as a model.

For discussion:

- If an urn contains two white and three red marbles, how would you show that ...

  ... in a draw of a single marble, the probability of drawing red is 3/5?

... if you draw (and keep) a red marble on a first trial, the probability of drawing a red marble on a second trial is 1/2?
... if you draw (and keep) a white marble on a first trial, the probability of drawing a red marble on a second trial is 3/4?
... if each drawn marble is returned to the urn after its drawing, the probability of drawing a red marble on the next draw is always 3/5?

Two relevant mathematical notations

You are almost ready to calculate some probabilities associated with hypergeometric experiments. First, however, there's no escaping the need to become acquainted (or, perhaps, re-acquainted) with a couple of useful mathematical notations.

factorial function

The first notation is that of the *factorial function* (sometimes called the *factorial operator*):

For any positive integer k, the expression $k!$ (read: "k factorial") is the product of all positive integers from 1 to k. Thus, you have the following values for the factorial function:

$$1! = 1$$
$$2! = 1 \cdot 2 = 2$$
$$3! = 1 \cdot 2 \cdot 3 = 6$$
$$4! = 1 \cdot 2 \cdot 3 \cdot 4 = 24$$
$$5! = 1 \cdot 2 \cdot 3 \cdot 4 \cdot 5 = 120$$

.
.
.

$$k! = 1 \cdot 2 \cdot 3 \cdot 4 \cdot 5 \cdots (k - 1)(k)$$
$$= k(k - 1)! = k(k - 1)(k - 2)! = \ldots .$$

The definition of the factorial function is extended to include zero-factorial (0!) with its value defined to be 1; i.e., $0! = 1$. At first glance, this extension may appear inconsistent with the main definition—even counter-intuitive. You will see the rationale in setting $0! = 1$ in the ensuing discussion.[3]

If you have a scientific calculator, this is a good time to try its factorial function, usually invoked with a key labeled with an exclamation mark (!). What do you get when you try 0!, 1!, 2.5!, and -3!? Try some larger numbers, say 10!, 50!, 70!, and 100!. But watch out: the factorial you ask for may be larger than any value the calculator's memory can carry. For example, $10! = 3{,}628{,}800$ and 70! is larger than 10^{100} (the integer 1 written with 100 trailing zeroes).

binomial coefficient

The second notation is that of the *binomial coefficient* which is written as two non-negative integers, one above the other, in parenthesis. It is evaluated using factorial functions according to the expression

$$\binom{a}{b} = \frac{a!}{b!(a-b)!},$$

which is read "*a* things taken *b* at a time" because it represents the number of ways that you may select ***b*** items out of a collection of ***a*** distinct items.[4] For example: Suppose you have 50 movies in your VCR collection. The number of double-feature evenings available to you is

[3] Because negative and fractional arguments for the factorial function are not needed in this discussion, we choose *not* to define it here for those values.

[4] The reason this expression is called the "binomial coefficient" will become apparent during the study of the binomial distribution in Chapter 17.

$$\binom{50}{2} = \frac{50!}{2!(50-2)!} = \frac{(50)(49)(48!)}{(2)(1)\times(48!)}$$
$$= \frac{(50)(49)}{(2)} = (25)(49)$$
$$= 1{,}225.$$

For discussion:

- The following argument provides plausibility to the definition 0! = 1. Starting arbitrarily with 4! and working our way down, we find:

  4! = (3!)(4)
  3! = (2!)(3)
  2! = (1!)(2)
  1! = (0!)(1).

  If you agree with the last relation, then 0! cannot be anything but 1. If you don't agree, we all have some pondering to do.

- What will you do to calculate, or to approximate, a large factorial (such as 70!) when your calculator cannot hold numbers larger than 10^{100}?

- In practice, you rarely have to calculate large factorials, because the binomial coefficient involves many terms that either cancel out or nearly cancel out. For example,

$$\begin{bmatrix} 70 \\ 2 \end{bmatrix} = \frac{70!}{2!\ 68!} = \frac{(70)(69)68!}{2!\ 68!} = \frac{(70)(69)}{2} = 2{,}415.$$

At last! The hypergeometric distribution stands up!

Finally! Armed with the factorial function and the binomial coefficient, you are ready to express and calculate simple probabilities associated with sampling from a finite population. Specifically, the probability of obtaining exactly h items with a specific attribute out of a sample of size n, given that the associated population has N items, exactly M of which have the attribute of interest, is

$$Pr\{h\} = \frac{\begin{bmatrix} M \\ h \end{bmatrix} \begin{bmatrix} N-M \\ n-h \end{bmatrix}}{\begin{bmatrix} N \\ n \end{bmatrix}}, \quad h = 0,\ 1,\ \ldots\ ,\ \min(M,\ n).$$

Examine $Pr\{h\}$ carefully. The denominator gives the number of ways a sample of n items can be selected from a population of size N. The numerator gives the number of ways of selecting a sample of size n that contains exactly h out of the M items with a given attribute *and* containing exactly $(n - h)$ out of the $(N - M)$ items in the population without the given attribute.

The expression for $Pr\{h\}$ can be rewritten by replacing each term by its factorial equivalence, resulting in a longer, yet possibly more practical, expression:

$$Pr\{h\} = \frac{M!}{h!(M-h)!} \; \frac{(N-M)!}{(n-h)!(N-M-n+h)!} \; \frac{n!(N-n)!}{N!}.$$

Limited tables of the hypergeometric distribution can be found in the literature. For example, Owen (1962, pp. 458-479), provides probabilities associated with various configurations of N, M, n, and h, up to $N = 21$. Beyer (1974, pp. 246-249) provides similar tables up to $N = 10$.

The mean of the hypergeometric distribution (that is, the average value of the random variable H in "the long run") is given by

$$Mean(H) = nM / N.$$

This expression, $Mean(H)$ for the mean value of H, is not surprising because M/N is the proportion of items with the attribute of interest in the population, and you would expect a similar proportion to apply to the sample.

The variance of the hypergeometric distribution is given by

$$Var(H) = \frac{nM}{N} \frac{N - M}{N} \frac{N - n}{N - 1}.$$

Development of the mean and variance of H is given in Mood, Graybill, and Boes (1974, p. 91). The standard deviation for H is obtained, of course, as the square root of $Var(H)$.

Probability calculations associated with the hypergeometric distribution are involved and lengthy. In many applications, however, the hypergeometric distribution can be approximated by other distributions if certain conditions are met. For example, if the ratio n/N is small, say $n/N < 0.10$, the effect of non-replacement diminishes, and the binomial distribution of Chapter 17 can be used to approximate the hypergeometric distribution with relative impunity. Hald (1952, p. 691) goes further, pointing out that the normal approximation can be used if the variance of H is larger than 9.

Applications

This section contains three applications of hypergeometric experiments: Example 16-1 demonstrates the calculation of probabilities, Example 16-2 describes hypothesis-testing and the effects of the distribution's being discrete, and Example 16-3 illustrates the normal approximation to the distribution.

Example 16-1: An inventory audit

In an inventory of 25 items, five (i.e., 20%) are claimed to be defective. Owing to the cost of inspection, you are faced with verifying the inventory with a sample of size $n = 10$. What is the probability that, if you randomly select 10 items, none will be defective? one? two? more than two?

You have $N = 25$, $M = (0.2)(25) = 5$, $n = 10$, and $h = 0, 1$, and 2, with the following probabilities calculated:

$$Pr\{0\} = \frac{5!}{0!(5-0)!} \; \frac{(25-5)!}{(10-0)!(25-5-10+0)!} \; \frac{10!(25-10)!}{25!} = 0.0565$$

$$Pr\{1\} = \frac{5!}{1!(5-1)!} \; \frac{(25-5)!}{(10-1)!(25-5-10+1)!} \; \frac{10!(25-10)!}{25!} = 0.2569$$

$$Pr\{2\} = \frac{5!}{2!(5-2)!} \; \frac{(25-5)!}{(10-2)!(25-5-10+2)!} \; \frac{10!(25-10)!}{25!} = 0.3854.$$

Thus, the probabilities that the sample of size $n = 10$ will contain 0, 1, or 2 defective items are 0.0565, 0.2569, and 0.3854, respectively. It follows that the probability of more than 2 defective items in the sample is

1.0000 - 0.0565 - 0. 0.2569 - 0.3854 = 0.3012.

The results of the probabilistic calculation are summarized and completed for $h = 3$, 4, and 5 in Table 16-1. Also given in this table is the cumulative probability for each value of h, $Pr\{H \leq h\}$, as well as its complement, $Pr\{H > h\} = 1 - Pr\{H \leq h\}$.

Table 16-1:
Summary of calculations for Example 16-1

Hypergeometric probabilities for $N = 25$, $M = 5$, and $n = 10$			
h	$Pr\{H = h\}$	Cumulative: $Pr\{H \leq h\}$	Complement: $Pr\{H > h\}$
0	0.0565	0.0565	0.9435
1	0.2569	0.3134	0.6866
2	0.3854	0.6988	0.3012
3	0.2372	0.9360	0.0640
4	0.0593	0.9953	0.0047
5	0.0047	1.0000	0.0000

Example 16-2:
Testing hypothesis for the inventory audit of Example 16-1

You can construct a test of a hypothesis about the number of items in a finite population with an attribute of interest using the hypothesis-testing processes introduced in Chapter 9. Although the focus here is on a one-sided test, the extension to two-sided alternative can be similarly developed. However, since the hypergeometric distribution deals with discrete events, the critical region is comprised of discrete points, rather than an interval.

Suppose that the claim is made that a population of size N contains at most M_0 items with a specific attribute. You wish to test whether that claim is correct, allowing a Type I error of size α_0. You set

H_0: $M = M_0$,
H_0: $M > M_0$, and
$\alpha = \alpha_0$.

The critical region is selected so that so that, if H (the number of items with the attribute in the sample) is "excessively large," the hypothesis is rejected. Formally, the critical region is made of values of h, such that

$$\sum_{h=h_0+1}^{n} Pr\{h|M_0\} \leq \alpha_0,$$

where $Pr\{h|M_0\}$is the probability of obtaining exactly h items with the specific attribute when the number of items with that attribute in the population is M_0. In a straightforward approach, a table such as Table 16-1 is constructed and the value of h_0 that bounds the critical region is found from the last column (the complement probability).

To illustrate, suppose you are promised that no more than five items of the 25-piece inventory are defective. You plan to collect a random sample of size $n = 10$ to test that claim with $\alpha = 0.05$. The critical region is determined from Table 16-1 by considering the last column and finding the largest probability that is smaller than $\alpha = 0.05$. In this case, that probability is 0.0047. The value of h associated with this probability is $h = 4$. Hence, you take $h_0 = 4$ as the critical value and reject the hypothesis if the number of defective items in the 10-item sample is four or larger.

Here is the rationale for selecting $h_0 = 4$ as the "beginning" (i.e., the critical value) of the critical region: You are seeking a value of h_0 such that exactly $\alpha = 0.05$ of the time you will observe h_0 or more defective items when the hypothesis H_0: $M = 5$ is true. But an *exact* solution is not possible in this particular case. Instead, you note that $h_0 = 3$ is not adequate because it and all larger values occur with probability 0.0640. You then try $h_0 = 4$ and find it and all larger values occur with probability 0.0047; thus, $0.0047 \leq \alpha = 0.05 \leq 0.0640$ and decide $h_0 = 4$ is the critical value that will assure your not exceeding 0.05 level of significance. Note especially that you would determine an identical protocol, with the same critical value, for any α between 0.0047 and 0.0640.

Example 16-3: Using the normal approximation to the hypergeometric distribution

Examples 16-1 and 16-2 should be sufficiently convincing that working with hypergeometric probabilities can be tedious, even if they can be calculated in a straightforward fashion. Relief is only a normal approximation away—as long as Hald's (1952, p. 691) criterion is met; that is, as long as you can determine that $Var(H) > 9$. If this condition is met, then the standardized statistic

$$Z = \frac{H - Mean(H)}{SD(H)}$$

is distributed approximately as a standardized normal variable. A test of an hypothesis can be exercised by placing an hypothesized value for M/N and comparing the resulting statistic to normal quantiles. This approximation is demonstrated with a cable-inspection example:

A collection of 360 cables is audited for proper shielding. The manufacturer claims that no more than 20% of the cables are improperly shielded. A cable-installing contractor wins the job by, among other things, echoing the manufacturer's claim. Suppose you have the resources to examine a sample of size $n = 100$. You find that 30 of the cables in the sample are improperly shielded. Do you have enough evidence to challenge the 20% limit claimed by the cable-installing contractor?

In this example, you have $N = 360$ and $n = 100$. The null and the alternative hypotheses, translated to integers, are written as H_0: $M = 72$, H_1: $M > 72$, and $\alpha = 0.05$ is chosen. To see if Hald's criterion for the use of the normal distribution is met, you compute the value of *Var*(*H*) under the null hypothesis and obtain:

$$var(H) = \frac{nM}{N} \frac{N-M}{N} \frac{N-n}{N-1}$$

$$= \frac{(100)(72)}{360} \frac{360-72}{360} \frac{360-100}{360-1}$$

$$= 11.59.$$

Because 11.59 is larger than 9, you can exploit Hald's criterion to complete your analysis with the normal approximation to the hypergeometric.

The mean and the standard deviation of H are calculated by

$$Mean(H) = nM/N = (100)(72)/360 = 20$$

and

$$SD(H) = \sqrt{Var(H)} = \sqrt{11.59} = 3.40.$$

The corresponding Z statistic is then calculated as

$$Z = \frac{30-20}{3.40} = 2.94.$$

Because Z is larger than 1.645 (the one-sided critical value for $\alpha = 0.05$), you reject the hypothesis that the proportion of improperly shielded cables is at most 20%.

For discussion:

- Reconsider Example 16-3. What would you conclude if 15 cables with improper shielding were found in the sample?

What to remember about Chapter 16

Chapter 16 displayed and discussed the following concepts connected with the hypergeometric distribution:

- *binary attributes*
- *sampling with and without replacement*
- *factorial function*
- *binomial coefficient*
- *hypergeometric distribution function.*

In addition, you were given the means to:

- *calculate specific hypergeometric probabilities for given population size, sample size, and number of items with a given attribute*
- *estimate the mean and variance for the number of attributes in the sample*
- *determine when the hypergeometric distribution can be approximated by the normal distribution.*

17

Binomial experiments

What to look for in Chapter 17

Chapter 17 is devoted to binomial experiments. These experiments involve a special, but not uncommon, type of discrete events. Among other things, this chapter shows how to calculate the probability of occurrence of such events.

Binomial and hypergeometic experiments have some similarities and often are confused with each other. Follow this chapter and you will be able to:

- *recognize sampling scenarios that can be treated as binomial experiments*
- *distinguish between and recognize hypergeometric and binomial sampling scenarios*
- *calculate the population mean and variance for the proportion of items of interest in binomial experiments*
- *estimate a population's proportion of items of interest when the binomial distribution is appropriate*

- *recognize when the binomial distribution can be approximated by the normal distribution*
- *know when the binomial distribution can**not** be approximated by the normal distribution*
- *recognize when you can construct confidence intervals for proportions using the normal approximation*
- *know what to do to construct confidence intervals for proportions when the normal approximation is not up to the task.*

Four requirements for a binomial experiment

As its name suggests, *binomial experiments*—like the hypergeometric experiments in Chapter 16—deal with outcomes that are binary. Examples of binary outcomes are {*good*, *bad*}, {*satisfactory*, *unsatisfactory*}, and {*OK*, *not-OK*}. The knowledge and understanding of when an experiment is qualified as a binomial experiment are essential to its meaningful and defensible statistical and probabilistic analysis. Just as with hypergeomtric experiments, if an outcome of a binomial experiment lands in a grey area (e.g., *maybe* or *almost*), then you must look for another approach (perhaps the *multi*nomial) to analyze your data. Alternatively, you may agree to classify a "gray-area" response, like "slightly pregnant," consistently and unambiguously as one or another. However, you must be sure that the disposition of such ambiguous outcomes be determined before data are collected; certainly, you must make a such decision *before* you have an opportunity to interject either your or your management's bias into the experiment's analysis.

If the preceding paragraph seems familiar, it ought to. You saw much the same discussion in Chapter 16. Because many features of binomial and hypergeometric experiments are similar, certain of these ideas bear repeating.

An experiment qualifies as a binomial experiment if the following four requirements are met:

Requirement 1: The experiment is conducted in $\underline{n}$ trials, all of which are conducted under identical conditions.[1]	**Requirement 1** says that the experiment is made up of exactly *n* trials *and* that all *n* trials are conducted under the same set of circumstances. This requirement for identical conditions does not have to be taken literally. In flipping a coin, for instance, you need not insist that the coin be flipped at the same time of the day and only on Monday. For that matter, you may even permit another coin to be flipped. The important principle to remember is that events that may affect the experiment's results must be the same. ***Example:*** *If you wish to assess the probability that pipe joints are welded properly, you must be sure that all the joints in your sample are welded by the same welder, or at least, by welders of similar training and certification.*

[1] A single trial is often called a Bernoulli trial, named after the Swiss theologian-cum-mathematician James (a.k.a. Jakob or Jaques) Bernoulli, (1654-1705). One of several generations of a distinguished family of mathematicians and physical scientists, he is credited with coining the term *integral*.

Requirement 2: The ith trial, $i = 1, \ldots, n$, results in either a *success* (recorded as $Y_i = 1$) or a *failure* (recorded as $Y_i = 0$). Note also that *success* in this chapter is akin to *attribute* in Chapter 16.	**Requirement 2** insists that the response must be binary (dichotomous, if it suits you). Whether you toss a coin or inspect a valve for compliance, the experiment must lead to a success or a failure. And, of course, when you define a success, you must have a frame of reference clearly in mind, since success to one player may be a failure to another. To handle the results of a binomial experiment mathematically, you denote the result of the ith trial by Y_i and assign $Y_i = 1$ if the trial yields a success; otherwise, you assign $Y_i = 0$. ***Example:*** *A properly welded pipe joint is scored as 1, while an improperly welded joint is scored as 0.*
Requirement 3: The probability of a success, denoted by π (lower-case Greek letter *pi*), is constant from trial to trial. As with all probabilities, the condition $0 \leq \pi \leq 1$ must be met.	**Requirement 3** states that the probability of success, π, is constant from trial to trial. If, for example, your experiment is the throw of a six-faced die, the probability of getting a 6 is presumed to be unchanged from one throw to the next. Indeed, a starting model for such an experiment could specify $Pr\{6\} = 1/6$. ***Example:*** *If the quality of the welding improves over time (perhaps owing to experience or to improved equipment), then this assumption of constant probability is violated. Similarly, if the quality of welding deteriorates over time (perhaps owing to fatigue, over-confidence, or mere sloppiness), then the binomial experiment is not applicable.*

Requirement 4: The n trials are independent.	**Requirement 4** states that the trials are independent of each other; that is, that the system has no "memory." If you win eight times in a row in a coin-tossing contest, Requirement 4 means that your chance of winning on the ninth trial is not different from that of any other toss. ***Example:*** *Each pipe-joint weld is performed independently of the others; i.e., whatever happens during any weld is neither influenced by nor has influence on what happens during the other welds.*

Your focus in binomial experiments is on either or both of two statistics:

(1) $B = \Sigma Y_i$, the number of successes in n trials

and

(2) $P = B/n$, the sample proportion of successes.

To repeat: it is extremely important to recognize whether your study meets the binomial experiment's requirements. And yet, when certain conditions prevail, some departure from some of these assumptions may be acceptable. Don't hesitate to ask your friendly statistician!

For discussion:

- Why shouldn't you simply look at the "grey-area" responses and then decide how to classify them? (Hint: How would you *like* "grey-area" responses to be classified?)

- Give an example from your own experience in which the binomial experiment is applicable. Show that, when one requirement is violated, other requirements may be violated, too.
- Whenever Team A plays basketball against Team B, the probability of Team A winning is 0.60. The two teams play n games and the number of winnings by each team is recorded. Would you consider this a binomial experiment?
- Describe a scenario and conditions where an inspection of special purpose pipes qualifies as a binomial experiment. Which of the four requirements is the most questionable?

Binomial probabilities

You are now ready to calculate some probabilities associated with binomial experiments. In the conduct of a binomial experiment involving n trials, the number of successes is one of the integers in the set $\{0, 1, 2, \ldots, n\}$. Before the experiment, you don't *know* which of these values will be obtained. But, if you know the binomial parameter π, you can calculate the probability that the random variable B will attain any of the $(n + 1)$ possible values in the set.

binomial density function

The *binomial density function* gives the probability that $B = b$, where $b = 0$ or 1 or 2 or ... or n successes in n trials, when the probability of a success in a single trial is π. The binomial probability function is written and calculated as

$$Pr\{B = b; n, \pi\} = \begin{bmatrix} n \\ b \end{bmatrix} \pi^b (1 - \pi)^{n-b}$$

$$= \frac{n!}{b!(n - b)!} \pi^b (1 - \pi)^{n-b},$$

where $0 \le \pi \le 1$ and $b = 0, 1, \ldots, n$. The binomial parameters are n and π, as indicated in the expression for its probability function. When there is little opportunity to be misunderstood, you may find it convenient to write $Pr\{b\}$ in place of the longer expression $Pr\{B = b; n, \pi\}$.

For discussion:

- Note that, in (what may be) the majority of textbooks on statistics, the population parameter which describes the proportion of characteristics of interest often is denoted by the English letter p. To maintain consistency in usage across this book, we elect to employ the lower-case Greek letter π to indicate the binomial parameter. Note further that we use the upper-case letter B to designate a random variable from a binomial experiment, whereas some other books denote it by X or Y. Once the outcome of the binomial experiment is known (no longer a random variable), it is denoted by the lower-case English letter b.

- To meet the binomial distribution's requirement for sampling with a constant probability π, you may treat the experiment as sampling with replacement. Discuss the pros and cons of sampling with replacement in each of the following four experiments:

  (1) Tossing the same coin repeatedly to establish whether heads and tails appear with equal frequency.

  (2) Rolling the same die repeatedly to investigate whether the die is fair.

(3) Sampling suburban residents to see how many would be willing to have their home's electricity generated by nuclear energy.

(4) Sampling employees for drug use.

Example 17-1:
A sequence of heads in a series of coin tosses

What is the probability of throwing 6 heads in a row with a fair coin?

You have $b = 6$, $n = 6$, and $\pi = 0.5$, from which you calculate

$$Pr\{6\} = \frac{6!}{6!0!}(0.5)^6(0.5)^0 = 0.015625.$$

Example 17-2:
Acceptance sampling of laser printer cartridges

In the testing of laser printer cartridges for length of service, a cartridge is defective if it fails to deliver at least 3,000 printed pages. Given that 20% of a typical line of reconditioned cartridges are defective, you can address the following kinds of questions:

(a) What is the probability of obtaining *exactly* two defective cartridges in a random sample of 50 cartridges?

You have $b = 2$, $n = 50$, and $\pi = 0.2$, from which

$$Pr\{2\} = \frac{50!}{2!\ 48!}(0.2)^2(0.80)^{48} = 0.001093.$$

(b) What is the probability of obtaining *at most* two defective cartridges in a random sample of $n = 50$?

To meet the "at most" requirement, the sample must yield exactly 0 or exactly 1 or exactly 2 defective cartridges. Since those events are mutually exclusive, the probabilities associated with these events are additive. You've already calculated $Pr\{2\}$. Now you need to calculate $Pr\{1\}$ and $Pr\{0\}$:

$$Pr\{1\} = \frac{50!}{1!49!}(0.2)^1(0.80)^{49} = 0.000178$$

and

$$Pr\{0\} = \frac{50!}{0!50!}(0.2)^0(0.80)^{50} = 0.000014.$$

Putting the pieces together, the probability of obtaining at most 2 defective cartridges is:

$$\begin{aligned} Pr\{B \leq 2\} &= Pr\{0\} + Pr\{1\} + Pr\{2\} \\ &= 0.001285. \end{aligned}$$

(c) Does this investigation qualify as a binomial experiment? If not, which requirements of the binomial experiment are violated?

(d) How would *you* sample cartridges from a production line? Or can you? Or: is that even the right question?

Example 17-3: Drug-testing

Each designated employee in a government agency is a candidate for drug-testing. The tests are administered on each of 10 randomly selected days in any given calendar year. For each of those 10 days, 10% of those employees are selected at random from the pool of eligible employees. The sample is selected independently of any previous sample. Hence, some employees may be selected more than once in a year, while others may not be selected at all.

What is the probability that an employee will not be selected during the year? What is the probability that an employee will be selected once, twice, or more than twice during the year?

The probability of being selected at a given test is $\pi = 0.1$. The number of trials is $n = 10$. From which you get

$$Pr\{0\} = \frac{10!}{0!\ 10!}(0.1)^0(0.9)^{10} = 0.3487$$

$$Pr\{1\} = \frac{10!}{1!\ 9!}(0.1)^1(0.9)^9 = 0.3874$$

$$Pr\{2\} = \frac{10!}{2!\ 8!}(0.1)^2(0.9)^8 = 0.1937.$$

Thus, the probabilities that an individual employee will be tested 0, 1, or 2 times during the year are 0.3487, 0.3487, and 0.1937, respectively. The probability that an individual will be tested more than twice is

$$1.0000 - 0.3487 -0.3874 - 0.1937 = 0.0702.$$

For discussion:

- The sampling scheme described in Example 17-3 is labeled a "100% sampling rate," a common but unfortunate term because it has the potential to mislead. The label is meant to describe a plan by which the total number of individual drug tests conducted in a specific year is equal to the number of employees in the pool. Discuss reasons why the label, "100% sampling rate," is potentially misleading. What impression does this label convey?

- Consider now a "50% sampling rate" scheme, in which the total number of individual drug tests administered during a year's testing is equal to one-half of the total number of employees in the pool. The tests are conducted ten times a year, so that the probability of being selected at each test is 0.05. Show that the probability of an employee's being tested zero, one, two, or more than two times in one year is Pr{0} = 0.5987, Pr{1} = 0.3151, Pr{2} = 0.0746, and Pr{more than 2} = 0.0116.

- How would *you* test employees for drug use and make sure that every employee is tested at least once each year?

A special exercise

Suppose that approximately 10% of the routine production runs of computer memory chips are defective. In order for your computer to run a new batch of software properly, you must endow it with eight additional memory chips. Your computer dealer carries those chips in a large bin. What is the probability that, if you randomly select 8 chips, all of them will work?

Assuming that the chips are randomly selected from the bin, the probability that all eight chips will function properly (i.e., that zero chips will fail) is given by (using $n = 8$ and $\pi = 0.1$) the expression

$$Pr\{0\} = \frac{8!}{0!8!}(0.1)^0(0.9)^8 = 0.4305.$$

This probability, slightly larger than 0.43, reflects a uncomfortable "likeliness" that your computer will be operational tonight. Thus, you might decide to buy nine chips to improve your "chances" of winding up with eight working chips. All you have to do is to calculate the probability that the number of bad chips in a sample of $n = 9$ is either zero or one, because either outcome is acceptable to the task of getting your computer to operate your new software.

But: How many memory chips *should* you buy? Even if you buy 50 chips, there still is a non-zero probability that at most 7 of them will work properly. This worrisome state is exactly where the probabilistic rubber meets the practical road.

For one thing, you can set your own criterion. Thus, suppose you decide to buy enough chips to provide some reasonable assurance, say 95%, that you will not have to return to the store for more chips that day. All you're faced with is a little calculating.

If you buy exactly nine randomly selected chips, then the probability that at least 8 of them will function is the probability than either zero or one of the chips will fail. The two outcomes are mutually exclusive, so you set out to calculate the probability that no failures appear in the 9 chips *and* the probability that exactly one failure appears in the 9 chips. Then all you do is add the two probabilities, and you have your result.

Your calculations should look something like these two expressions:

$$Pr\{0\} = \frac{9!}{0!9!}(0.1)^0(0.9)^9 = 0.3874$$

and

$$Pr\{1\} = \frac{9!}{1!8!}(0.1)^1(0.9)^8 = 0.3874.$$

(Oh, my! What happened? Did you anticipate the result that $Pr\{0\} = Pr\{1\}$? What explanation can you come up with?)

Your assurance of not having to return to the store is now measured by the sum of these probabilities; that is, $0.3874 + 0.3874 = 0.7748$. Although this is nearly twice the assurance given by trying to get by with the purchase of eight chips, it still doesn't reach the level of 95% that you promised yourself. So you consider the purchase of 10 chips. You find that the probability of obtaining at least eight good chips in a random sample of 10 chips is $Pr\{0\} + Pr\{1\} + Pr\{2\} = 0.3487 + 3874 + 0.1937 = 0.9298$, *still* not enough to meet your self-established 95% criterion.

One thing you *could* do: Simply lower your standard from 95% to, say, 90%. Now you're in business with $n = 10$ because 0.90 is less than 0.9298. But a small twinge of conscience may be lurking close by. You're not sticking to your original criterion. You're letting the numbers modify your intention. What to do?

Well, you could continue the search for the sample size that satisfies the 95% criterion. And that's just what this special exercise is all about.

Consider next a sample of 11 chips. Will *it* be large enough? You find out by calculating the following terms for $n = 11$:

$Pr\{0\}$ = ______

Pr{1} = ______ $Pr\{0\} + Pr\{1\}$ = ______

$Pr\{2\}$ = ______ $Pr\{0\} + Pr\{0\} + Pr\{2\}$ = ______

$Pr\{3\}$ = ______ $Pr\{0\} + Pr\{1\} + Pr\{2\} + Pr\{3\}$ = ______

Do you now have the required assurance? If *yes*, congratulations! If *no*, double check your calculations—or repeat the process with $n = 12$ and $n = 13$ and ... and so on.

> *If you did your calculations correctly, you found that the probability of obtaining at least eight functioning chips when $n = 11$ is 0.9814, approximately 0.98.*

For discussion:

- Reconsider the special exercise involving the purchase of computer chips. Was the analysis based on a binomial experiment? Why do you think so? If you think it wasn't, what conditions would have rendered it a binomial experiment?

- In the special exercise, your assurance criterion of 95% was set arbitrarily. You may also argue that, in forcing the purchase of an

11th chip, the *arbitrary* selection of 95% was not justified in light of the very little extra assurance (98% versus 93%) you bought for the price of the last chip. That's certainly one way of looking at the problem. But look at it from the other direction. The probability of having to return to the store when you buy 11 chips is 1.0000 - 0.9814 = 0.0186, while for 10 chips the probability is 1.0000 - 0.9298 = 0.0702, nearly four times as high as the probability of drawing at least eight good chips in a sample of size 11.

To repeat: The decision of selecting the sample size ultimately rests with the user. Financial, time, and other considerations certainly enter into your considerations. Here's a thought:

> Do recognize and articulate constraints and limitations early in your considerations. Once your planned long-range goals and processes are decided, stick to them. As a rule, adhering to your guideline(s) and to your established decision criteria will keep you out of the trouble that's waiting whenever you fall back upon *ad hoc*, spur-of-the-moment, short-term processes. Successful investors and professional gamblers seem to have an innate understanding of this principle.

Measures of statistics derived from the binomial probability function

If you find B defective switches when you inspect a random sample of size n from a shipment of switches, then the sample proportion of defective switches is B/n. This sample proportion is a prime candidate as an estimator of the population proportion, namely π—in this case, denoting the probability of a switch being defective.

Naturally, both B (the count of items of interest) and B/n (the sample proportion of items of interest) are statistics of interest. Whereas many statistics books concentrate on the statistic B for mathematical reasons, we prefer to emphasize the statistic B/n for practical reasons. For notational and mnemonic convenience, we set $P = B/n$.

Because they are statistics, B and $P = B/n$ each has a mean, a variance, and a standard deviation. These three performance measures are displayed in Table 17-1.

Table 17-1:
Measures of statistics derived from the binomial probability function

Measure	Statistic when binomial parameter is π	
	B	$P = B/n$
Mean	$n\pi$	π
Variance	$n\pi(1 - \pi)$	$\pi(1 - \pi)/n$
Standard deviation	$\sqrt{n\pi(1-\pi)}$	$\sqrt{\pi(1-\pi)/n}$

On the calculation of binomial probabilities

Although the formula for calculating a binomial probability, namely

$$Pr\{B = b; n, \pi\} = \frac{n!}{b!(n-b)!}\pi^b(1 - \pi)^{n-b}$$

is not especially intimidating or difficult to evaluate, repeated evaluations for many different sets of the three arguments can be boring and tedious. Fortunately, you have several "outs."

First, you have Table T-7. It gives, to four decimal places, the binomial probabilities involving combinations of $\pi = 0.01, 0.05, 0.10, 0.25$, and 0.50 and $b = 0, 1, \ldots, 9$ for $n = 1, 2, \ldots, 40$.

Second, you may have access to a variety of published binomial tables. Among them are: a pamphlet issued by the Ordnance Corps (1952), a book by Romig (1953), a compilation by the staff of Harvard University's Computation Laboratory (1955), and Beyer (1974, pp. 182-193).

Third, you may have a hand-held calculator with the binomial functions built in. Then it's a matter of pressing buttons and reading displays.

Fourth, you may have a spreadsheet program in your desktop computer. Some spreadsheets have the binomial function built in. For others, you can set up a spreadsheet that makes use of a binomial recursion formula such as

$$Pr\{b + 1\} = Pr\{b\} \frac{\pi}{1 - \pi} \frac{n - b}{b + 1},$$

in which the parameters π and n are fixed and the recursion begins with the evaluation of $Pr\{0\}$ and progresses through $Pr\{1\}$, $Pr\{2\}$, and so on until $Pr\{n\}$ is evaluated.

Fifth, if certain conditions pertain to the binomial's parameters, you can use the normal approximation, which is discussed in the following section.

Sixth, if certain conditions pertain the binomial parameters, you can use the Poisson approximation, which is discussed in Chapter 18.

The normal approximation to the binomial distribution

Although B always is an integer in binomial experiments, certain conditions allow you to *approximate* the distribution of B and, consequently, the distribution of $P = B/n$ by the ubiquitous *normal distribution*. This approximation simplifies the term-by-term calculations inherent in the binomial by employing the easy-to-use standard normal table (such as Table T-1) and allowing you to solve many problems without having to dive into an extensive enumeration exercise. The conditions usually are stated, by such writers as Dixon and Massey (1983, p. 170) as:

$$n\pi > 5 \text{ and } n(1 - \pi) > 5.$$

Some writers suggest that, if the binomial parameter π is unknown, you use $P = B/n$ to approximate π, yielding the following modified conditions:

$$n(B/n) > 5 \text{ and } n(1 - B/n) > 5,$$

which in turn simplify to:

$$B > 5 \text{ and } (n - B) > 5.$$

When these conditions are satisfied, you can call on all the procedures involved in normal estimation and hypothesis-testing processes. Thus, you may test whether π is equal to a given value (using either a one- or a two-sided test), or you can produce a confidence interval on the binomial parameter.

Thus, a general null hypothesis for the binomial parameter may be stated as H_0: $\pi = \pi_0$. You write the test statistic as:

$$Z = \frac{B/n - \pi_0}{\sqrt{\frac{\pi_0 (1 - \pi_0)}{n}}}.$$

Example 17-4:
Binomial hypothesis-testing using the normal approximation

In a random sample of 63 buckets of water from a lake near a manufacturing facility, 16 were found to be contaminated. The plant manager claims that no more than 10% of the water is contaminated. The estimate of the contamination rate is $p = b/n = 16/63 = 0.254$. Would you reject the manager's claim?

You have H_0: $\pi = 0.1$, H_1: $\pi > 0.1$, and $\alpha = 0.05$.

First note that $n\pi = 63(0.1) = 6.3 > 5$ *and* that $n(1 - \pi) = 63(0.9) = 65.7 > 5$, so that the normal approximation to the binomial is appropriate. You proceed by constructing a standard normalized test statistic and reject the null hypothesis, H_0: $\pi = 0.1$, if the calculated statistic is larger than $z_{0.950} = 1.645$. For this example, the value of the statistic is

$$z = \frac{16/63 - 0.1}{\sqrt{\frac{0.1\,(1.0-0.1)}{63}}} = 4.07.$$

Because the test statistic $z = 4.07$ is larger than 1.645, you reject H_0, claiming statistical evidence that the level of contamination is larger than 10%.

We all are grateful for the normal approximation to the binomial distribution. If it hadn't been for this (or some other) approximation,[2] we would have had to shift into a full enumeration mode and find, separately, the probability of obtaining 0, 1, 2, ... contaminated buckets of water out of 63 drawn.

Confidence intervals for the binomial parameter: When the normal approximation suffices

When the normal approximation to the binomial distribution suffices, the construction of a confidence interval about the binomial parameter π is simple. If the normal approximation is not applicable, the section following gives you some hints. In such cases, construction of the confidence interval is tedious and often is best left to a professional statistician while you go about other business.

When the normal approximation applies, the construction of the confidence interval follows the procedure used for the population mean in Chapter 8, with a small difference, you decide what it is.

Remember that, when σ is known and you have a normal distribution, a 95% two-sided confidence interval about the population mean μ is given by

(sample mean) $\pm$ 1.960 (standard error of the mean).

If a one-sided confidence interval is desired, you substitute 1.645 for 1.960 and use either end according to your needs.

By analogy, the 95% two-sided confidence interval for π is given by

[2] Generally is credited to J. Bernoulli who was mentioned in the first footnote in this chapter.

$B/n \pm 1.960 \times$ (estimated standard deviation of B/n)

$$= B/n \pm 1.960\sqrt{\frac{B/n(1 - B/n)}{n}}$$

$$= P \pm 1.960\sqrt{\frac{P(1 - P)}{n}}.$$

Example 17-5:
Do computer cables meet standard specifications?

If seven of 100 computer cables are found not to meet standard specifications, then a 95% upper confidence interval on the fraction of out-of-specification cables is given by:

$$\frac{7}{100} + 1.645\sqrt{\frac{(\frac{7}{100})(\frac{93}{100})}{100}} = 0.07 + 0.042$$

$$= 0.112.$$

This sets the 95% upper limit for the fraction of cables not meeting standard specifications at slightly more than 11%.

Confidence intervals for the binomial parameter: When the normal approximation does *not* suffice

Recall that the normal approximation to the binomial can be used when these conditions are met:

B >5 *and* (n - B) > 5.

However, failure to meet these conditions does not mean that you cannot find confidence limits for the measure you're interested in. This problem arises in numerous situations, primarily those in which the event of interest is "rare," such as those involving extremely high-quality products or the occurrence of disasters like floods and earthquakes and collapsing highway bridges.

Note that there is a flip-side to "rare" events: Because the binomial parameter π and its complement $(1 - \pi)$ play symmetric roles in these considerations, what is said for "rare" events applies equally to "common" events.

The following discussion is adapted from Mood and Graybill (1963, pp. 260-262). Their notation is modified to conform to this text.

To state the problem: You have a sample of size n with an observed number of successes, say B. You wish to place a 95% confidence interval on the binomial parameter π.

The 95% two-sided confidence *upper* limit for π, call it π_U, is the value of π for which

$$\sum_{k=0}^{B} \binom{n}{k} \pi^k (1 - \pi)^{n-k} = 0.025,$$

and the 95% two-sided confidence *lower* limit for π, call it π_L, is the value of π for which

$$\sum_{k=B}^{n} \binom{n}{k} \pi^k (1 - \pi)^{n-k} = 0.025.$$

If you have $B = n$, set $\pi_U = 1$; similarly, if you have $B = 0$, set $\pi_L = 0$.

Solving either of these equations can be daunting. No matter what values of n and B you have, you must solve a polynomial in the parameter π. Fortunately, as shown by Bowker and Lieberman (1972, pp. 466-467), values of π_U

and π_L can be expressed in terms of quantiles of the F distribution.

Let n be the number of trials and b be the number of occurrences of the event of interest. Then upper and lower 100(1 - α) confidence limits are given by

$$p_U = \frac{(b+1)f_{(1-\alpha/2)}(2(b+1),\ 2(n-b))}{(n-b) + (b+1)f_{(1-\alpha/2)}(2(b+1),\ 2(n-b))}$$

$$p_L = \frac{b}{b + (n-b+1)f_{(1-\alpha/2)}(2(n-b+1),\ 2b)}.$$

Bowker and Lieberman (1972, p. 467) illustrate with this calculation (which here reflects this book's notation):

> For example, suppose that 4 defects are observed in sample of 25. The ... estimate [of π] is $p = 0.16$. To find a 95% confidence interval for π with $\alpha = 0.05$, $n = 25$, $b = 4$,
>
> $$f_{0.0975}(10,\ 42) = 2.37$$
> $$f_{0.0975}(44,\ 8) = 3.82.$$
>
> The interval is then given by
>
> $$p_U = \frac{5(2.37)}{21 + 5(2.37)} = 0.361$$
>
> $$p_L = \frac{4}{4 + 22(3.82)} = 0.045.$$

Because the case of observing $b = 0$ in a binomial sample of size n occurs often enough to warrant explication, we work through the details for a specific $n = 25$. First, the lower limit is $\pi_L = 0$. Because the observed $b = 0$, the equation for π_U reduces to finding the value of π for which

$$\binom{25}{0} \pi^0(1-\pi)^{25-0} = (1 - \pi)^{25} = 0.025.$$

To solve this equation, you resort to logarithms (here we use natural logarithm) and obtain

$$25 \times \ln(1 - \pi) = \ln(0.025) = -3.68888,$$

from which

$$\ln(1 - \pi) = -3.68888/25 = -0.1475552.$$

Taking exponentials to the base e of both sides of this equation yields a simple linear equation in π; i.e.,

$$(1 - \pi) = e^{-0.1475552} = 0.86281481$$

whose solution is the desired upper limit $\pi_U = 1 - 0.8628 = 0.1372$. Thus, when $n = 25$ and $b = 0$, you can state that the true value of the binomial parameter π lies in the interval (0, 0.1372) with 95% confidence.

For discussion:

- Does the water contamination problem in Example 17-4 describe a binomial experiment?
- How would *you* measure water contamination if faced with a similar situation?
- Give your own interpretation of the results of Example 17-4.
- What do you call the plant manager? Unlucky?

- Suppose that in reality the level of contamination is not higher than 10%, but because of the random nature of the problem you drew 16 contaminated buckets in a sample of 63. Your sample indirectly leads to the firing of the plant manager. Are you sorry? Have you suffered a loss of credibility? What can you say to restore your credibility? Indeed, what is the nature of one's credibility in matters that involve hypothesis testing?

- How would you draw a random sample of 63 buckets from a body of water?

- Suppose eight out of eighty buckets were contaminated? What would you calculate, and what would you conclude?

- Does the number 63 in Example 17-4 sound strange to you? Who would propose such a sample size? Why not a nice round number like 10 or 50 or 100?

- How is it possible to obtain a positive confidence limit when you observe *no successes* in a binomial experiment? Is 0 data? Is 0 different from *no data*?

Now for a somewhat different discussion: Considerations on political polling

- This topic definitely is not energy-related, but it's of broad interest to all of us, at least in the period before Election Day. Beyond that general interest, however, it helps us tie together two powerful concepts: that of sample size determination and that of the normal approximation to the binomial distribution. Let's begin by simplifying the polling problem.

  Suppose we need to estimate the sample size required to predict a forthcoming gubernatorial election. We assume that two candidates are running neck-and-neck. (If one candidate were clearly ahead of the other, you wouldn't have to sample. Or would you?)

The assumptions in sampling the population are stated below:

> There are only two candidates, say, A and B.
>
> The sample is selected at random.
>
> Every person in the sample will be voting.
>
> Every person interviewed is cooperative and honest.
>
> The opinions expressed today are not going to change between today and Election Day.

We may simplify the problem and ask each interviewee whether he/she will vote for Candidate *A*; Candidate *B*'s tally is obtained by subtraction. The number of expressions of support, n_A, recorded for Candidate *A*, divided by sample size, *n*, provides an estimate of π, the proportion of the population's votes that will go to Candidate *A* on Election Day.

In determining the sample size required to obtain a meaningful (i.e., defensible from a polling standpoint) estimate of π, we place a *bound* on the magnitude of the *error* of our estimate. You may rightly ask: What do you mean, *error*? The answer lies in recalling that, in statistical discussion, an error is the difference between an observation and "the truth."

Typically, we see π converted to a percentage, so that we'll couch some of this discussion in terms of a number that looks like $100\pi\%$; i.e., we will report a statistic that looks like $100p\%$. Suppose, further, that we wish our estimate to be "within $\pm 3\%$" of the true, yet unknown, value $100\pi\%$. And suppose, even further, that we wish to have "95% confidence" to that effect. (How large a sample would we need to provide "100% confidence"? After all, 100% is only a little bit larger than 95%, isn't it?)

Using the methodology for sample-size determination given in Chapter 8, we set the largest possible half-width of the confidence interval about π to .03, yielding:

$$1.96\sqrt{\frac{0.5)(0.5)}{n}} = 0.03.$$

Solving this equation for n yields $n = 1{,}067.11$, which is rounded up to 1,068 to obtain the next largest integer that satisfies the conditions.

You may have noticed that, when you view a pre-election report that includes a statement about a 3% "margin of error," you may also be told that the results are based on a sample size of $n = 1{,}100$. What you may not be told directly is that the pollster and the client have a specific confidence level in mind when designing and reporting the study; often, it indeed is 95%—but, unless you are given concomitant information about the poll, you may never know for certain. From a different viewpoint, if you can tolerate results reported with a larger "margin of error," say, 5%, then a sample size of less than 400 is sufficient—assuming, of course, that you hold at 95% confidence.

For even further discussion:

If you were a candidate for office, and therefore a candidate to be a pollster's client, how might you evaluate the pollster's ideas?

- What sample size would satisfy *your* "need to know"?

- Now suppose each interview costs $5.00 to conduct. Does that fact affect your acceptable sample size?

- Does it matter to you if the poll is carried out by standing at the legendary corner of 5th and Main and stopping "random" pedestrians

... or if it's conducted by telephone ... or if it collects its data by knocking on doors ... or if it is performed by placing an ad in a newspaper and asking for write-in indications ... or if it is ... ?

- Try to fill in those final ellipses with some other polling procedures. Don't worry about their being plausible.

- Discuss. Or *is* there anything to discuss?

The binomial approximation to the hypergeometric distribution

Binomial probabilities can be used to approximate hypergeometric probabilities—under certain conditions that are relatively easy to meet. Brownlee (1965, p. 167) says: "A rough criterion for the validity of the approximation is that $n/N < 0.1$; i.e., for the sample size to be less than 10 per cent of the population size." The principal advantages to the binomial over the hypergeometric are that binomial probabilities are easier to calculate and that there are more tables available for this task.

In the following demonstration, you compare probabilities calculated for the two distributions for a sample size $n = 5$ from a population of size $N = 60$ with the number of items with a specific attribute $M = 25$. Table 17-2 contains the probabilities for $h = 0, 1, 2, 3, 4, 5$ for both distributions. Recall that h is the number of items in the sample with the attribute of interest, in parallel with the b successes for the binomial distribution.

Table 17-2:
An example comparing hypergeometric and binomial probabilities

Number of attributes/ successes in sample of size $n = 5$ (h or b)	Distribution		Difference (relative difference)
	Hypergeometric: $Pr\{H = h; N = 60, M = 25, n = 5\}$	Binomial: $Pr\{B = b; n = 5, \pi = 25/60\}$	
0	0.0594	0.0675	-0.0081 (-13.6%)
1	0.2397	0.2412	-0.0015 (-0.6%)
2	0.3595	0.3446	0.0149 (4.1%)
3	0.2506	0.2462	0.0044 (1.8%)
4	0.0811	0.0879	-0.0068 (-8.4%)
5	0.0097	0.0126	-.0029 (-29.9%)

Thus, for this example, the two sets of probabilities agree to within one or two digits in the second digit. But their relative agreements may be deceiving, depending upon their application. You will find generally that the agreement improves as N gets larger and/or the sampling fraction n/N gets smaller.

You also may wish to compare the mean and variance of the hypergeometric and the binomial distributions to see how close they are when the binomial parameter π is set equal to M/N and when the sampling fraction n/N is small.

What to remember about Chapter 17

Chapter 17 discussed and expanded upon binomial experiments; i.e., those experiments to whose outcomes the binomial distribution can be applied. It addressed a special, but very common, type of discrete event and showed you how to calculate the probability of occurrence of such events.

By following the ideas in this chapter, you will be able to:

- *recognize sampling scenarios that can be treated as binomial experiments*
- *distinguish between recognize hypergeometric and binomial sampling scenarios*
- *calculate the population mean and variance for the proportion of items of interest*
- *estimate a population's proportion of items of interest*
- *recognize when the binomial distribution can be approximated by the normal distribution*
- *approximate the binomial distribution by the normal distribution*
- *construct confidence intervals for proportions using the normal approximation.*

Poisson experiments

What to look for in Chapter 18

Chapter 18 introduces the Poisson distribution[1] which is often used as a model for calculating probabilities of events associated with time and space. For example, with the help of the Poisson distribution function, you will be able to calculate the probabilities of 0 or 1 or 2 or more failures of a component with a known constant failure rate, provided you can satisfy certain assumptions akin to the requirements for the binomial experiment in Chapter 17. Chapter 18 shows you how to:

- *recognize sampling scenarios that can be treated as Poisson experiments*
- *recognize when the binomial distribution can be approximated by the Poisson distribution*

[1] After Siméon Denis Poisson (1781-1840), a French mathematician.

- *compare selected binomial probabilities to their Poisson counterparts*
- *estimate the population mean, variance, and standard deviation for the number of rare events in a fixed time interval*
- *construct a confidence interval for the mean of the number of rare events in a fixed unit of time.*

Why the Poisson distribution?

Two major areas in statistics and probability bring the Poisson distribution into play:

(1) the calculation of probabilities of rare events[2] and

(2) the approximation of certain binomial probabilities (where the binomial calculations can be lengthy and cumbersome and tedious).

These two areas are examined in the following sections.

Probabilities of rare events

The Poisson distribution often is used as a model for calculating probabilities associated with the occurrence of rare events in a fixed time interval or in a fixed unit of space. Indeed, every study with non-negative countable responses is a legitimate candidate for modelling by the Poisson distribution. Some examples of data that can be modelled by the Poisson distribution are the count of:

[2] *Rare events* refer to happenings that are uncommon (in either time or space) in a particular domain of discourse. What is a "rare" event in one situation may not be unusual in another, even though the essence of the event itself remains unchanged.

- trucks overturned on Washington, DC's Beltway in a given month
- babies born in a cab on the way to the hospital in a specific week
- flaws in 100 continuous yards of fabric
- kinks in 1,000 yards of stretched copper wire
- lightning hits within a one-kilometer circle around a particular power station during a particular year
- ducks hit by motorists in Montgomery Village, MD, on May 7, 1993
- birds sitting on your TV antenna next Thursday at 2 p.m.
- alpha particles emitted from a specific source in a 15-second time interval.

Often, in statistical literature, the Poisson distribution is developed as a limiting form of the binomial. However, as indicated by the examples above, there are numerous situations in which the Poisson distribution is valuable in its own right. In either case, a fully rigorous mathematical discourse on the subject is beyond our intentions here. Instead, we spin off a plausibility argument—borrowing freely from Rosner (1989, pp. 88-90), given here with some modification to match this book's notation. The gist of the argument lies in two assumptions and a direct consequence. Although the development is given in terms of events in time, application to events in space is straightforward.

Consider any small subinterval of a time period t and denote it by Δt.

Assumption 1: Assume that

(a) The probability of observing 1 event is approximately directly proportional to the length of the subinterval Δt. That is, $Pr\{1 \text{ event in } \Delta t\}$ is approximately equal to $\theta\Delta t$ for some constant θ.

(b) The probability of observing 0 events over the subinterval Δt is approximately $1 - \theta\Delta t$.

(c) The probability of observing more than one event in the subinterval Δt is negligible.

(d) The approximations in (a), (b), and (c) improve as the length of the subinterval Δt approaches zero.

Assumption 2: If an event occurs in one subinterval Δt, it has no bearing on the probability of an event in any other subinterval.

Consequence: The expected number of events per unit time is the same throughout the entire time interval t. Thus, an increase—or a decrease—or any other type of change—in the anticipated incidence of the event as time goes on within the time period t would violate this assumption. Note that t should not be overly long, since this assumption is less likely to hold as t increases.

The random variable of interest is P (after Poisson) which is the total number of occurrences of the event in a total time interval t. If the stated assumptions are met, P is a Poisson variable. The probability that the Poisson variable $P = p$ (where $p = 0, 1, 2, \ldots$) is given by

$$Pr\{P = p\} = \frac{e^{-\theta t}(\theta t)^p}{p!}, \quad p = 0, 1, 2, \ldots .$$

When there is no ambiguity, you may choose to write the simpler expression $Pr\{P = p\} = Pr\{p\}$.

It can be shown (*cf.* Mood, Graybill, and Boes, 1974, pp. 93-94) that the mean of the random variable P is θt, itself often denoted by the lower-case Greek letter lambda, λ; that is, you write $\lambda = \theta t$. This result ought not to be surprising, because P is the simple summation of events, each of which occurs with probability $\theta\Delta t$. What *may* come as a surprise to you, however, is that the variance of P also is λ. It therefore follows that the standard deviation for P is $\sqrt{\lambda}$. Stated in terms of λ, the probability of p events occurring during the total time period t is

$$Pr\{P = p\} = \frac{e^{-\lambda}\lambda^p}{p!}, \qquad p = 0, 1, 2, \dots .$$

Thus, in this formulation, the Poisson density function depends upon only the single parameter λ.

The probabilities associated with various values of λ and p are evaluated easily on a hand-held scientific calculator (just don't overflow the calculator's memory!). But you may find it simpler to consult a source like Table T-9, where the probabilities already are calculated for you for selected λ and p.

Application of the Poisson distribution is demonstrated in two examples. Example 18-1 applies to Poisson events in time (errors in message transmission), while Example 18-2 applies to Poisson events in space (failures of urethane coating adhesions).

Example 18-1: Errors in message transmission

A status report of plant operation is continuously updated and transmitted to the plant manager via a modem. The probability of error transmission in any one-hour segment is believed to be $\theta = 0.01$. If a segment of 10 one-hour periods is selected at random, what is the probability that during that period no errors will be found? exactly one error? exactly two errors? two or more errors?

The average number of errors in a 10-hour segment is $\lambda = n\theta = (10)(0.01) = 0.10$, so that $e^{-\lambda} = 0.9048$. The probabilities of zero, one, or two errors can be obtained from Table T-9. However, they are calculated here for edification purposes. Verify these calculations, as well as the equivalent table entries, and be edified.

$Pr\{0\} = (e^{-0.1})(0.1)^0/0! = (0.9048)(0.1)^0/\ 0! = 0.9048,$

$Pr\{1\} = (0.9048)(0.1)^1/1! = 0.0905,$

$Pr\{2\} = (0.9048)(0.1)^2/2! = 0.0045,$ and

$$\begin{aligned} Pr\{\text{more than } 2\} &= 1 - Pr\{0\} - Pr\{1\} - Pr\{2\} \\ &= 1 - 0.9048 - 0.0905 - 0.0045 \\ &= 0.0002. \end{aligned}$$

Thus, the probability of two or more errors in a 10-hour segment is 0.0045 + 0.0002 = 0.0047.

Although you can find the values for this example in Table T-9, you ought not expect tabled values for *every* combination of λ and p. You may have to interpolate in Table T-9—or do the calculations yourself.

Example 18-2: Failures of urethane coating adhesions

The adhesion of urethane coating to concrete surfaces may be tested by a "scratch test" applied to a "block" of material that is one square foot in surface area. The manufacturer of a urethane coating claims a 95% adhesion rate (meaning that 95% of the scratch tests will be "successful" in that none of the coating is affected by the test). If 12 one-square-foot blocks of concrete are selected at random,what is the probability that at least 11 out of the 12 will pass?

The probability of failing the test is 1.00 - 0.95 = 0.05, which is "reasonably small" and thus meets Poisson assumptions. You reword the question to ask: What is the probability of failing none or one of the scratch tests?

From $\theta = 0.05$ and $n = 12$, you have $\lambda = (0.05)(12) = 0.60$; you may obtain the required Poisson probabilities from Table T-9 or calculate the answers directly as:

$Pr\{0\} = (e^{-0.60})(0.60^0)/0! = 0.5488$,

$Pr\{1\} = (e^{-0.60})(0.60^1)/1! = 0.3293$, and

$1 - Pr\{0\} - Pr\{1\} = 1 - 0.5488 - 0.3293 = 0.1219$.

Thus, the probability of at least 11 acceptable blocks is about 0.12.

For discussion:

- Are all the assumptions required for a Poisson experiment met in Examples 18-1 and 18-2?

- Example 18-1 considered a set of 10 uninterrupted 60-minute time periods for the purpose of counting transmission errors. In reality and in theory, the 10 time segments do not have to be consecutive. As a matter of fact, non-consecutive time segments may be considered desirable. Why?

- In Example 18-2, the 12 concrete blocks were selected at random. Is there any advantage to selecting contiguous blocks? Any disadvantage?

- Which of the following examples qualifies as a Poisson variable?

  (a) the number of failures of an instructor to come to class
  (b) the number of births of 10-pound babies in a given hospital during a given month

(c) the number of times the word "statistics" is mispronounced in class
(d) the number of grammatic and/or spelling errors in this book.

- Now it's your turn. Describe a Poisson experiment that is applicable to your work environment.

- If the probability of an elevator failure is said to be 0.014 in an arbitrary 24-hour period, what is the average number of elevator failures and the standard deviation for a 5-day work week? For a random set of 20 days in the summer? For a random set of 20 days in the winter?

- Reconsider Example 18-1. If you were told that the probability of a single error in a one-hour interval is 0.40, too large for the Poisson experiment, what would you do to force the problem into the framework of a Poisson experiment? Is such forcing justified?

Using the Poisson distribution as an approximation to the binomial distribution

The second major area where the Poisson distribution is used as an approximation of the binomial distribution. Recall from Chapter 17 that if $n\pi$ is larger than 5, then the normal distribution provides a good approximation to the binomial. This is good news. More good news is that, if $n\pi$ is smaller than 5, a good approximation to the binomial distribution is provided by the Poisson distribution. A mathematical proof of the approximation may be found in Hoel (1971, pp. 64-65).

Example 18-3 illustrates the agreement between the binomial and the Poisson distributions. And remember, the approximation improves as $n\pi$ decreases.

Example 18-3:
Errors in a printed document

A large official publication has misprint errors on 5% of its pages. If a random sample of 40 pages is examined, what is the probability that no pages in error will be found? Exactly one page with at least one error be found? Exactly two pages with at least one error be found?

Although the calculation of these probabilities can be easily done on a hand-held calculator, let's use Tables T-8 (for $n = 40$ and $\pi = 0.05$) and T-9 (for $\lambda = n\pi = 40 \times 0.05 = 2.0$) to obtain the corresponding binomial and Poisson probabilities. These are summarized in Table 18-1. In the ranges of the values of interest, the two distributions agree to within 0.01 of each other. In general, the approximation improves as n increases and/or as π decreases.

Table 18-1:
Comparing binomial and Poisson probabilities for $n = 40$, $\pi = 0.05$, and $\lambda = n\pi = 2.0$

Required probabilities	Exact binomial probabilities ($n = 40$, $\pi = 0.05$)	Corresponding Poisson probabilities ($\lambda = n\pi = 2.0$)
Pr{0}	0.1285	0.1353
Pr{1}	0.2706	0.2707
Pr(2}	0.2777	0.2707

Constructing a confidence interval for a Poisson parameter

Once a Poisson experiment is conducted, it's reasonable to make an inference about the "long-run average number of successes" produced in a similar setting, denoted here by λ. A popular approach to making this inference is the construction of a confidence interval for λ. Of course, if a test of hypothesis is desired, you can determine whether the confidence interval contains λ_0 (the hypothesized value of λ), and, if not, reject that hypothesis.

As shown by Brownlee (1965, pp. 172-174), you can form such confidence intervals with the use of the chi-squared distribution, pertinent values of which appear in this book's Table T-2. Here are the steps required to form a $100(1 - \alpha)\%$ confidence interval on λ:

(1) You observe P events in a total time interval of length t.

(2) Set two different degrees of freedom: $DF_L = 2P$ and $DF_U = 2(P + 1)$.

(3) The lower confidence limit is given by

$$\lambda_L = \frac{1}{2}\chi^2_{(\alpha/2)}(DF_L),$$

and the upper confidence limit is given by

$$\lambda_U = \frac{1}{2}\chi^2_{(1-\alpha/2)}(DF_U).$$

Note that, if $P = 0$, you immediately set $\lambda_L = 0$.

(4) You are then $100(1 - \alpha)\%$ confident that the true value of λ lies between λ_L and λ_U.

(4') Given that $\lambda = t\theta$, you can use the results of step (4) to create a $100(1 - \alpha)\%$ confidence interval on θ. Writing $\theta = \lambda/t$, you have $\theta_L = \lambda_L/t$ and $\theta_U = \lambda_U/t$.

By way of illustration, here are the steps required to form a 95% confidence interval on λ:

(1) You observe $p = 0$ events in a total time interval of length $t = 60$.

(2) Set the degrees of freedom $df_L = 0$ and $df_U = 2(0 + 1) = 2$.

(3) The lower confidence limit is given by

$$\lambda_L = 0,$$

and the upper confidence limit is given by

$$\lambda_U = \frac{1}{2}\chi^2_{(.975)}(2) = \frac{1}{2}(7.38) = 3.69.$$

(4) You are then 95% confident that the true value of λ lies between $\lambda_L = 0$ and $\lambda_U = 3.69$.

(4') Given that $\lambda = t\theta$, you can use the results of step (4) to create a 95% confidence interval on θ. Recall that $t = 60$. Writing $\theta = \lambda/t$, you have $\theta_L = \lambda_L/t = 0/60 = 0$ and $\theta_U = \lambda_U/t = 3.69/60 = 0.0615$.

Note that, in keeping with the convention adopted in Chapter 17 for binomial experiments, observing a value of zero in a Poisson experiment *also* leads to the immediate conclusion that the lower end of a two-sided confidence interval is zero.

Table T-10 removes much of your computational angst; it gives the confidence limits for the Poisson parameter λ for selected values of p from 0 to 50 and for several choices of the level of significance. The table is easy to use, as Example 18-4 illustrates. The only information you need is the count, p, of occurrence of the events of interest. Of course, you must decide the confidence level and whether a one- or a two-sided confidence interval will be used.

Example 18-4: On the failure of motors to start

Assume that the number of failures of a motor to start on demand during a one-month period is treated as a Poisson random variable. Suppose that, in a specific month, a particular motor fails twice in 100 trials. Based on these data, you seek a one-sided upper 95% limit for the average number of monthly failures per 100 trials for this motor.

Using Table T-10 with $p = 2$ and $(1 - \alpha) = 0.95$, you find the upper limit to be 6.30. Thus, given these data, any hypothesized value for λ larger than 6.30 is rejected.

What to remember about Chapter 18

Chapter 18 introduced the Poisson distribution which is often used as a model for calculating probabilities of events associated with time and space. With the help of the Poisson distribution function, you learned how to calculate the probabilities of 0 or 1 or 2 or more failures of a component with a known constant failure rate, provided certain assumptions are met. Chapter 18 showed you how to:

- *recognize sampling scenarios that can be treated as Poisson experiments*
- *recognize when the binomial distribution can be approximated by the Poisson distribution*
- *compare selected binomial probabilities to their Poisson counterparts*
- *estimate the population mean, variance, and standard deviation for the number of rare events in a fixed time interval*
- *construct a confidence interval for the mean of the number of rare events in a fixed unit of time.*

19

Quality assurance

What to look for in Chapter 19

Chapter 19 focuses on those concepts that result in procedures designed and constructed to further *the pursuit and the achievement of quality*. Perhaps those procedures alert a manufacturer that a particular product may not be meeting the product's specifications. Or they may call a service provider's attention to a mis-operating servicing system. Whatever the specific application, the concepts—and the procedures that spin from them—are collected under a general rubric called *quality assurance*. In connection with these quality-driven procedures, you will encounter a number of specialized terms and concepts:

- *quality control*
- *process control*
- *acceptance sampling by variables and by attributes*.

Chapter 19 is deliberately short; it sets the scene, gives some rationale and insight into quality assurance processes, and serves to introduce process control and acceptance sampling, which are discussed in Chapters 20 and 21, respectively.

What is quality assurance?

quality assurance

In its broadest terms, *quality assurance* encompasses all the things that you and your organization do to make sure that your products/services provide the value and the performance expected of them by their recipients, your customers. Quality assurance efforts tend to be associated with products, from sampling a day's production of automobiles for fuel efficiency to proofreading responses to customer complaints to funding research into advanced technology. But the basics of quality assurance are equally applicable to the treatment of patients in a hospital or to plumbing service in your home.

Part 50, Appendix B, *Title 10 (Energy) of the* U.S. Code of Federal Regulations, 1993, states:

quality control

> "Quality Assurance" comprises all those planned and systematic actions necessary to provide adequate confidence that a structure, system, or component will perform satisfactorily in service. Quality assurance includes *quality control* (italics added) which comprises those quality assurance actions related to the physical characteristics of a material, structure, component, or system which provides a means to control the quality of the material, structure, component, or system to predetermined requirements.

To augment this definition of quality assurance, here is Kendall and Buckland's (1971, p. 121) definition of quality control:

> **Quality control** The statistical analysis of process inspection data for the purpose of controlling the quality of a manufactured product which is produced in mass. It aims at tracing and eliminating systematic variations in quality, or reducing them to an acceptable level,

leaving the remaining variation to chance. The process is said then to be statistically under control.

Quality assurance is a two-way street—all products and services have both producers (senders) and customers (receivers). At times, the producer and the customer are one and the same. Quality assurance criteria are constructed for specific purposes. Some criteria are designed to protect the consumer from a "raw deal" or from a health risk or from shoddy merchandise. Others are designed to protect the producer from excessive waste of time and material and overcrowded inventory. Still others are designed with both the consumer and the producer in mind, although sampling designs with such a balanced intention usually require great effort and resources and cooperation among the affected parties.

Do these criteria sound contradictory? They are not if both consumer and producer have the same quality assurance objectives in mind. Consider the impacts of Japanese products on the American economy in the 1970s and 1980s. Quality was a major driving force: consumers wanted high-quality autos and electronics, and producers found ways to assure that quality. Japanese industrialists understand quality assurance quite clearly. The enduring irony is that they learned it from W. Edwards Deming, a well-known American statistician, who explained this philosophy to American industry during the 1940s and 1950s, with his message falling on proverbial deaf ears.

Quality assurance procedures need data: in its name, you apply a variety of statistical techniques to a variety of types of data. The two main data types you are likely to encounter are continuous (such as weights and diameters) and discrete binary data (such as yes/no and in-compliance/out-of-compliance). You have met both types in the preceding chapters.

As you might expect, however, the practice of quality assurance has it own vocabulary. The treatment of

sampling by variables, by attributes

continuous variables often is called *sampling by variables*, while the treatment of discrete binary data often is called *sampling by attributes*.

Process control: Building quality in

process control

The claim that a product meets its promised specifications can be investigated during production, during post-production, or during both. The testing of product quality during production is called *process control*. Process control means that items are routinely sampled and checked to determine whether the production specifications are maintained during the product's manufacture. If the specifications are not met, then production is said to be *out of control*, and corrective actions are called for. These corrective actions can include an intensive search for an assignable cause, an adjustment to the production process, and a thorough retesting of the process before it is restarted. However, even though corrective actions are taken to tighten (or adjust) the quality of the ongoing production, these actions do not necessarily include discarding all or part of the sampled product.

Although process control is designed to guard against the production of poor quality, it may also serve to indicate when the actual product is superior to that which is claimed by the manufacturer. This sometimes provides the manufacturer an opportunity to restate the product's specifications and to improve its competitive edge. At other times, the manufacturer may choose to relax some too-tight process controls, resulting in time and material savings, while still maintaining the claimed product specification.

Some aspects of process control are discussed in Chapter 20.

Acceptance sampling: Verifying quality

acceptance sampling, lot

Statistical quality assurance activities conducted after items are out of production (indeed, often after delivery is made to the consumer) fall into the domain of *acceptance sampling*. The basic idea of acceptance sampling is simple: A collection of items, usually called a *lot*, is examined, and the decision then is made to either accept or reject the entire collection—the lot—on the basis of that examination. It is the practice of acceptance sampling that's difficult: Convincing a group of managers that a lot of material must be rejected often requires the sharpest kind of diplomatic skills. Acceptance sampling follows an explicit and strict protocol that culminates in the acceptance or the rejection of an entire lot of items.

Some aspects of acceptance sampling are discussed in Chapter 21.

For discussion

- Discuss the sentence used earlier in the chapter: "At times, the producer and the customer are one and the same."

- What are the main differences between process control and acceptance sampling?

What to remember about Chapter 19

Chapter 19 focused on the rationale and the philosophy of quality assurance, a collection of procedures primarily designed to warn producers when the quality of their products has deteriorated to a point where corrective actions are necessary or might be necessary soon. The discussion centered on two main topics:

- *an appreciation for the needs and purposes of the quality assurance discipline*

- *recognition of the differences in intent and procedures associated with process control (the subject of Chapter 20) and acceptance sampling (the subject of Chapter 21).*

20

Quality assurance through process control

What to look for in Chapter 20

Chapter 20 builds on Chapter 19's discourse on quality assurance, emphasizing the specific topic of *process control*. It looks particularly at the construction and use of control charts, one of the most firmly established of all quality assurance techniques. Three types of charts are discussed and illustrated:

- *control charts for means*
- *control charts for standard deviations*
- *control charts for proportions*.

Process control and control charts for means

control charts

Control of a production process is often monitored through *control charts*, graphic constructs that are designed to alert a quality control manager when a process is out of control—or, at least, appears to be heading in that direction. Duncan (1986, p. 417) summarizes the purposes and the consequences of using control charts:

> A control chart is a statistical device principally used for the study and control of repetitive processes. Dr. Walter A. Shewhart, its originator, suggests that the control chart may serve, first, to define the goal or standard for a process that the management might strive to attain; second, it may be used as an instrument for attaining that goal; and, third, it may serve as means of judging whether the goal has been reached. It is thus an instrument to be used in specification, production, and inspection and, when so used, brings these three phases of industry into an interdependent whole.

Each control chart focuses on one target quantity—Shewhart's "goal or standard"—such as the claimed mean, the advertised standard deviation or range, or the promised maximum proportion of defective items. A control chart for the mean, for instance, is constructed around the target quantity, which we designate by μ_{target}. The sample means of *consecutively produced samples* are plotted on the control chart with *control limits* constructed above and below target value. These ideas are illustrated in Example 20-1.

Example 20-1: Control chart for percentage of uranium in UO_2 powder

Jaech (1973, pp. 68-69) gives an example involving the construction of a control chart for the average percent uranium in batches of UO_2 powder for means of 19 batches, reproduced here in Table 20-1. Jaech assumes that the standard factor (i.e., the target mean, μ_{target}) is 87.60% and uses $\sigma = 0.06\%$ (absolute percent).

Table 20-1:
Average percent uranium in batches of UO_2 powder

Batch	1	2	3	4	5	6	7
Mean	87.54	87.56	87.50	87.47	87.64	87.56	87.71
Batch	8	9	10	11	12	13	14
Mean	87.61	87.60	87.60	87.47	87.60	87.69	87.78
Batch	15	16	17	18	19		
Mean	87.69	87.72	87.77	87.79	87.78		

upper, lower control limits

The associated control chart is given in Figure 20-1. The heavy horizontal center line is plotted at the target value, μ_{target} = 87.60%. The *upper control limit* (*UCL*) is plotted at $\mu_{target} + 3\sigma = 87.60 + 3(0.06) = 87.78$ and the *lower control limit* (*LCL*) at $\mu_{target} - 3\sigma = 87.60 - 3(0.06) = 87.42$.

The individual means of the 19 batches of UO_2 powder are plotted from left-to-right, in the order in which they are produced. The importance of ordering, usually in time, in control charts cannot be over-emphasized. As you will see, evidence of malcontrol is revealed by that ordering.

Figure 20-1:
A control chart for batches of UO_2 powder recorded in Table 20-1

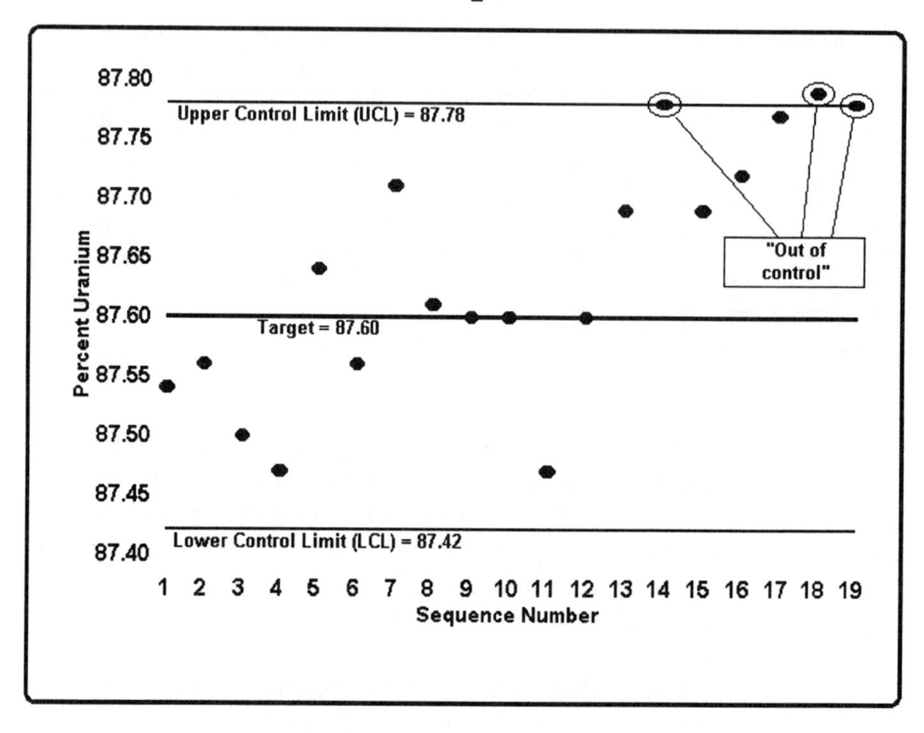

It is important to note that, if the data values in Table 20-1 were based on multiple readings (i.e., on sample size $n > 1$), the control limits would be written as $87.60 \pm 3\sigma/\sqrt{n}$, where σ designates the standard deviation associated with individual readings.

action limits

The selection of the value 3 as the coefficient of σ is consistent with long-standing practice in the field of quality control. Indeed, control charts were introduced as "3-sigma control charts" in Shewhart's (1931) pioneering work on quality control; consequently, you may encounter references to 3-sigma control charts as *Shewhart charts*. Sometimes you will find "3-sigma limits" called *action limits*, a term designed to convey the urgency implied when observations fall beyond them.

warning limits

As control chart practice developed in the United States, especially during and immediately after World War II, a seemingly endless procession of modified control limits appeared. For instance, *warning limits* (sometimes called *alarm limits* or "2-sigma limits") may be set at $\mu_{target} \pm 2\sigma/\sqrt{n}$, with the implication of "heads up"—but with somewhat less urgency than action limits. The generic terminology, of course, refers to "k-sigma limits" which are set at $\mu_{target} \pm k\sigma/\sqrt{n}$.

"Run"-ning with control charts

When you see a point lying outside the control limits—that is, the 3-sigma limits (the action limits)—the time has come to take action. But how should you react to a point lying outside the 2-sigma limits, the "warning limits"? Do you take "warning"? What do you do next? Do you call in the boss? Do you have a committee meeting? The answer, depending on the production process, may be all or none or some of the above.

The most important thing to do is to watch what happens on the chart as successive points are plotted. Do they fall between the warning limits, or do they continue to fall outside? If the latter, how do you react to points falling between the warning and the action limits?

run theory

Particularly powerful guidance, validated by decades of practice, can be found in a special area of statistics: *run theory*. Run theory finds application in many quantitative problems. Early references include Stevens (1939), Wald and Wolfowitz (1940), and Swed and Eisenhart (1943). Attention here, however, is paid to its application to control charts. Brownlee (1965, pp. 224-232) and Duncan (1986, pp. 417-435) make the necessary connections and guide the following discussion.

run

Duncan (1986, p. 328) says: "A *run* [italics added] is a succession of items of the same class." Thus, suppose you have a collection of nine items: three of class *A*, two of class *B*, and four of class *C*. If you draw them at random, one at a time, without replacement, you might get the sequence {*BAACCCABC*}, thus providing two runs of *A* (one of length 2 and one of length 1), two runs of *B* (each of length 1), and two runs of *C* (one of length 3 and one of length 1).

With respect to control charts, you may look for several types of runs, such as outside certain *k*-sigma limits or above and/or below the target line. Indeed, in some situations, you might be on guard against "too many runs," a situation that can develop when an inspector is helping you make "things average out."

Here is a paraphrase of Duncan's (1986, pp. 434-345) criteria for suggesting an out-of-control condition in your process:

1. One or more points outside the control limits.

2. One or more points in the vicinity of a warning limit. A recommended step is the immediate taking and analysis of additional data to determine the actual level of production.

3. A run of 7 or more points. Such a run may simply be above or below the target line. Or it may be a "run up" (a succession of increases in value) or a "run down" (a succession decreases in value).

4. Cycles or other non-random patterns. Because of their infinite variety, such cycles or patterns are difficult to specify. Watch your charts—you'll know them when you see them.

5. A run of 2 or 3 points outside the 2-sigma limits.

6. A run of 4 or 5 points outside the 1-sigma limits.

To make these matters explicit, consider Figure 20-2 which is a re-rendering of the control chart in Figure 20-1.

Figure 20-2:
Control chart reproduced from Figure 20-2 with 1-sigma and 2-sigma limits added

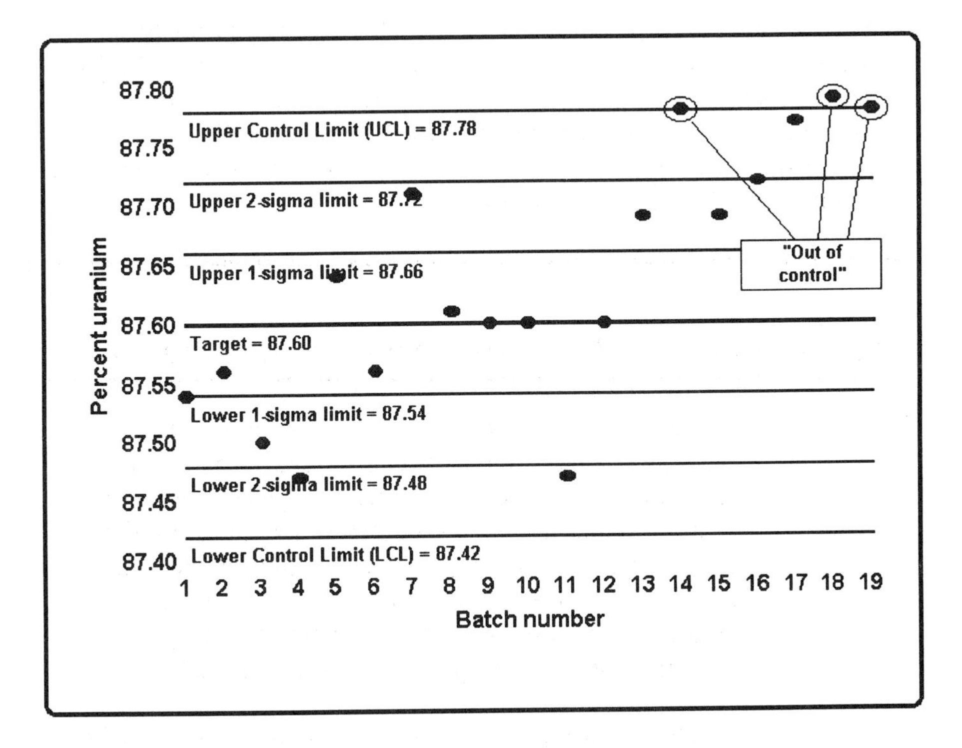

Starting with Batch 1, you have an "run up" of length 1, immediately followed by a "run down" of length 2. Only short runs occur until Batch 11, after which you see a "run up" of length 3. Indeed, Batch 14 is right on the upper control limit. Moreover, Batches 13 through 16 or 17 provide the alerting 4 or 5 points outside the 1-sigma limits. So you can see that, irrespective of Batch 14's being on the UCL, beginning with Batch 13, an out-of-control story is being told.

This type of application of run theory applies to all types of control charts, not just the means charts, one of which is used here for illustration.

For discussion:

- Some control chart practitioners, particularly those in Great Britain, use multipliers of 3.09 for action limits and 1.96 for warning limits. The numbers 1.96, 2, 3, and 3.09 are closely associated with probabilities associated with the normal distribution. What are the pros and cons of, say, choosing 3.09 instead of 3 when you determine action limits for a production process?

Process control and control charts for dispersion

As indicated, most points plotted on a control chart are statistics derived from a sample[1] of size $n > 1$. Each sample can therefore yield a measure (indeed, measures) of its own dispersion. Among these measures are the usual suspects: the sample variance, the sample standard deviation, and the sample range. Each of these is used by various quality assurance practitioners according to the particular needs of the process being monitored.

To claim that one has achieved "good production quality" requires not only a good average performance but also good consistency, or small variability, among the items being produced. To this end, some measure of within-sample dispersion variability of the samples is subjected to

[1] According to Duncan (1986, p. 427): *It is important to note that the samples on a control chart should represent subgroups of output that are as homogeneous as possible.... Nothing is more important in the setting up of a control chart than the careful determination of subgroups.* [Duncan's italics]

control charts and examined for its being "in control" or "out of control," just as means are.

Suppose you decide to apply control chart procedures to the standard deviations of the samples you're using for controlling the mean. Suppose further that you have chosen a target measure of variability, denoted σ_{target}. You then construct a control chart for the sample variances that is "centered" on the target standard deviation, σ_{target}.

Your next task is the quantification of the upper and lower control limits. Because you're dealing with variances of samples and because decades of practice have shown its efficacy, those control limits are based on the fact that the statistic $(n - 1)S^2/\sigma^2$ is distributed as a chi-squared random variable with $(n - 1)$ degrees of freedom (see Chapter 10). Thus, if you want upper and lower control limits that will sound a false alarm with probability α, you do the following:

(1) From Table T-2, which provides quantiles for the chi-squared distribution, you obtain the value $\chi^2_{(1-\alpha/2)}(n-1)$.

(2) You equate this quantile to $(n-1)S^2/\sigma^2_{target}$ and solve for S^2 in terms of σ^2_{target}, where σ^2_{target} is the square of the target measure of variability; this yields the upper control limit, say σ^2_{UCL}, for variances.

(3) To obtain the lower control limit, say σ^2_{LCL}, you find the quantile $\chi^2_{(\alpha/2)}(n-1)$ and solve the resulting equation for σ^2_{LCL}.

(4) You then move from working with variances to working with standard deviations by taking the square roots of σ^2_{UCL} and σ^2_{LCL}.

These four steps are illustrated in Example 20-2.

Example 20-2: Control limits for σ

Suppose you wish to monitor the claim that the standard deviation of the process, declared in Example 20-1 to be 0.06%, is still in effect. Construct a 99.5% upper control limit, assuming that daily samples of size n = 10 are available for the calculation of S.

(1) From Table T-2, which provides quantiles for the chi-squared distribution, you obtain $\chi^2_{(0.995)}(9) = 23.6$.

(2) You equate this quantile to $(n-1)S^2/\sigma^2_{target} = 9S^2/(0.06)^2$; i.e, $23.6 = 9S^2/(0.06)^2$. Solving for S^2, you obtain $S^2 = (23.6)(0.0036)/9 = 0.0094$. This is the upper control limit for σ^2; that is, $\sigma^2_{UCL} = 0.0094$.

(3) To obtain the lower control limit, say σ^2_{LCL}, you find the quantile $\chi^2_{(\alpha/2)}(n-1)$ and solve the resulting equation for σ^2_{LCL}. In the best academic tradition, this last calculation is an exercise left to the reader.

(4) You then move from working with variances to working with standard deviations by taking square roots of the upper and lower control limits for σ^2. For example, $\sigma_{UCL} = \sqrt{0.0094} = 0.097$.

On occasion, you may find it considerably easier to calculate the sample range than the sample standard deviation, especially with small samples. The process control of variability can be accomplished by comparing the sample ranges to established control limits for the range. Although the comparative calculational advantage of the range over the standard deviation has faded in the later part of the 20th century with increasingly available microcomputers, control charts for ranges remain ubiquitous in production facilities. Control limits for the range are functions of σ, the standard deviation, and n, the sample size. When the population specification for the standard deviation σ is given, Bowen and Bennett (1988, pp. 218-223) provide multipliers of σ to obtain 3-standard-deviation upper and lower control limits for the range for samples of size 10 or smaller. For samples of size up to 25, the appropriate limits may be obtained through a simple manipulation of tabled values given in Beyer (1974, Section XI).

Control charts for attributes

Consider next a process control where the specification of proportion of defective items in the lot/population is set at $\pi = \pi_{target}$. If the sample size is large enough that both $n\pi$ and $n(1 - \pi)$ are larger than 5, then a normal approximation to the distribution of the sample proportion is reasonable, as discussed in Chapter 17. The control chart is then constructed with a center line at π and upper and lower control limits set at $\pi \pm 3\sqrt{\pi(1 - \pi)/n}$, where $\pi(1 - \pi)/n$ is the variance of $\hat{\pi} = B/n$, the sample proportion of attributes of interest, as used in Chapter 17.

As an example, if the target value of π is 0.278 and $n = 100$, then control limits for π are

$$0.278 \pm 3\sqrt{\frac{(0.278)(0.722)}{100}}$$

from which the $UCL = 0.278 + 0.134 = 0.412$ and the $LCL = 0.278 - 0.134 = 0.144$.

Further reading about process control in general, and on control charts in particular, may be found in Burr (1976, 1979) and Duncan (1986).

For discussion

- Consult Table T-1 to show that, when $z = 3$, the probability associated with either UCL and LCL is 0.0013.

- Why, when σ is known, do we use $z = 1.96$ ($z = 2$, approximately) as a multiplier of σ for hypothesis-testing, whereas in constructing control limits the multiplier of σ is 3? Can you say that hypothesis-testing is more (or less) demanding than process control?

- What are the similarities between control limits and confidence intervals? In what ways are they different?

- Why are alarm limits used in addition to control limits?

- As presented in Figure 20-1, the two control limits (and the two alarm limits, if used, as in Figure 20-2) are equi-distant from μ_{target}. Are there situations when the limits should *not* be equally distanced from μ_{target}?

- Why is it easier to calculate the sample's range than its standard deviation?

- As described in Example 20-1, the control chart for the mean is built with a known σ. What can you do if σ is not known?

What to remember about Chapter 20

Chapter 20 built on Chapter 19's discussion of quality assurance and focused on the topic of *process control.* It looked particularly at the construction and use of control charts, one of the most powerful of all quality assurance techniques. Three types of charts were discussed and illustrated:

- *control charts for means*
- *control charts for standard deviations*
- *control charts for proportions.*

21

Quality assurance through the 95/95 acceptance criterion

What to look for in Chapter 21

Chapter 21 concludes the three-chapter overview of quality assurance. It is devoted to *acceptance sampling*, a collection of processes by which a lot of production items is judged by its attributes *after* it is produced. Special attention is paid to *the 95/95 acceptance criterion*, primarily because of its common use in the Nuclear Regulatory Commission's inspection programs. This chapter develops this methodology, presents some of the alternative sampling plans within a 95/95 acceptance criterion, and shows how easy it is to misuse the criterion. In the course of constructing 95/95 sampling plans, you will be reminded of such concepts and procedures as:

- *calculation of binomial probabilities*
- *individual and cumulative probabilities*
- *mutually exclusive and independent events.*

You will be also shown

- *how to construct 95/95 sampling plans*
- *how the methodology can be extended to construction of other sampling plans, such as 90/95 or 90/80*
- *why it is easy to misuse and misinterpret the 95/95 acceptance criterion.*

Acceptance sampling interpreted

acceptance sampling

Acceptance sampling covers those procedures that determine a course of action *after* a lot[1] of items has been produced. As its name implies, acceptance sampling is a sampling procedure with explicit criteria for the determination of whether a lot (i.e., a population) of items is acceptable. If a lot is not acceptable, it is rejected. The consequences of such rejection—scrapping the product, reworking the product to improve it, bankruptcy of the manufacturer—are too multifarious to explore here. Suffice to say, rejection of any manufactured product is seldom taken lightly.

The value of acceptance sampling to the entire production process is captured in these phrases from Duncan (1986, pp. 161-162):

> It is to be emphasized that the purpose of acceptance sampling is to determine a course of action, not to estimate lot quality....
>
> It is also to be emphasized that acceptance sampling is not an attempt to "control" quality. The latter is the purpose of control charts....

[1] In most acceptance-sampling literature, a *lot* is defined in terms similar to those used by Kendall and Buckland (1975, p. 88) as: " ... a group of units of a product produced under similar conditions and therefore, in a sense, of homogeneous origin; e.g., a set of screws produced by a lathe or a set of light bulbs produced by a number of similar machines. It is sometimes implicit that a lot is for inspection." Thus, *lot* fits the meaning of *population* as used in this book.

> The indirect effects of acceptance sampling on quality are likely to be much more important that the direct effects.... Acceptance sampling ... indirectly improves quality ... through its encouragement of good quality by a high rate of acceptance and its discouragement of poor quality by a high rate of rejection.
>
> Furthermore, if acceptance sampling is used ... at various stages of production, it may have beneficial effects in general on the quality of production.... Production personnel will ... become quality conscious and there will be an interest in quality on the part of both inspection and production [personnel]. The rule will be: Make it right the first time." These psychological aspects of acceptance sampling are of major importance.

Duncan concludes his discourse by indicating the conditions under which acceptance sampling is likely to be used:

> 1. When the cost of inspection is high and the loss from the passing of a defective item is not great. It is possible in some cases that no inspection at all will be the cheapest plan.
>
> 2. When 100 percent inspection is fatiguing and a carefully worked-out sampling plan will produce as good or better results. [Because] ... 100 percent inspection may not mean 100 percent perfect quality, ... the percentage of defective items passed may be higher than under a scientifically designed sampling plan.
>
> 3. When inspection is destructive. In this case sampling must be employed.

Just as process control (cf. Chapter 20) differentiates between sampling by attributes and sampling by variables, so does acceptance sampling. But, in either kind of sampling, acceptance sampling applied to a single lot is essentially a test of a hypothesis, in which a sample statistic is compared against a claimed product's characteristic (such as a mean or a standard deviation). The important driving feature of acceptance sampling, however, is that there is a greater concern about protecting the consumer by maintaining the *consumer's risk* below a given level. The concern with consumer's risk, typically, translates into a small probability of a Type II error. That is, you wish to make sure that the probability of accepting poor quality material is kept small. In this context, the Type II error is made if you don't reject a "bad" lot. For further reading on sampling for variables in the framework of acceptance sampling, refer to Duncan (1974) or Burr (1976) or Schilling (1982).

This chapter concentrates on acceptance sampling by attributes; that is, on procedures used when the items in the lot are inspected for the presence of an attribute of interest. To focus the discussion, assume that the attribute is undesirable, such as an item's being defective, broken, scratched, or not in compliance for some reason. Suppose further that the product's specification claims the proportion of items in the lot with the attribute is not larger than π_{spec}.

Clearly, to be absolutely certain sure that the fraction of the lot with the attribute of concern does not exceed π_{spec}, you need to examine every item in the lot. But such an effort, mounted for every shipment of every kind of material is an unacceptable burden. For some products, like sealed systems or single-use items, such inspection is impossible because the item's integrity is destroyed by the very act of inspection.

A statistical sampling, however, might ease your burden, requiring you to inspect only a fraction of each submitted lot, while providing some assurance—in lieu of an absolute guarantee—that the proportion of items with the attribute is in accord with the stated lot specification.

More specifically, statements based the *95/95 acceptance criterion* about a lot sampled for attributes convey a sense similar to these:

> *We are 95% assured that at least 95% of the items in the lot are in compliance.*

> *We are 95% confident that at least 95% of the cables in a bundle are traceable to their source.*

Although the chapter concentrates on 95% assurance of 95% quality, you can easily extend the methodology and lay a foundation for statements like the following:

> *We are 99% confident that at least 99% of the pipe supports are properly welded.*

> *We are 90% certain that at least 80% of the invoices were paid within 30 days of their receipt.*

But beware! The last statement is *not* equivalent to:

> *We are 80% sure that at least 90% of the invoices were paid within 30 days of their receipt.*

The assurance-to-quality criterion

Suppose you are given a lot of reinforcing bars (rebars) which was subjected to statistical sampling for meeting strength specifications. The quality statement issued for your review says:

> *We are 95% confident that at least 90% of the rebars in the lot are acceptable.*

This means that, based on the inspectors' data, no more than 10% of the rebars are unacceptable and that the inspectors are willing to back that statement with no less than a 19-to-1 bet.

This specific criterion is often given the shorthand notation of *95/90*. You read this as meaning that *you are 95% confident that at least 90% of the product is satisfactory for the purpose it was intended*. Thus, the first number (95) designates the *assurance* and the second number (90) designates the *quality*.

As a mnemonic for keeping track of these two values, consider this as being an *assurance-to-quality* statement and denote it by the symbol *A/Q*. In many inspection programs in general, and in NRC programs in particular, the *A/Q statement is set at 95/95*. Thus, the focus here is on the 95/95 criterion to illustrate the technique with this criterion, and to include "the 95/95 criterion" in the chapter's title.

The rules of the game

The general *A/Q* criterion is accompanied by a strict set of rules. Stick to the rules and stay out of distracting debate. Changes to the rules should be considered only on the conservative side and by compelling arguments. These rules (some of which reiterate the previous discussion) are:

- The *A/Q* criterion is a design criterion. You determine the sample size in advance and agree in advance that a one-time decision will be made to either accept the entire lot or to reject the entire lot, depending on the results of the collected sample.

- The lot must be unequivocally defined in the sampling plan. It is inappropriate to redefine the lot once the inspection begins. Note, however, that the lot size does not necessarily have to be finite or known.

- The lot is made up of similar items that are treated alike; i.e., they are interchangeable in terms of their intended use.

- If the lot about which an assurance statement is desired is made up of several sub-lots, you must address each sub-lot separately. Avoid mixing apples and oranges or you'll wind up making a quality statement about "fruit salad."

- The statistical assurance is expressed as two numbers (e.g., 95/95, 90/95, 95/99). The first number (*A*) is the assurance that the acceptable proportion is met. The second number (*Q*) is the acceptable percentage of "good" items.

- The assurance level (*A*) gives you a "comfort index." This index is the probability (expressed as percentage) that the lot will *not* be accepted when the lot's quality is less than *Q*.

- The *A/Q* provides assurance to the consumer—but it does not, in and of itself, provide assurance to the producer. Thus, at this point, you do not necessarily know the probability that the lot will be rejected when in reality the quality of the lot is at least as good as the product specification, *Q*. To examine this issue, you turn to the operating characteristics (OCs) of the plans under consideration. These OCs provide information similar to the operating characteristic curves for hypothesis-testing that were discussed in Chapter 9. See, for example, Duncan (1986, pp. 163-368) or Schilling (1982).

- Once you select a sampling plan and begin sampling, do not switch plans. You will be shown, by way of example at the end of this chapter, how changing horses in the middle of the stream is likely to cause a deterioration of the promised assurance.

The calculation of probabilities associated with the *A/Q* process is based on the binomial distribution, the requirements of which are detailed in chapter 17 and are repeated below:

Requirement 1: The experiment is conducted in n trials, or items, all of which are conducted under identical conditions.

Requirement 2: The ith trial ($i = 1, \ldots, n$) results either in a *success* (i.e., the attribute of interest is identified in the ith item) and recorded as a success with $Y_i = 1$ or in a *failure* with $Y_i = 0$.

Requirement 3: The probability of success, denoted by the parameter π, is constant from trial to trial.

Requirement 4: The n trials are independent.

If the listed requirements cannot be met, or at least, accounted for, other approaches need to be considered. For example, if the lot size is small, the hypergeometric distribution (Chapter 16) may be the appropriate vehicle for the probability calculation. In this case, you may wish to consult Sherr (1973) for the required sampling plan.

Clearly, since a lot comprises a finite number of individual items, you do not have a constant probability of drawing an item with an attribute. However, if the lot is large compared to the sample size (say, $n < N/10$), then, for most applications, there is very little difference between basing the probabilistic considerations on the binomial distribution or on the hypergeometric distribution (that adjusts for the lot's finiteness). Moreover, operating under the binomial assumptions makes the sample larger than need be for a finite lot; hence, the assurance actually is higher than claimed.

acceptance number, c

The sampling plan determines the sample size, denoted by n, and the maximum number of items with the attributes of interest that can be tolerated in the sample. The last number is called the *acceptance number*, denoted by c. If the sample yields more than c defective items, the entire lot is rejected.

In the single-sampling context, there is no "second chance." The lot is either accepted or rejected. However, sampling schemes may be designed (usually at the expense of a larger initial sample) to go through double- or multiple-sampling plans. You may read about those special plans in Duncan (1986, pp. 184-213) and Schilling (1982, pp. 127-153).

The *A/Q* is best illustrated by an example. So, armed with the understanding of probability, independent events, mutually exclusive events, and individual and cumulative probabilities, you move gently—or is it *gingerly*?— into the next section for a detailed example.

Calculating probabilities associated with the *A/Q* = 95/95 criterion

Focus here is on the *A/Q* = 95/95 criterion for quality assurance; used routinely by the U.S. Nuclear Regulatory Commission, it has a special role throughout the nuclear power industry.

A 95/95 assurance-to-quality plan for attribute sampling is not unique. In fact, you are about to encounter three different acceptance sampling plans (these, by the way, are by no means exhaustive), each of which meets the desired 95/95 criteria. For ease of comparison, the three plans are displayed side-by-side in Table 21-1, cleverly labeled as **Plan A**, **Plan B**, and **Plan C**, followed by a discussion of their respective features.

Table 21-1:
Three different sampling plans that meet the 95/95 assurance criterion

Plan A	Plan B	Plan C
Select n_A = 59 items at random from the lot.	Select n_B = 93 items at random from the lot.	Select n_C = 124 items at random from the lot.
If 0 items fail, accept the lot.	If 0 or 1 item fails, accept the lot.	If 0, 1, or 2 items fail, accept the lot.
If even 1 unacceptable item is found, reject the lot.	If 2 or more unacceptable items are found, reject the lot	If 3 or more unacceptable items are found, reject the lot.
For this plan, the acceptance number is $c = 0$.	For this plan, the acceptance number is $c = 1$.	For this plan, the acceptance number is $c = 2$.

To show that each of the three candidate plans provides the desired A/Q= 95/95 criterion, you will need to perform some calculations of binomial probabilities for $\pi = \pi_{spec} = 0.05$. Because Table T-8 is restricted to probabilities for $n \leq 40$, Table 21-2 was constructed to give the selected probabilities needed for your immediate task. You may easily verify Table 21-2's values with the methodology developed in Chapter 17.

Table 21-2:
Selected binomial probabilities, π = 0.05

	Individual probabilities			Cumulative probabilities		
n	$b = 0$	$b = 1$	$b = 2$	$b \leq 0$	$b \leq 1$	$b \leq 2$
58	0.0510	0.1558	0.2337	0.0510	0.2069	0.4406
59	0.0485	0.1506	0.2298	0.0485	0.1991	0.4289
92	0.0089	0.0432	0.1035	0.0089	0.0521	0.1556
93	0.0085	0.0415	0.1005	0.0085	0.0500	0.1504
123	0.0018	0.0118	0.0378	0.0018	0.0136	0.0514
124	0.0017	0.0113	0.0365	0.0017	0.0130	0.0495

To show that each of the three plans meets the 95/95 criterion, consider:

The quality aspect that 95% of the items are in compliance is conveniently translated to the complementary aspect that the proportion of items out of compliance is no larger than 0.05; hence, the associated probabilities are calculated for $\pi = 0.05$.

The 95% assurance aspect means that, if the promised quality is not present in the lot (i.e., that $\pi > 0.05$), the procedure would lead to rejection of the lot with at least 0.95 probability. (This translates to a probability of less than 0.05 that a "bad" lot will be accepted.

Now examine **Plan A**, where the acceptance number is $c = 0$. Look at Table 21-2 under $b \leq 0$ under

Cumulative probabilities (even though for $c = 0$, the individual and the cumulative probabilities are identical). You find that, if sample size is $n = 58$, the probability of accepting a lot with 5% defective items is 0.0510. This is slightly larger than the promised value of 0.05. When $n = 59$, the probability of acceptance of the lot is smaller than 0.05, which does satisfy the 95/95 criterion.

Plan B allows 0 or 1 items out of compliance. For $b \leq 1$, Table 21-2 shows that $n = 93$ gives you exactly a value of 0.05 for the probability of accepting the lot when $\pi = 0.05$. Hence, the 95/95 criterion is satisfied for $n = 93$ and $c = 1$.

Following a similar argument, you find that **Plan C**, for which $n = 124$ and $c = 2$, satisfies the 95/95 criterion.

As already indicated, there are numerous other sampling plans that are capable of meeting the 95/95 criterion. To limit the discussion, focus now on **Plans A**, **B**, and **C**. Which of the three plans do you like best? And, of course, why?

As a consumer, you don't really care, since all three plans give you the same assurance. As a producer, however, you believe that you deserve a fair shake from the consumer's quality agents. If only these three plans are offered to you, you must make the choice based on how good your product really is and on a number of economic considerations as well.

As the producer, you prefer as small a sample as possible, and you opt for **Plan A**. However, if **Plan A** is chosen, then how will you react to the fact that the entire lot is rejected if even a single item out of 59 (that's less than 2% of the sample!) is found out of compliance? If you are very confident that the quality of your manufactured product is very high, you may choose to stay with **Plan A**. On the other hand, to safeguard against what you surely view as an unlikely event, you may prefer

Plan C, with its larger sample size, because, as your intuition and analysis tell you, it is "easier to pass" a test with 2 defective items out of 124 than 1 out of 93, and certainly easier than 0 out of 59.

For discussion:

- Refresh your memory of Duncan's list of conditions under which acceptance sampling "is likely to be used." Which of them apply to your work?
- What is the difference between a 95/90 and a 90/95 sampling plan?
- Suppose your friend needs to confirm that a shipment of bolts contains no more than 5% defective bolts. To that end, he collects a sample of size 20 and finds no defective bolts. The sample percentage of defective bolts is less than 5% defective, and your friend concludes that the shipment is acceptable. Is anything wrong with your friend's logic? Discuss.
- If you were a producer and were given a choice between a 95/90 plan and a 90/95 plan, which would you choose? What data and calculations do you need to help you make that decision?
- Why do you use Table 21-2 with $\pi = 0.05$ when you are worried $\pi > 0.05$?
- Show an example of a finite lot where it does not make sense to use either Plan A, Plan B, or Plan C.
- Suppose you wish to construct a 90/80 sampling plan with acceptance number $c = 0$. Describe what probabilities you need and how you would go about calculating the sample size required to meet the stated criterion.

What's wrong with this picture?

As emphasized earlier, once you select a plan, you must stick with it. This section discusses the common mistake of plan-switching and recalculates the assurance for the sampling effort to show that the assurance is smaller than believed. For this illustration, consider a hypothetical producer who agrees at the outset that each of the three sampling plans described earlier meets the required 95/95 criterion. The three sampling plans are summarized in Table 21-3.

Table 21-3:
Three 95/95 sampling plans summarized from Table 21-1

Plan A:	**Plan B:**	**Plan C:**
Take a sample of 59 items.	Take a sample of 93 items.	Take a sample of 124 items.
If 0 defective items are found, accept the lot.	If 0 or 1 defective items are found, accept the lot.	If 0, 1, or 2 defective items are found, accept the lot.
Otherwise, reject the lot.	Otherwise, reject the lot.	Otherwise, reject the lot.

multiple-sampling plan

The producer, not necessarily a villain, mind you, sees nothing wrong with the following *multiple-sampling plan* which he believes still meets the 95/95 assurance criterion. This plan would lead to the acceptance of the lot if *any* of the following four events occurs:

Event 1: Use the approved **Plan A**. Collect 59 items. If zero defective items are found, stop the sampling and accept the lot.

Event 2: Use the approved **Plan A**. Collect 59 items. If exactly one defective item is found, collect 34 additional items. This brings the total number of inspected items to 93, as required for the approved **Plan B**. If no defective items are found among the additional 34 items, stop the sampling and accept the lot.

Event 3: Use the approved **Plan A**. Collect 59 items. If exactly two defective items are found, collect 65 additional items. This brings the total number of inspected items to 124, as required by the approved **Plan C**. If no defective items are found among the additional 65 additional items, stop the sampling and accept the lot.

Event 4: Use the approved **Plan A**. Collect 59 items. If exactly one defective item is found, collect 34 additional items, bringing the total number of items inspected to 93, as required for the approved **Plan B**. If exactly one defective item is found among the 34 items, collect 31 additional items, bringing the total number of items inspected to 124, as required for the approved **Plan C**. If no additional defective items are found, stop the sampling and accept the lot.

At first glance, the producer's multiple-sampling strategy appears reasonable and acceptable since each event satisfies one of the approved plans. But it always pays to revisit first principles and calculate the assurance associated with this sampling plan. To calculate that assurance, you start with some intermediate probability calculations. And, although you could obtain at least some of those probabilities from a table, it is instructive to calculate them and present all the necessary calculations together. Using Chapter 17 methodology, recall that

$$Pr\{B,n|\pi=0.05\}=\frac{n!}{B!(n-B)!}(0.05)^B\ (0.95)^{n-B},$$

where $Pr\{B,\ n|\pi\ =\ 0.05\}$ denotes the probability that B defective items will be found in a sample of size n, given that the proportion of defective items in the lot is $\pi\ =\ 0.05$. Because the binomial parameter $\pi\ =\ 0.05$ throughout this discourse, the shorthand notation is used in the following evaluations of individual binomial probabilities:

$$Pr\{0,31\} = \frac{31!}{0!31!}(0.05)^0(0.95)^{31} = 0.2039$$
$$Pr\{0,34\} = \frac{34!}{0!34!}(0.05)^0(0.95)^{34} = 0.1748$$
$$Pr\{0,59\} = \frac{59!}{0!59!}(0.05)^0(0.95)^{59} = 0.0485$$
$$Pr\{0,65\} = \frac{65!}{0!65!}(0.05)^0(0.95)^{65} = 0.0356$$
$$Pr\{1,31\} = \frac{31!}{1!30!}(0.05)^1(0.95)^{30} = 0.3327$$
$$Pr\{1,34\} = \frac{34!}{1!33!}(0.05)^1(0.95)^{33} = 0.3128$$
$$Pr\{1,59\} = \frac{59!}{1!58!}(0.05)^1(0.95)^{58} = 0.1506$$
$$Pr\{2,59\} = \frac{59!}{2!57!}(0.05)^2(0.95)^{57} = 0.2298.$$

Next, calculate the probability of occurrence of each of the four events listed earlier:

$$\begin{aligned} Pr\{\text{Event 1}\} &= Pr\{b = 0,\ n = 59\} \\ &= Pr\{0,\ 59\} = 0.0485 \end{aligned}$$

$$\begin{aligned} Pr\{\text{Event 2}\} &= P\{(b = 1,\ n = 59\ \textit{and}\ b = 0,\ n = 34\} \\ &= Pr\{1,\ 59\} \times Pr\{0,\ 34\} \\ &= (0.1506)(0.1748) \\ &= 0.0263 \end{aligned}$$

$$
\begin{aligned}
Pr\{\text{Event 3}\} &= Pr\{b = 2,\ n = 59 \text{ and } b = 0,\ n = 65\} \\
&= Pr\{2,59) \times Pr\{0,65) \\
&= (0.2298)(0.0356) = 0.0082
\end{aligned}
$$

$$
\begin{aligned}
Pr\{\text{Event 4}) &= Pr\{b = 1,\ n = 59 \text{ and } b = 1,\ n = 34, \\
&\qquad \text{and } b = 0,\ n = 31) \\
&= Pr\{1,59) \times Pr\{1,34) \times Pr\{0,31) \\
&= (0.1506)(0.3128)(0.2039) \\
&= 0.0096.
\end{aligned}
$$

Finally, put the pieces together. These four events are mutually exclusive; therefore, the probability of accepting the lot is the sum of the four probabilities. This is calculated as:

$$
\begin{aligned}
&Pr\{\text{lot accepted}\} \\
&\quad = 0.0485 + 0.0263 + 0.00829 + 0.0096 \\
&\quad = 0.0926.
\end{aligned}
$$

Hence, with 5% of the items being defective, the probability of accepting the lot is 0.0926—almost twice as large as the 0.05 that was intended. The probability of rejecting the lot is 1.0000 - 0.0096 = 0.9074.

THE ASSURANCE IS ONLY 90.7%, NOT THE CLAIMED 95%.

If the sampling scheme permits additional sampling (e.g., continue sampling and, if necessary, switch to a plan which allows 3 defective items out of 153), the composite assurance deteriorates even further.

Of course, legitimate multiple-sampling plans can be constructed so that the plan assurance will be 95/95—but, of course, such plans are by no means unique. You will find a discussion of multiple-sampling plans in, for example, Duncan (1986, pp. 204-213).

Finally, keep in mind that this chapter's acceptance sampling plans are based on infinite, or at least, "large," lots where the binomial distribution is used as the theoretical base. Small lots, however, are better served by sampling plans that are based on the hypergeometric distribution (see Chapter 16). The design of sampling plans for lots governed by the hypergeometric distribution is far from trivial. For more details, consider Sherr (1972).

For discussion

- The last section illustrated that a particular multiple-sampling scheme gives 91/95, rather than 95/95, assurance. But, then, what's wrong with a 91/95 assurance?
- Why do you expect the sample size to be smaller if a plan is based on the hypergeometric rather than on the binomial distribution?
- Many sampling tasks require the sample size to be a percentage of the lot of interest. Is this type of requirement justified? Is a *minimum* sampling percentage ever justified?

Sampling plans to meet other quality-to-assurance specifications

You can specify any pair of numbers—with each member of the pair bounded by 0 and 100, of course—to form an assurance-to-quality (*A/Q*) criterion. Conversely, any acceptance sampling plan can be assigned an *A/Q*.

For any given situation *A/Q criterion*, you have two principal courses of action:

(1) Search the quality control literature for sampling plans that satisfy your particular *A/Q* criterion. Suggested sources include Duncan (1986), Burr (1979), and Schilling (1982).

(2) Make your own calculations. Follow the general process laid out in this chapter. Although tables and hand-held calculators are better tools than pencil and paper, it's best to work with an interactive computer program that lets you tweak the parameters (primarily the sample size n and the acceptance number c) as you explore possible solutions.

What to remember about Chapter 21

Chapter 21 discussed features of quality assurance that are associated with acceptance sampling of a lot, or population, which is judged by its attributes. More particularly, special attention was paid to *the 95/95 acceptance criterion*, primarily because of its common use in the Nuclear Regulatory Commission's inspection programs. The chapter showed how this methodology is developed, presented some of the alternatives within a 95/95 acceptance criterion, and showed how easy it is to misuse the criterion. In the course of constructing 95/95 sampling plans, you were reminded of such ideas as:

- *calculation of binomial probabilities*
- *individual and cumulative probabilities*
- *mutually exclusive and independent events.*

You were also shown:

- *how to construct 95/95 sampling plans*
- *how the methodology can be extended to construction of other sampling plans, such as 90/95 or 90/80*
- *why it is easy to misuse and misinterpret the 95/95 acceptance criterion.*

In the course of this chapter you learned to:

- *appreciate the need for quality assurance inspection before a lot is accepted*
- *calculate the assurance associated with a sampling plan for a "large" lot*
- *calculate the assurance associated with simple cases of multiple sampling*
- *recognize when a sampling plan's assurance is not as good as it seems and when professional statistical support is helpful, perhaps even imperative.*

Afterword: Know what thou art missing

This book's scope provides neither the depth nor the breadth to contain an exhaustive treatment of any specific topic. Furthermore, many specialty fields in statistics—or offshoots of statistics—do not receive so much as an honorable mention in the text. As part of an atonement process for both of these types of shortfalls, this *Afterword* lists a number of these specialties. Each is described in a short phrase and given at least one reference to encourage your own exploration and discovery.[1]

Bon voyage!

[1] As statisticians carrying the torch of our profession, we *were* going to list these topics at random. Bowing to both external and internal pressures, we compromised on alphabetical listing. The descriptions themselves are drawn from many sources, principal among them are Kendall and Buckland (1971) and Kruskal and Tanur (1978), as well as our own experiences and viewpoints.

Topic	Description	Reference(s)
Bayesian statistics	Inference based on subjective probability.	Press (1988)
Biostatistics	Statistical procedures applied to biological and medical sciences.	Armitage (1971) Daniel (1974)
Categorial data analysis	Methods of reporting and interpreting multi-dimensional cross-tabulated data	Agresti (1990) Upton (1978)
Computation and algorithms	Research into and efficient use of computing methods specific to statistical applications.	Kennedy and Gentle (1980)
Covariance analysis	See **Linear models**	
Decision theory	Study of strategies for selecting one action from a set of available actions.	Chernoff and Moses (1959) Ferguson (1967)
Demography	Quantitative study of human populations.	Barclay (1958)
Discriminant analysis	Simultaneous study of differences among two or more groups with respect to several variables.	Klecka (1980)
Distribution-free methods	See **Nonparametric statistics.**	
Dynamic mathematical modeling	Techniques and applications useful in studying systemic changes over time.	Sandefur (1993)
Econometrics	Formulation of economics in mathematical and statistical terms.	Goldberger (1964) Johnston (1984)
Epidemiology	Specialized statistical-modeling, data-analytic, and inferential methods applied to epidemics.	Lilienfeld and Lilienfeld (1980)
Experimental design	Theory and practice of efficient experimentation.	Davies (1978) Mason, *et al.* (1989)

Topic	Description	Reference(s)
Exploratory data analysis	Manipulation, summarization, and display of data to improve their comprehensibility and to uncover underlying structure and to detect important departures from that structure.	Hoaglin, Mosteller, and Tukey (1983, 1985)
Geostatistics	Statistical study of events that fluctuate in space and time.	Isaaks and Srivastava (1990)
Linear models	Application and analysis of linear relationships among variable, both random and deterministic.	Graybill (1961) Neter, *et al.* (1990)
Multivariate analysis	Study of multidimensional distributions and samples from those distributions.	Morrison (1976) Tabachnick and Fidell (1989) Tatsuoka (1971)
Nonparametric statistics	Methods which do not depend upon the form of an underlying distribution.	Gibbons and Chakraborti (1992) Hollander and Wolfe (1973)
Outliers	An *outlier* is one of a set of values that lies unexpectedly distant from most of the other members of the set.	Barnett and Lewis (1984) Beckman and Cook (1983)
Perception of graphics	Selecting among many candidate graphical displays to convey correctly the information contained in a set of data.	Cleveland (1985, 1993) Tufte (1983, 1990)
Probability	See Chapters 3 and 15 in this book.	Feller (1957) Parzen (1960)
Psychometrics	Measurement of psychological variables and related research methodologies.	Ghiselli (1964) Guilford (1954)
Regression analysis	See **Linear Models**	

Topic	Description	Reference(s)
Robust statistics	Methods of estimation designed to reduce unwanted effects of inappropriate assumptions.	Huber (1981)
Sample size determination	Finding efficient sample sizes to meet experimental and resource requirements.	Desu and Raghavarao, (1990) Odeh and Fox (1975)
Sample surveys	Design and practice of surveys.	Cochran (1977) Levy and Lemeshow (1991)
Simulation methods	Study of systems too complicated for explicit analytical solution (e.g., political, physical, sociological, economic) through computer modeling.	Kennedy and Gentle (1980) Whicker and Sigelman (1991)
Theory of stochastic processes	The "dynamic" part of probability theory, in which a collection of random variables (called a *random process*) with respect to their interdependence and limiting behavior.	Parzen (1962)
Time series	A sequence of numerical data in which each item is associated with a particular instant in time.	Nelson (1973) Pankratz (1983)

"Old statisticians never die ... they just get broken down by age and sex."

S. Smith, quoted in Madansky (1988, p. 252).

Bibliography

Note: These entries point to authors who and material that we have found helpful, useful, insightful, and/or meaningful in our own pursuits of statistical quality and quantitative literacy. We believe that each of them, even if not mentioned specifically in the text itself, offers special rewards.

DL & RHM

Agresti, A. 1990, *Categorical Data Analysis*, John Wiley & Sons, Inc., New York, NY.

Armitage, P., 1971, *Statistical Methods in Medical Research*, John Wiley & Sons, Inc., New York, NY.

Babbie, E., 1990, *Survey Research Methods*, 2nd Ed., Wadsworth Publishing Company, Belmont, CA.

Barclay, G.W., 1958, *Techniques of Population Analysis*, John Wiley & Sons, Inc., New York, NY.

Barnett, V., 1991, *Sample Survey Principles and Methods*, Oxford University Press, New York, NY.

Barnett, V., and T. Lewis, 1984, *Outliers in Statistical Data*, 2nd Ed., John Wiley & Sons, Inc., New York, NY.

Bates, D. M., and D. G. Watts, 1988, *Nonlinear Regression Analysis and Its Applications*, John Wiley & Sons, Inc., New York, NY.

Beckman, R. J., and R. D. Cook, 1983, Outlier.......s, *Technometrics*, **25**, 119-149.

Beyer, W. H., 1974, *Handbook of Tables for Probability and Statistics*, 2nd Ed., The Chemical Rubber Co., Cleveland, OH.

Bickel, P. J., E. A. Hammel, and J. W. O'Conner, 1975, Sex Bias in Graduate Admission: Data from Berkeley, *Science*, **Vol. 187**, No. 4175, 398-404.

Bowen, W. M., and C. A. Bennett (Eds.), 1988, *Statistical Methods for Nuclear Material Management*, NUREG/CR-4604, U.S. Nuclear Regulatory Commission, U.S. Government Printing Office, Washington, DC.

Bowker, A. H., and G. J. Lieberman, 1972, *Engineering Statistics*, Prentice-Hall, Inc., Englewood Cliffs, NJ.

Box, G. E. P., and N. R. Draper, 1987, *Empirical Model-Building and Response Surfaces*, John Wiley & Sons, Inc., New York, NY.

Box, G. E. P., W. G. Hunter, and J. S. Hunter, 1978, *Statistics for Experimenters: An Introduction to Design, Data Analysis, and Model Building*, John Wiley & Sons, Inc., New York, NY.

Box, G. E. P, and G. M. Jenkins, 1976, Rev. Ed., *Time Series Analysis Forecast and Control*, Holden-Day, Inc., San Francisco, CA.

Brownlee, K. A., 1965, *Statistical Theory and Methodology in Science and Engineering*, 2nd Ed., John Wiley & Sons, Inc., New York, NY.

Burr, I. W., 1976, *Statistical Quality Control Methods*, Marcel Dekker, Inc., New York, NY.

Burr, I. W., 1979, *Elementary Statistical Quality Control*, Marcel Dekker, Inc., New York, NY.

Campbell, S. K., 1974, *Flaws and Fallacies in Statistical Thinking*, Prentice-Hall, Englewood Cliffs, NJ.

Chambers, J. M., W. S. Cleveland, B. Kleiner, and P. A. Tukey, 1983, *Graphical Methods for Data Analysis*, Duxbury Press, Boston, MA.

Chapman, D. G. 1951, Some Properties of the Hypergeometric Distribution with Application to Zoological Sample Censuses, *University of California Publications in Statistics,* **1**, University of California Press, Berkeley, CA, 131-160.

Chatfield, C., 1991, Avoiding Statistical Pitfalls, *Statistical Science*, **6**, 240-268.

Chernoff, H., and L. E. Moses, 1959, *Elementary Decision Theory*, John Wiley & Sons, Inc., New York, NY.

Cleveland, W. S., 1985, *The Elements of Graphing Data*, Wadsworth Advanced Book Program, Monterey, CA.

Cleveland, W. S., 1993, *Visualizing Data*, Hobart Press, Summit, NJ.

Cleveland, W. S., and R. McGill, 1984, Graphical Perception: Theory, Experimentation, and Application to the Development of Graphical Methods, *Journal of the American Statistical Association*, **79**, 531-554.

Cochran, W. G., 1952, The χ^2 Test of Goodness of Fit, *Annals of Mathematical Statistics*, **23**, 315-345.

Cochran, W.G., 1954, Some Methods for Strengthening the Common χ^2 Tests, *Biometrics*, **10**, 417-451.

Cochran, W. G., 1977, *Sampling Techniques*, 3rd Ed., John Wiley & Sons, Inc., New York, NY.

Cochran, W. G., and G. M. Cox, 1957, *Experimental design*, 2nd Ed., John Wiley & Sons, Inc., New York, NY.

Coleman, D. E., 1985, Measuring Measurements, *RCA Engineer*, **30**-3, 16-23.

Cramér, H., 1958, *Mathematical Methods of Statistics*, Princeton University Press, Princeton, NJ.

Daniel, C., and F. S. Wood, with the assistance of J. W. Gorman, 1980, *Fitting Equations to Data: Computer Analysis of Multifactor Data*, 2nd Ed., John Wiley & Sons, Inc., New York, NY.

Daniel, W. W., 1974, *Biostatistics: A Foundation for Analysis in the Health Sciences*, John Wiley & Sons, Inc., New York, NY.

Davies, O. L. (Ed.), 1978, *The Design and Analysis of Industrial Experiments*, 2nd Ed., Hafner Publishing Co., New York, NY.

Davies, O. L., and P. L. Goldsmith (Eds.), 1972, *Statistical Methods in Research and Production, with Special Reference to the Chemical Industry*, 4th Ed., Hafner Publishing Co., New York, NY.

Deming, W. E., 1982, Quality, Productivity, and Competitive Position, MIT Center for Advanced Engineering Study, Cambridge, MA.

Desu, M. and D. Raghavarao, 1990, *Sample Size Methodology*, Academic Press, Inc., San Diego, CA.

Dixon, W. J., and F. J. Massey, Jr., 1983, *Introduction to Statistical Analysis*, 4th Ed., McGraw-Hill Book Company, Inc., New York, NY.

Draper, N. R., and H. Smith, 1981, *Applied Regression Analysis*, 2nd Ed., John Wiley & Sons, Inc., New York, NY.

Duncan, A. J., 1986, *Quality Control and Industrial Statistics*, 5th Ed., Irwin, Homewood, IL.

Dunn, O. J., and V. A. Clark, 1987, *Applied Statistics: Analysis of Variance and Regression,* 2nd Ed., John Wiley & Sons, Inc., New York, NY.

Eadie, W. T., D. Dryard, F. E. James, M. Roos, and B. Sadoulet, 1971, *Statistical Methods in Experimental Physics*, American Elsevier Publishing Company, Inc., New York, NY.

Eisenhart, C., M. W. Hastay, and W. A. Wallis, 1947, *Techniques of Statistical Analysis*, McGraw-Hill Book Company, Inc., New York, NY.

Feller, W., 1957, *An Introduction to Probability Theory and Its Applications, Volume I*, 2nd Ed., John Wiley & Sons, Inc., New York, NY.

Ferguson, T. S., 1967, *Mathematical Statistics: A Decision Theoretic Approach*, Academic Press, New York, NY.

Fisher, R. A., 1970, *Statistical Methods for Research Workers*, 14th Ed., Oliver and Boyd, Edinburgh, Scotland.

Fleiss, J. L., 1981, *Statistical Methods for Rates and Proportions*, 2nd Ed., John Wiley & Sons, Inc., New York, NY.

Frodesen, A. G., O. Skjeggestad, and H. Tofte, 1979, *Probability and Statistics in Particle Physics*, Universitetsforlagat, Bergen-Oslo-Tromso, Norway.

Ghiselli, E. E. 1964, *Theory of Psychological Measurement*, McGraw-Hill Book Company, Inc., New York, NY.

Gibbons, J. D., and S. Chakraborti, 3rd Ed., 1992, *Nonparametric Statistical Inference*, Marcel Dekker, Inc., New York, NY.

Gilbert, N., 1976, *Statistics*, W. B. Saunders Co., Philadelphia, PA.

Goldberger, A. S., 1964, *Econometric Theory*, John Wiley, New York, NY.

Graybill, F. A., 1961, *An Introduction to Linear Statistical Models, Volume I*, McGraw-Hill Book Company, Inc., New York, NY.

Guenther, W. C., 1964, *Analysis of Variance*, Prentice-Hall, Inc., Englewood Cliffs, NJ.

Guilford, J. P., 1954, *Psychometric Methods*, McGraw-Hill Book Company, Inc., New York, NY.

Hahn, G. J., 1984, Experimental Design in the Complex World, *Technometrics*, **26**, 19-22.

Hahn, G. J., and W. Q. Meeker, 1991, *Statistical Intervals: A Guide for Practitioners*, John Wiley, New York, NY.

Hahn, G. J., and S. S. Shapiro, 1967, *Statistical Models in Engineering*, John Wiley & Sons, Inc., New York, NY.

Hald, A., 1952, *Statistical Tables and Formulas*, John Wiley & Sons, Inc., New York, NY.

Hald, A., 1952, *Statistical Theory with Engineering Applications*, John Wiley & Sons, Inc., New York, NY.

Hammersly, J. M., and D. C. Hanscomb, 1964, *Monte Carlo Methods*, Methuen, London.

Harter, H. L.,1974-76, The Method of Least Squares and Some Alternatives, *International Statistical Review* Part I, **42:** 147-174; Part II, **42:** 235-264; Part III, **43:** 1-44; Part IV, **43:** 125-190; Part V, **43:** 269-278; Part VI Subject and Author Indexes, **44:** 113-159.

Hastie, T. J., and R. J. Tibshirani, 1990, *Generalized Additive Models,* Chapman and Hall, New York, NY.

Hastings, N. A. J., and W. B. Peacock, 1975, *Statistical Distributions: A Handbook for Students and Practitioners*, A Halsted Press Book, John Wiley & Sons, Inc., New York, NY.

Hicks, C. R., 1982, *Fundamental Concepts in the Design of Experiments*, 3rd Ed., Holt, Rinehart, and Winston, New York, NY.

Hoaglin, D. C., F. Mosteller, and J. W. Tukey, (Eds.), 1983, *Understanding Robust and Exploratory Data Analysis*, John Wiley & Sons, Inc., New York, NY.

Hoaglin, D. C., F. Mosteller, and J. W. Tukey, (Eds.), 1985, *Exploring Data Tables, Trends, and Shapes*, John Wiley & Sons, Inc., New York, NY.

Hoel, P. G., 1971, *Introduction to Mathematical Statistics*, 4th Ed., John Wiley & Sons, Inc., New York, NY.

Hogan, H., 1992, The Post-Enumeration Survey: An Overview, *The American Statistician*, **46**, No. 4, 261-269.

Hollander, M., and D. A. Wolfe, 1973, *Nonparametric Statistical Methods*, John Wiley & Sons, Inc., New York, NY.

Hooke, R., 1983, *How to Tell the Liars from the Statisticians*, Marcel Dekker, Inc., New York, NY.

Huber, P. J., 1981, *Robust Statistics*, John Wiley & Sons, Inc., New York, NY.

Huff, D., 1954, *How to Lie with Statistics*, W. W. Norton & Company, New York, NY.

Hunter, J. S., 1975, The Technology of Quality, *RCA Engineer*, **30**-3, 8-15.

Isaaks, E. H., and R. M. Srivastava, 1990, *An Introduction to Applied Geostatistics*, Oxford University Press, Cary, NC.

Jaech, J. L., 1973, *Statistical Methods in Nuclear Material Control*, TID-26298, NTIS, Springfield, VA.

James, G., and R. C. James, 1951, *Mathematics Dictionary*, D. Van Nostrand Co., Inc., Princeton, NJ.

Johnson, M. E., 1987, *Multivariate Statistical Simulation: A Guide to Selecting and Generating Continuous Multivariate Distributions*, John Wiley & Sons, Inc., New York, NY.

Johnston, J., 1984, *Econometric Methods*, 3rd Ed., McGraw-Hill Book Company, New York, NY.

Kasprzyk, D., G. Duncan, G. Kalton, and M. P. Singh (Eds.), 1989, *Panel Surveys*, John Wiley & Sons, Inc., New York, NY.

Kempthorne, O., 1952, *The Design and Analysis of Experiments*, John Wiley & Sons, Inc., New York, NY.

Kendall, M. G., and W. B. Buckland, 1971, *A Dictionary of Statistical Terms*, 3rd Ed., Hafner Publishing Co., New York, NY.

Kendall, M., A. Stuart, and J. K. Ord, 1989, *Kendall's Advanced Theory of Statistics*, 5th Ed., Volume 1: Distribution Theory, Volume 2: Classical Inference and Relationship, Oxford University Press, New York, NY.

Kendall, M., A. Stuart, and J. K. Ord, 1987, *Kendall's Advanced Theory of Statistics*, 4th Ed., Volume 3: Design and Analysis, and Time-Series, Oxford University Press, New York, NY.

Kennedy, W. J., Jr., and J. E. Gentle, 1980, *Statistical Computing*, Marcel Dekker, Inc., New York, NY.

Klecka, W. R., 1980, *Discriminant Analysis*, Sage Publications, Newbury Park, CA.

Kotz, S., and N. L. Johnson (Eds.), with C.B. Read, 1988, *Encyclopedia of Statistical Sciences* (in nine volumes), John Wiley & Sons, Inc., New York, NY.

Kruskal, W. H., and J. M. Tanur (Eds.), 1978, *International Encyclopedia of Statistics*, Volumes 1 and 2, The Free Press, New York, NY.

Ku, H. H. (Ed.), 1969, *Precision Measurement and Calibration, Selected NBS Papers on Statistical Concepts and Procedures*, NBS Special Publication 300 — Volume I, National Bureau of Standards, U.S. Government Printing Office, Washington, DC.

Lancaster, H. O., 1969, *The Chi-Squared Distribution*, John Wiley & Sons, Inc., New York, NY.

Large, D. W., and P. Michie, 1981, Proving that the Strength of the British Navy Depends on the Number of Old Maids in England: A Comparison of Scientific with Legal Proof, *Environmental Law*, **11**, 557-638.

Levy, P. S., 1991, *Sampling of Populations—Methods and Applications*, John Wiley & Sons, Inc., New York, NY.

Levy, P. S., and S. Lemeshow, 1991, *Sampling of Populations: Methods and Applications*, John Wiley & Sons, Inc., New York, NY.

Lieberman, G. J., and D. B. Owen, 1961, *Tables of the Hypergeometric Probability Distribution*, Stanford University Press, Stanford, CA.

Lilienfeld, A. M., and D. E. Lilienfeld, 1980, *Foundations of Epidemiology*, Oxford University Press, Cary, NC.

Lilliefors, H. W., 1967, On the Kolmogorov-Smirnov Test for Normality with Mean and Variance Unknown, *Journal of the American Statistical Association, **62**: 399-402.*

Little, R. J. A., and D. B. Rubin, 1987, *Statistical Analysis with Missing Data*, John Wiley & Sons, Inc., New York, NY.

Lwanga, S. K., and S. Lemeshow, 1991, *Sample Size Determination in Health Studies: A Practical Manual*, World Health Organization, Geneva, Switzerland.

Madansky, A., 1988, *Prescriptions for Working Statisticians*, Springer-Verlag, New York, NY.

Mason, R. L., R. F. Gunst, and J. L. Hess, 1989, *Statistical Design and Analysis of Experiments with Applications to Engineering and Science,* John Wiley & Sons, Inc., New York, NY.

McCuen, R. H., 1985, *Statistical Methods for Engineers*, Prentice-Hall, Inc., Englewood Cliffs, NJ.

McNemar, Q., 1947, Note on the Sampling Error of the Difference between Correlated Proportions or Percentages, *Psychometrika*, **12**, 153-157.

Meyer, P. L., 1970, *Introductory Probability and Statistical Applications*, 2nd Ed., Addison-Wesley Publishing Company, Reading MA.

Meyer, S. L., 1975, *Data Analysis for Scientists and Engineers*, John Wiley & Sons, Inc., New York, NY.

Miller, I., Freund, J. E., and Johnson, R. A., 1990, *Probability and Statistics for Engineers*, Prentice-Hall, Inc., Englewood Cliffs, NJ.

Mood, A. M., and F. A. Graybill, 1963, *Introduction to the Theory of Statistics*, 2nd Ed., McGraw-Hill Book Company, Inc, New York, NY.

Mood, A. M., F. A. Graybill, and D. C. Boes, 1974, *Introduction to the Theory of Statistics*, 3rd Ed., McGraw-Hill Book Company, Inc, New York, NY.

Mosteller, F., and J. W. Tukey, 1971, *Data Analysis and Regression*, Addison-Wesley Publishing Company, Reading MA.

Natrella, M. G., 1966, *Experimental Statistics*, National Bureau of Standards Handbook #91, U. S. Government Printing Office, Washington, DC.

Nelson, C. R., 1973, *Applied Time Series Analysis for Managerial Forecasting*, Holden-Day, Inc., San Francisco, CA.

Neter, J., W. Wasserman, and M. H. Kutner, 1990, *Applied Linear Statistical Models: Regression, Analysis of Variance, and Experimental Designs*, 3rd Ed., Irwin, Homewood, IL.

NRC, 1989, *Nuclear Regulatory Commission Information Digest, 1992 Edition*, NUREG-1350, Volume 4, U.S. Government Printing Office, Washington, DC.

Odeh, R. R., and M. Fox, 1975, *Sample Size Choice: Charts for Experiments with Linear Models*, Marcel Dekker, Inc., New York, NY.

Odeh, R. E., and D. B. Owen, 1980, *Tables for Normal Tolerance Limits, Sampling Plans, and Screening*, Marcel Dekker, Inc., New York, NY.

Ordnance Corps, 1952, *Tables of the Cumulative Binomial Probabilities*, Ordnance Corps Pamphlet ORDP 20-1, US. Government Printing Office, Washington, DC.

Ostle, B., and R. W. Mensing, 1979, *Statistics in Research*, 3rd Ed., Iowa State University Press, Ames, IA.

Ott, L., and W. Mendenhall, 1984, *Understanding Statistics*, 4th Ed., Duxbury Press, Boston, MA.

Owen, D. B., 1962, *Statistical Tables*, Addison-Wesley Publishing Co., Reading, MA.

Pankratz, A., 1983, *Forecasting with Univariate Box-Jenkins Models: Concepts and Cases*, John Wiley & Sons, Inc., New York, NY.

Pankratz, A., 1991, *Forecasting with Dynamic Regression Models*, John Wiley & Sons, Inc., New York, NY.

Parzen, E., 1960, *Modern Probability Theory and Its Applications*, John Wiley & Sons, Inc., New York, NY.

Parzen, E., 1962, *Stochastic Processes*, Holden-Day, Inc., San Francisco, CA.

Pearson, E. S. and H. O. Hartley, 1970 (Volume I) and 1976 (Volume II), *Biometrika Tables for Statisticians*, Cambridge University Press, Lowe and Brydone (Printers) Ltd., Thetford, Norfolk, England.

Plackett, R. L., 1981, *The Analysis of Categorical Data*, 2nd Ed., Macmillan Publishing Co., Inc., New York, NY.

Press, S. J., 1988, *Bayesian Statistics: Principles, Models, and Applications*, John Wiley & Sons, Inc., New York, NY.

Reichmann, W. J., 1971, *Use and Abuse of Statistics*, Oxford University Press, New York, NY.

Romig, H. G., 1953, *50-100 Binomial Tables*, John Wiley & Sons, Inc., New York, NY.

Rosner, B., 1989, *Fundamentals of Biostatistics*, 3rd Ed., P.W.S. Kent, Boston, MA.

Rossi, P. H., J. D. Wright, and A. B. Anderson (Eds.), 1983, *Handbook of Survey Research*, Academic Press, Inc., Orlando, FL.

Rubin, D. B., 1987, *Multiple Imputation for Nonresponse in Surveys*, John Wiley & Sons, Inc., New York, NY.

Sandefur, J. T., 1993, *Discrete Dynamical Modeling*, Oxford University Press, Cary, NC.

Satterthwaite, F. E., 1946, An Approximate Distribution of Estimates of Variance Components, *Biometrics Bulletin*, **2**, 110-114.

Schilling, E. G., 1982, *Acceptance Sampling in Quality Control*, Marcel Dekker, Inc., New York, NY.

Searle, S. R., 1987, *Linear Models for Unbalanced Data*, John Wiley & Sons, Inc., New York, NY.

Shafer, G., 1990, The Unity and Diversity of Probability, *Statistical Science*, **5**, No. 4, 435-462.

Shapiro, S. S., and A. J. Gross, 1981, *Statistical Modeling Techniques*, Marcel Dekker, Inc., New York, NY.

Shapiro, S. S., and M. B. Wilk, 1965, An Analysis of Variance Test for Normality (complete samples), *Biometrika*, **52**, 591-611.

Sherr, T. S., 1972, *Attribute Sampling Inspection Procedure Based on the Hypergeometric Distribution*, WASH-1210, U.S. Atomic Energy Commission. U.S. Government Printing Office, Washington, DC.

Shewhart, W. A., 1931, *Economic Control of Quality of Manufactured Product*, D. Van Nostrand Co., Inc., Princeton, NJ.

Shooman, M. L., 1990, *Probabilistic Reliability: An Engineering Approach*, 2nd Ed., McGraw-Hill Book Company, Inc., New York, NY.

Shreider, Y. A., 1966, *The Monte Carlo Method*, translated by G. J. Tee, Pergammon Press, New York, NY.

Siegel, S., 1956, *Nonparametric Statistics for the Behavioral Sciences*, McGraw-Hill Book Company, Inc., New York, NY.

Snedecor, G. W., and W. G. Cochran, 1989, *Statistical Methods*, 8th Ed., Iowa State University Press, Ames, IA.

Staff of the Computational Laboratory, 1955, *Tables of the Cumulative Binomial Probability Distribution*, Harvard University Press, Cambridge, MA.

Steel, R. G. D., and J. H. Torrie, 1980, *Principles and Procedures of Statistics: A Biometrical Approach*, 2nd Ed., McGraw-Hill Book Company, Inc, New York, NY.

Stein, P., D. Coleman, and B. Gunter, 1986, Graphics for Display of Statistical Data, Supplement to *RCA Engineer*, **30**-3.

Stevens, S. S., 1946, On the Theory of Scales of Measurement, *Science*, **103**, 677-680.

Stevens, W. L., 1939, Distribution of Groups in a Sequence of Alternatives, *Annals of Eugenics*, **9**, 10-17.

Swed, F. S., and C. Eisenhart, 1943, Tables for Testing Randomness of Grouping in a Sequence of Alternatives, *Annals of Mathematical Statistics,* **14**, 66-87.

Tabachnick, B. G., and L. S. Fidell, 1989, *Using Multivariate Statistics*, 2nd Ed., Harper & Row, Publishers, New York, NY.

Tanur, J., F. Mosteller, W. H. Kruskal, R. F. Link, R. S. Pieters, and G. R. Rising (Eds.), 1989, *Statistics: A Guide to the Unknown*, 3rd Ed., Wadsworth & Brooks/Cole Advanced Books & Software, Pacific Grove, CA.

Toothaker, L. E., 1991, *Multiple Comparisons for Researchers*, Sage Publications, Inc., Newbury Park, CA.

Tufte, E. R., 1983, *The Visual Display of Quantitative Information*, Graphics Press, Cheshire, CT.

Tufte, E. R., 1990, *Envisioning Information*, Graphics Press, Cheshire, CT.

Tukey, J. W., 1977, *Exploratory Data Analysis*, Addison-Wesley Publishing Co., Reading, MA.

Tukey, J.W., 1993, Graphic Comparisons of Several Aspects: Alternatives and Suggested Principles, *Journal of Computational and Graphical Statistics*, with comments by H. Wainer, discussion by R. A. Becker and W. S. Cleveland, and rejoinder by J. W. Tukey, **2**, 1, 1-49.

Upton, G. J. G., 1978, *The Analysis of Cross-Tabulated Data*, John Wiley & Sons, Inc., New York, NY.

U.S. Code of Federal Regulations, 1993, Title 10 (Energy), U.S. Government Printing Office, Washington, DC..

Velleman, P. F., and D. C. Hoaglin, 1981, *Applications, Basics, and Computing of Exploratory Data Analysis*, Duxbury Press, Boston, MA.

Velleman, P. F., and L. Wilkinson, 1993, Nominal, Ordinal, Interval, and Ratio Typologies are Misleading, *The American Statistician*, **47**, 1, 65-72.

Wadsworth, H. M. (Ed.), 1989, *Handbook of Statistical Methods for Engineers and Scientists*, McGraw-Hill Book Company, Inc., New York, NY.

Wald, A., and J. Wolfowitz, 1940, On a Test Whether Two Samples are from the Same Population, *Annals of Mathematical Statistics*, **11**, 147-162.

Weinberg, G. H., and J. H. Shumaker, 1969, *Statistics, An Intuitive Approach*, 2nd Ed., Brooks/Cole Publishing Co., Monterey, CA.

Weisberg, S., 1985, *Applied Linear Regression*, 2nd Ed., John Wiley & Sons, Inc., New York, NY.

Welch, B. L., 1937, The Significance of the Difference between Two Means when the Population Variances are Unequal, *Biometrika,* **29**, 350-362.

Welch, B. L. 1947, The Generalization of "Student's" Problem when Several Different Variances are Involved, *Biometrika*, **34**, 28-35.

Whicker, M. L, and L. Sigelman, 1991, *Computer Simulation Applications: An Introduction*, Sage Publications, Inc., Newbury Park, CA.

Williams, B., 1978, *A Sampler on Sampling*, John Wiley & Sons, Inc., New York, NY.

Winer, B. J., 1971, *Statistical Principles in Experimental Design*, 2nd Ed., McGraw-Hill Book Company, Inc., New York, NY.

Statistical tables

Table T-1: The cumulative standardized normal distribution (with selected quantiles)

A random normal variable has a density function given by

$$n(y;\ \mu,\ \sigma) = \frac{1}{\sigma\sqrt{2\pi}}\ e^{-\frac{(y\ -\ \mu)^2}{2\sigma^2}}$$

$$-\infty < y < \infty; \quad -\infty < \mu < \infty, \quad 0 < \sigma.$$

Its two parameters are μ (the mean) and σ (the standard deviation). When $\mu = 0$ and $\sigma = 1$, the normal variable is said to be standardized. A normal variable Y becomes a standardized normal variable, Z, by the transformation $Z = (Y - \mu)/\sigma$.

Table T-1 gives values of the standardized normal distribution, $N(z;\ 0,\ 1)$, often written as $N(z)$, for selected values of z ($0 \leq z \leq 3.49$). For $z < 0$, set $z' = -z$ and use $N(z) = 1 - N(z')$. To use Table T-1:

- Find the two-decimal-place value of z = a.bc by locating the row containing a.b in the first column and the column containing 0.0c in the first row.

- Read $N(z)$ at the intersection of that row and that column.

Example 1: To find $N(1.45)$, locate the intersection of the 15th row and the 6th column in the body of Table T-1 and read the value $N(1.45) = 0.9265$. This is displayed on the graph of the standardized normal density function.

Example 2: To find $N(-1.45)$, write $N(-1.45)$
$= 1 - N(1.45)$
$= 1 - 0.9265$
$= 0.0735$.

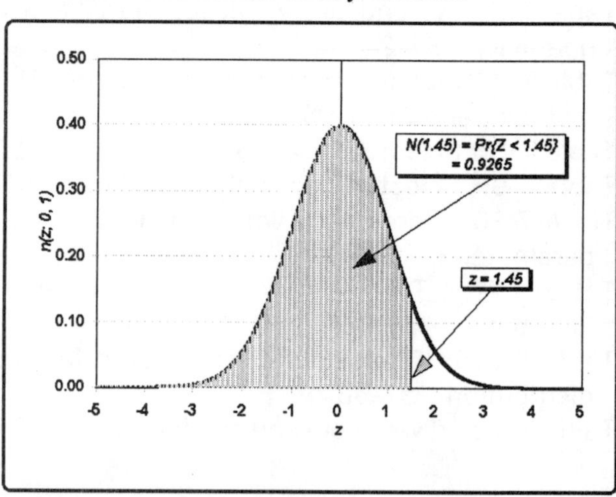

Table T-1: The cumulative standardized normal distribution (with selected quantiles)*

z	0.00	0.01	0.02	0.03	0.04	0.05	0.06	0.07	0.08	0.09
0.0	0.5000	0.5040	0.5080	0.5120	0.5160	0.5199	0.5239	0.5279	0.5319	0.5359
0.1	0.5398	0.5438	0.5478	0.5517	0.5557	0.5596	0.5636	0.5675	0.5714	0.5753
0.2	0.5793	0.5832	0.5871	0.5910	0.5948	0.5987	0.6026	0.6064	0.6103	0.6141
0.3	0.6179	0.6217	0.6255	0.6293	0.6331	0.6368	0.6406	0.6443	0.6480	0.6517
0.4	0.6554	0.6591	0.6628	0.6664	0.6700	0.6736	0.6772	0.6808	0.6844	0.6879
0.5	0.6915	0.6950	0.6985	0.7019	0.7054	0.7088	0.7123	0.7157	0.7190	0.7224
0.6	0.7257	0.7291	0.7324	0.7357	0.7389	0.7422	0.7454	0.7486	0.7517	0.7549
0.7	0.7580	0.7611	0.7642	0.7673	0.7704	0.7734	0.7764	0.7794	0.7823	0.7852
0.8	0.7881	0.7910	0.7939	0.7967	0.7995	0.8023	0.8051	0.8078	0.8106	0.8133
0.9	0.8159	0.8186	0.8212	0.8238	0.8264	0.8289	0.8315	0.8340	0.8365	0.8389
1.0	0.8413	0.8438	0.8461	0.8485	0.8508	0.8531	0.8554	0.8577	0.8599	0.8621
1.1	0.8643	0.8665	0.8686	0.8708	0.8729	0.8749	0.8770	0.8790	0.8810	0.8830
1.2	0.8849	0.8869	0.8888	0.8907	0.8925	0.8944	0.8962	0.8980	0.8997	0.9015
1.3	0.9032	0.9049	0.9066	0.9082	0.9099	0.9115	0.9131	0.9147	0.9162	0.9177
1.4	0.9192	0.9207	0.9222	0.9236	0.9251	0.9265	0.9279	0.9292	0.9306	0.9319
1.5	0.9332	0.9345	0.9357	0.9370	0.9382	0.9394	0.9406	0.9418	0.9429	0.9441
1.6	0.9452	0.9463	0.9474	0.9484	0.9495	0.9505	0.9515	0.9525	0.9535	0.9545
1.7	0.9554	0.9564	0.9573	0.9582	0.9591	0.9599	0.9608	0.9616	0.9625	0.9633
1.8	0.9641	0.9649	0.9656	0.9664	0.9671	0.9678	0.9686	0.9693	0.9699	0.9706
1.9	0.9713	0.9719	0.9726	0.9732	0.9738	0.9744	0.9750	0.9756	0.9761	0.9767
2.0	0.9772	0.9778	0.9783	0.9788	0.9793	0.9798	0.9803	0.9808	0.9812	0.9817
2.1	0.9821	0.9826	0.9830	0.9834	0.9838	0.9842	0.9846	0.9850	0.9854	0.9857
2.2	0.9861	0.9864	0.9868	0.9871	0.9875	0.9878	0.9881	0.9884	0.9887	0.9890
2.3	0.9893	0.9896	0.9898	0.9901	0.9904	0.9906	0.9909	0.9911	0.9913	0.9916
2.4	0.9918	0.9920	0.9922	0.9925	0.9927	0.9929	0.9931	0.9932	0.9934	0.9936
2.5	0.9938	0.9940	0.9941	0.9943	0.9945	0.9946	0.9948	0.9949	0.9951	0.9952
2.6	0.9953	0.9955	0.9956	0.9957	0.9959	0.9960	0.9961	0.9962	0.9963	0.9964
2.7	0.9965	0.9966	0.9967	0.9968	0.9969	0.9970	0.9971	0.9972	0.9973	0.9974
2.8	0.9974	0.9975	0.9976	0.9977	0.9977	0.9978	0.9979	0.9979	0.9980	0.9981
2.9	0.9981	0.9982	0.9982	0.9983	0.9984	0.9984	0.9985	0.9985	0.9986	0.9986
3.0	0.9987	0.9987	0.9987	0.9988	0.9988	0.9989	0.9989	0.9989	0.9990	0.9990
3.1	0.9990	0.9991	0.9991	0.9991	0.9992	0.9992	0.9992	0.9992	0.9993	0.9993
3.2	0.9993	0.9993	0.9994	0.9994	0.9994	0.9994	0.9994	0.9995	0.9995	0.9995
3.3	0.9995	0.9995	0.9995	0.9996	0.9996	0.9996	0.9996	0.9996	0.9996	0.9997
3.4	0.9997	0.9997	0.9997	0.9997	0.9997	0.9997	0.9997	0.9997	0.9997	0.9998

Selected quantile										
q:	**0.5**	**0.75**	**0.9**	**0.95**	**0.975**	**0.98**	**0.99**	**0.995**	**0.999**
z_q:	0.000	0.674	1.282	1.645	1.960	2.054	2.326	2.576	3.090

* Prepared by the authors.

Table T-2: Quantiles, $\chi^2_q(df)$, for the chi-squared distribution with *df* degrees of freedom

A chi-squared random variable has a density function given by

$$f(y) = \frac{y^{(df-2)/2}\, e^{-y/2}}{2^{df/2}\,[(df-2)/2]!}$$

$y > 0, \quad df = 1, 2, \ldots .$

Its one parameter is *df* (called *degrees of freedom*); it is a positive integer.

Table T-2 gives quantiles, $\chi^2_q(df)$, for selected values of q (where $0 < q < 1$) and *df*. To use Table T-2:

- Find the row corresponding to the degrees of freedom, *df*.
- Find the column corresponding to the value of q.
- Find the desired quantile at the intersection of that row and that column.

Example: The 0.95 quantile of the chi-squared distribution with 10 degrees of freedom is $\chi^2_{0.95}(10) = 18.3$. This is displayed on the graph of the chi-squared density function for $df = 10$.

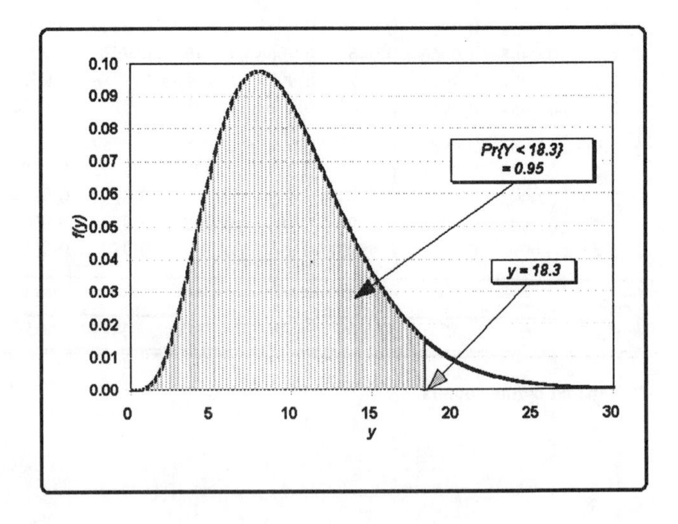

Table T-2: Quantiles, $\chi^2_q(df)$, for the chi-squared distribution with *df* degrees of freedom*

df	*q* 0.005	0.010	0.025	0.050	0.100	0.900	0.950	0.975	0.990	0.995
1	0.000039	0.000157	0.000982	0.00393	0.0158	2.71	3.84	5.02	6.63	7.88
2	0.0100	0.0201	0.0506	0.103	0.211	4.61	5.99	7.38	9.21	10.6
3	0.072	0.115	0.216	0.352	0.584	6.25	7.81	9.35	11.3	12.8
4	0.207	0.297	0.484	0.711	1.064	7.78	9.49	11.1	13.3	14.9
5	0.412	0.554	0.831	1.145	1.61	9.24	11.1	12.8	15.1	16.7
6	0.676	0.872	1.24	1.64	2.20	10.6	12.6	14.4	16.8	18.5
7	0.989	1.24	1.69	2.17	2.83	12.0	14.1	16.0	18.5	20.3
8	1.34	1.65	2.18	2.73	3.49	13.4	15.5	17.5	20.1	22.0
9	1.73	2.09	2.70	3.33	4.17	14.7	16.9	19.0	21.7	23.6
10	2.16	2.56	3.25	3.94	4.87	16.0	18.3	20.5	23.2	25.2
11	2.60	3.05	3.82	4.57	5.58	17.3	19.7	21.9	24.7	26.8
12	3.07	3.57	4.40	5.23	6.30	18.5	21.0	23.3	26.2	28.3
13	3.57	4.11	5.01	5.89	7.04	19.8	22.4	24.7	27.7	29.8
14	4.07	4.66	5.63	6.57	7.79	21.1	23.7	26.1	29.1	31.3
15	4.60	5.23	6.26	7.26	8.55	22.3	25.0	27.5	30.6	32.8
16	5.14	5.81	6.91	7.96	9.31	23.5	26.3	28.8	32.0	34.3
17	5.70	6.41	7.56	8.67	10.1	24.8	27.6	30.2	33.4	35.7
18	6.26	7.01	8.23	9.39	10.9	26.0	28.9	31.5	34.8	37.2
19	6.84	7.63	8.91	10.1	11.7	27.2	30.1	32.9	36.2	38.6
20	7.43	8.26	9.59	10.9	12.4	28.4	31.4	34.2	37.6	40.0
21	8.03	8.90	10.3	11.6	13.2	29.6	32.7	35.5	38.9	41.4
22	8.64	9.54	11.0	12.3	14.0	30.8	33.9	36.8	40.3	42.8
23	9.26	10.2	11.7	13.1	14.8	32.0	35.2	38.1	41.6	44.2
24	9.89	10.9	12.4	13.8	15.7	33.2	36.4	39.4	43.0	45.6
25	10.5	11.5	13.1	14.6	16.5	34.4	37.7	40.6	44.3	46.9
26	11.2	12.2	13.8	15.4	17.3	35.6	38.9	41.9	45.6	48.3
27	11.8	12.9	14.6	16.2	18.1	36.7	40.1	43.2	47.0	49.6
28	12.5	13.6	15.3	16.9	18.9	37.9	41.3	44.5	48.3	51.0
29	13.1	14.3	16.0	17.7	19.8	39.1	42.6	45.7	49.6	52.3
30	13.8	15.0	16.8	18.5	20.6	40.3	43.8	47.0	50.9	53.7
40	20.7	22.2	24.4	26.5	29.1	51.8	55.8	59.3	63.7	66.8
50	28.0	29.7	32.4	34.8	37.7	63.2	67.5	71.4	76.2	79.5
60	35.5	37.5	40.5	43.2	46.5	74.4	79.1	83.3	88.4	92.0
70	43.3	45.4	48.8	51.7	55.3	85.5	90.5	95.0	100	104
80	51.2	53.5	57.2	60.4	64.3	96.6	102	107	112	116
90	59.2	61.8	65.6	69.1	73.3	108	113	118	124	128
100	67.3	70.1	74.2	77.9	82.4	118	124	130	136	140

*Prepared by the authors.

Table T-3: Quantiles, $t_q(df)$, for Student's T distribution with df degrees of freedom

A Student's T random variable has a density function given by

$$f(y) = \frac{[(df - 1)/2]!}{(\pi df)^{1/2}(df/2 - 1)!}[1 + (y^2/df)]^{-(df + 1)/2}$$

$y > 0, \quad df = 1, 2, \ldots .$

Its one parameter is df (called degrees of freedom); it is a positive integer.

Table T-3 gives quantiles, $t_q(df)$, for selected values of q (where $0 < q < 1$) and df. To use Table T-3:

- Find the row corresponding to the degrees of freedom, df.
- Find the column corresponding to the value of q.
- Find the desired quantile at the intersection of that row and that column.

Example: The 0.95 quantile of Student's T distribution with 2 degrees of freedom is $t_{0.95}(2) = 2.92$. This is displayed on the graph of Student's T density function for $df = 2$.

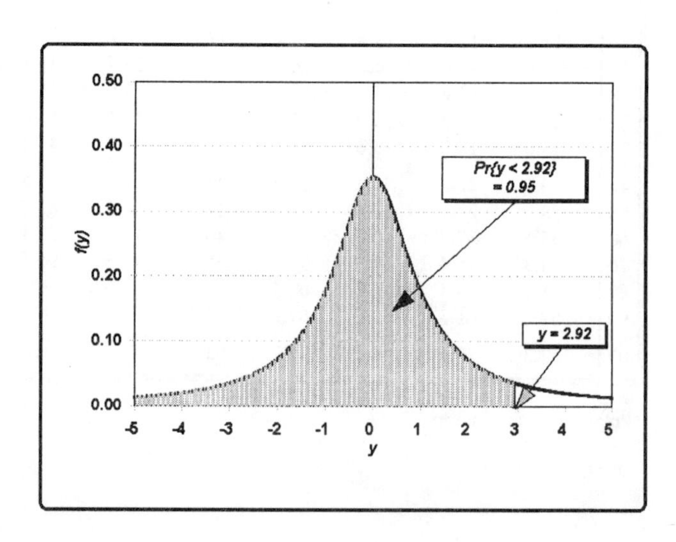

Table T-3: Quantiles, $t_q(df)$, for Student's *T* distribution with *df* degrees of freedom*

	q									
df	**0.50**	**0.60**	**0.70**	**0.75**	**0.80**	**0.90**	**0.95**	**0.975**	**0.99**	**0.995**
1	0.000	0.325	0.727	1.00	1.38	3.08	6.31	12.7	31.8	63.7
2	0.000	0.289	0.617	0.816	1.06	1.89	2.92	4.30	6.96	9.92
3	0.000	0.277	0.584	0.765	0.978	1.64	2.35	3.18	4.54	5.84
4	0.000	0.271	0.569	0.741	0.941	1.53	2.13	2.78	3.75	4.60
5	0.000	0.267	0.559	0.727	0.920	1.48	2.02	2.57	3.36	4.03
6	0.000	0.265	0.553	0.718	0.906	1.44	1.94	2.45	3.14	3.71
7	0.000	0.263	0.549	0.711	0.896	1.41	1.89	2.36	3.00	3.50
8	0.000	0.262	0.546	0.706	0.889	1.40	1.86	2.31	2.90	3.36
9	0.000	0.261	0.543	0.703	0.883	1.38	1.83	2.26	2.82	3.25
10	0.000	0.260	0.542	0.700	0.879	1.37	1.81	2.23	2.76	3.17
11	0.000	0.260	0.540	0.697	0.876	1.36	1.80	2.20	2.72	3.11
12	0.000	0.259	0.539	0.695	0.873	1.36	1.78	2.18	2.68	3.05
13	0.000	0.259	0.538	0.694	0.870	1.35	1.77	2.16	2.65	3.01
14	0.000	0.258	0.537	0.692	0.868	1.35	1.76	2.14	2.62	2.98
15	0.000	0.258	0.536	0.691	0.866	1.34	1.75	2.13	2.60	2.95
16	0.000	0.258	0.535	0.690	0.865	1.34	1.75	2.12	2.58	2.92
17	0.000	0.257	0.534	0.689	0.863	1.33	1.74	2.11	2.57	2.90
18	0.000	0.257	0.534	0.688	0.862	1.33	1.73	2.10	2.55	2.88
19	0.000	0.257	0.533	0.688	0.861	1.33	1.73	2.09	2.54	2.86
20	0.000	0.257	0.533	0.687	0.860	1.33	1.72	2.09	2.53	2.85
21	0.000	0.257	0.532	0.686	0.859	1.32	1.72	2.08	2.52	2.83
22	0.000	0.256	0.532	0.686	0.858	1.32	1.72	2.07	2.51	2.82
23	0.000	0.256	0.532	0.685	0.858	1.32	1.71	2.07	2.50	2.81
24	0.000	0.256	0.531	0.685	0.857	1.32	1.71	2.06	2.49	2.80
25	0.000	0.256	0.531	0.684	0.856	1.32	1.71	2.06	2.49	2.79
26	0.000	0.256	0.531	0.684	0.856	1.31	1.71	2.06	2.48	2.78
27	0.000	0.256	0.531	0.684	0.855	1.31	1.70	2.05	2.47	2.77
28	0.000	0.256	0.530	0.683	0.855	1.31	1.70	2.05	2.47	2.76
29	0.000	0.256	0.530	0.683	0.854	1.31	1.70	2.05	2.46	2.76
30	0.000	0.256	0.530	0.683	0.854	1.31	1.70	2.04	2.46	2.75
40	0.000	0.255	0.529	0.681	0.851	1.30	1.68	2.02	2.42	2.70
60	0.000	0.254	0.527	0.679	0.848	1.30	1.67	2.00	2.39	2.66
∞	0.000	0.253	0.524	0.674	0.842	1.28	1.645	1.96	2.33	2.58

* Prepared by the authors.

Table T-4: Quantiles, $f_q(df_1, df_2)$, for the F distribution with df_1 degrees of freedom in the numerator and df_2 degrees of freedom in the denominator

An F variable has a density function given by

$$f(y) = \frac{df_1^{(df_1/2)}\ df_2^{(df_2/2)}\ [(df_1 + df_2 - 2)/2]!}{[(df_1 - 2)/2]!\ [(df_2 - 2)/2]!} \times \frac{y^{(df_1-2)/2}}{(df_2 + df_1 y)^{(df_1+df_2)/2}}$$

$$y > 0, \quad df_1, df_2 = 1, 2, \ldots .$$

Its two parameters are df_1 (the *numerator degrees of freedom*) and df_2 (the *denominator degrees of freedom*); they both are positive integers.

Table T-4 gives quantiles, $f_q(df_1, df_2)$, for selected values of q (where $0 < q < 1$) and df_1 and df_2. To use Table T-4:

- Find the row corresponding to the denominator degrees of freedom, df_1.
- Find the sub-row corresponding to the value of q.
- The desired quantile is at the intersection of that sub-row and the column corresponding to the numerator degrees of freedom, df_2.

A particular identity is useful when you need smaller quantiles for the F distribution and you have the larger quantiles at hand (as in Table T-4):

$$f_{(1-q)}(df_1, df_2) = \frac{1}{f_q(df_2, df_1)}.$$

Table T-4: Quantiles, $f_q(df_1, df_2)$, for the F distribution with df_1 degrees of freedom in the numerator and df_2 degrees of freedom in the denominator (continued)

Example 1: The 0.90 quantile of the F distribution with 6 degrees of freedom in the numerator and 4 degrees of freedom in the denominator is $f_{0.90}(6, 4) = 4.01$. This is displayed on the graph of the F density function for $df_1 = 6$ and $df_2 = 4$.

Example 2: The 0.10 quantile of the F distribution with 6 numerator degrees of freedom and 4 denominator degrees of freedom is

$$f_{0.10}(6, 4) = \frac{1}{f_{0.90}(4, 6)} = \frac{1}{4.01} = 0.249.$$

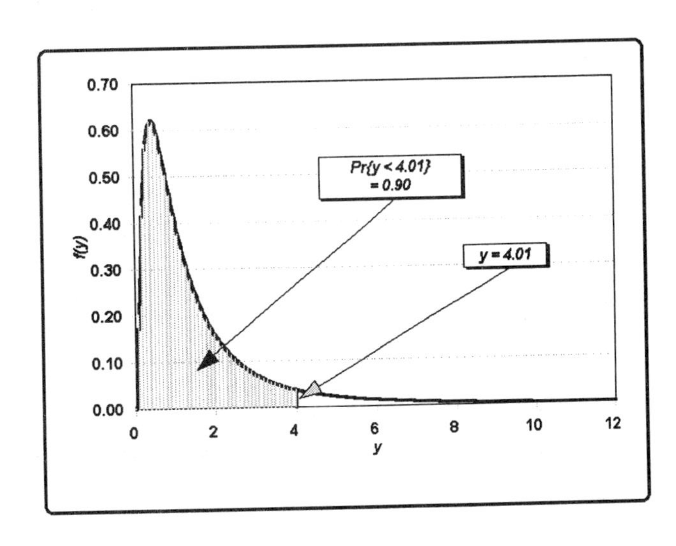

Table T-4: Quantiles, $f_q(df_1, df_2)$, for the F distribution with df_1 degrees of freedom in the numerator and df_2 degrees of freedom in the denominator*

df_2	q	df_1 = 1	2	3	4	5	6	7	8	9
1	**0.90**	39.9	49.5	53.6	55.8	57.2	58.2	58.9	59.4	59.9
	0.95	161	199	216	225	230	234	237	239	241
	0.975	648	799	864	900	922	937	948	957	963
	0.99	4050	5000	5400	5620	5760	5860	5930	5980	6020
	0.995	16200	20000	21600	22500	23100	23400	23700	23900	24100
2	**0.90**	8.53	9.00	9.16	9.24	9.29	9.33	9.35	9.37	9.38
	0.95	18.5	19.0	19.2	19.2	19.3	19.3	19.4	19.4	19.4
	0.975	38.5	39.0	39.2	39.2	39.3	39.3	39.4	39.4	39.4
	0.99	98.5	99.0	99.2	99.3	99.3	99.3	99.4	99.4	99.4
	0.995	199	199	199	199	199	199	199	199	199
3	**0.90**	5.54	5.46	5.39	5.34	5.31	5.28	5.27	5.25	5.24
	0.95	10.1	9.55	9.28	9.12	9.01	8.94	8.89	8.85	8.81
	0.975	17.4	16.0	15.4	15.1	14.9	14.7	14.6	14.5	14.5
	0.99	34.1	30.8	29.5	28.7	28.2	27.9	27.7	27.5	27.3
	0.995	55.6	49.8	47.5	46.2	45.4	44.8	44.4	44.1	43.9
4	**0.90**	4.54	4.32	4.19	4.11	4.05	4.01	3.98	3.95	3.94
	0.95	7.71	6.94	6.59	6.39	6.26	6.16	6.09	6.04	6.00
	0.975	12.2	10.6	9.98	9.60	9.36	9.20	9.07	8.98	8.90
	0.99	21.2	18.0	16.7	16.0	15.5	15.2	15.0	14.8	14.7
	0.995	31.3	26.3	24.3	23.2	22.5	22.0	21.6	21.4	21.1
5	**0.90**	4.06	3.78	3.62	3.52	3.45	3.40	3.37	3.34	3.32
	0.95	6.61	5.79	5.41	5.19	5.05	4.95	4.88	4.82	4.77
	0.975	10.0	8.43	7.76	7.39	7.15	6.98	6.85	6.76	6.68
	0.99	16.3	13.3	12.1	11.4	11.0	10.7	10.5	10.3	10.2
	0.995	22.8	18.3	16.5	15.6	14.9	14.5	14.2	14.0	13.8
6	**0.90**	3.78	3.46	3.29	3.18	3.11	3.05	3.01	2.98	2.96
	0.95	5.99	5.14	4.76	4.53	4.39	4.28	4.21	4.15	4.10
	0.975	8.81	7.26	6.60	6.23	5.99	5.82	5.70	5.60	5.52
	0.99	13.7	10.9	9.78	9.15	8.75	8.47	8.26	8.10	7.98
	0.995	18.6	14.5	12.9	12.0	11.5	11.1	10.8	10.6	10.4
7	**0.90**	3.59	3.26	3.07	2.96	2.88	2.83	2.78	2.75	2.72
	0.95	5.59	4.74	4.35	4.12	3.97	3.87	3.79	3.73	3.68
	0.975	8.07	6.54	5.89	5.52	5.29	5.12	4.99	4.90	4.82
	0.99	12.2	9.55	8.45	7.85	7.46	7.19	6.99	6.84	6.72
	0.995	16.2	12.4	10.9	10.1	9.52	9.16	8.89	8.68	8.51

* Prepared by the authors.

Table T-4: Quantiles, $f_q(df_1, df_2)$, for the F distribution with df_1 degrees of freedom in the numerator and df_2 degrees of freedom in the denominator (continued)

		df_1								
df_2	q	10	12	15	20	24	30	60	120	∞
1	0.90	60.2	60.7	61.2	61.7	62.0	62.3	62.8	63.1	63.3
	0.95	242	244	246	248	249	250	252	253	254
	0.975	969	977	985	993	997	1000	1010	1010	1020
	0.99	6060	6110	6160	6210	6230	6260	6310	6340	6370
	0.995	24200	24400	24600	24800	24900	25000	25300	25400	25500
2	0.90	9.39	9.41	9.42	9.44	9.45	9.46	9.47	9.48	9.49
	0.95	19.4	19.4	19.4	19.4	19.5	19.5	19.5	19.5	19.5
	0.975	39.4	39.4	39.4	39.4	39.5	39.5	39.5	39.5	39.5
	0.99	99.4	99.4	99.4	99.4	99.5	99.5	99.5	99.5	99.5
	0.995	199	199	199	199	199	199	199	199	200
3	0.90	5.23	5.22	5.20	5.18	5.18	5.17	5.15	5.14	5.13
	0.95	8.79	8.74	8.70	8.66	8.64	8.62	8.57	8.55	8.53
	0.975	14.4	14.3	14.3	14.2	14.1	14.1	14.0	13.9	13.9
	0.99	27.2	27.1	26.9	26.7	26.6	26.5	26.3	26.2	26.1
	0.995	43.7	43.4	43.1	42.8	42.6	42.5	42.1	42.0	41.8
4	0.90	3.92	3.90	3.87	3.84	3.83	3.82	3.79	3.78	3.76
	0.95	5.96	5.91	5.86	5.80	5.77	5.75	5.69	5.66	5.63
	0.975	8.84	8.75	8.66	8.56	8.51	8.46	8.36	8.31	8.26
	0.99	14.5	14.4	14.2	14.0	13.9	13.8	13.7	13.6	13.5
	0.995	21.0	20.7	20.4	20.2	20.0	19.9	19.6	19.5	19.3
5	0.90	3.30	3.27	3.24	3.21	3.19	3.17	3.14	3.12	3.11
	0.95	4.74	4.68	4.62	4.56	4.53	4.50	4.43	4.40	4.37
	0.975	6.62	6.52	6.43	6.33	6.28	6.23	6.12	6.07	6.02
	0.99	10.1	9.89	9.72	9.55	9.47	9.38	9.20	9.11	9.02
	0.995	13.6	13.4	13.1	12.9	12.8	12.7	12.4	12.3	12.1
6	0.90	2.94	2.90	2.87	2.84	2.82	2.80	2.76	2.74	2.72
	0.95	4.06	4.00	3.94	3.87	3.84	3.81	3.74	3.70	3.67
	0.975	5.46	5.37	5.27	5.17	5.12	5.07	4.96	4.90	4.85
	0.99	7.87	7.72	7.56	7.40	7.31	7.23	7.06	6.97	6.88
	0.995	10.3	10.0	9.81	9.59	9.47	9.36	9.12	9.00	8.88
7	0.90	2.70	2.67	2.63	2.59	2.58	2.56	2.51	2.49	2.47
	0.95	3.64	3.57	3.51	3.44	3.41	3.38	3.30	3.27	3.23
	0.975	4.76	4.67	4.57	4.47	4.41	4.36	4.25	4.20	4.14
	0.99	6.62	6.47	6.31	6.16	6.07	5.99	5.82	5.74	5.65
	0.995	8.38	8.18	7.97	7.75	7.64	7.53	7.31	7.19	7.08

Table T-4: Quantiles, $f_q(df_1, df_2)$, for the F distribution with df_1 degrees of freedom in the numerator and df_2 degrees of freedom in the denominator (continued)

		df_1								
df_2	q	1	2	3	4	5	6	7	8	9
8	0.90	3.46	3.11	2.92	2.81	2.73	2.67	2.62	2.59	2.56
	0.95	5.32	4.46	4.07	3.84	3.69	3.58	3.50	3.44	3.39
	0.975	7.57	6.06	5.42	5.05	4.82	4.65	4.53	4.43	4.36
	0.99	11.3	8.65	7.59	7.01	6.63	6.37	6.18	6.03	5.91
	0.995	14.7	11.0	9.60	8.81	8.30	7.95	7.69	7.50	7.34
9	0.90	3.36	3.01	2.81	2.69	2.61	2.55	2.51	2.47	2.44
	0.95	5.12	4.26	3.86	3.63	3.48	3.37	3.29	3.23	3.18
	0.975	7.21	5.71	5.08	4.72	4.48	4.32	4.20	4.10	4.03
	0.99	10.6	8.02	6.99	6.42	6.06	5.80	5.61	5.47	5.35
	0.995	13.6	10.11	8.72	7.96	7.47	7.13	6.88	6.69	6.54
10	0.90	3.29	2.92	2.73	2.61	2.52	2.46	2.41	2.38	2.35
	0.95	4.96	4.10	3.71	3.48	3.33	3.22	3.14	3.07	3.02
	0.975	6.94	5.46	4.83	4.47	4.24	4.07	3.95	3.85	3.78
	0.99	10.04	7.56	6.55	5.99	5.64	5.39	5.20	5.06	4.94
	0.995	12.8	9.43	8.08	7.34	6.87	6.54	6.30	6.12	5.97
12	0.90	3.18	2.81	2.61	2.48	2.39	2.33	2.28	2.24	2.21
	0.95	4.75	3.89	3.49	3.26	3.11	3.00	2.91	2.85	2.80
	0.975	6.55	5.10	4.47	4.12	3.89	3.73	3.61	3.51	3.44
	0.99	9.33	6.93	5.95	5.41	5.06	4.82	4.64	4.50	4.39
	0.995	11.8	8.51	7.23	6.52	6.07	5.76	5.52	5.35	5.20
15	0.90	3.07	2.70	2.49	2.36	2.27	2.21	2.16	2.12	2.09
	0.95	4.54	3.68	3.29	3.06	2.90	2.79	2.71	2.64	2.59
	0.975	6.20	4.77	4.15	3.80	3.58	3.41	3.29	3.20	3.12
	0.99	8.68	6.36	5.42	4.89	4.56	4.32	4.14	4.00	3.89
	0.995	10.8	7.70	6.48	5.80	5.37	5.07	4.85	4.67	4.54
20	0.90	2.97	2.59	2.38	2.25	2.16	2.09	2.04	2.00	1.96
	0.95	4.35	3.49	3.10	2.87	2.71	2.60	2.51	2.45	2.39
	0.975	5.87	4.46	3.86	3.51	3.29	3.13	3.01	2.91	2.84
	0.99	8.10	5.85	4.94	4.43	4.10	3.87	3.70	3.56	3.46
	0.995	9.9	6.99	5.82	5.17	4.76	4.47	4.26	4.09	3.96
24	0.90	2.93	2.54	2.33	2.19	2.10	2.04	1.98	1.94	1.91
	0.95	4.26	3.40	3.01	2.78	2.62	2.51	2.42	2.36	2.30
	0.975	5.72	4.32	3.72	3.38	3.15	2.99	2.87	2.78	2.70
	0.99	7.82	5.61	4.72	4.22	3.90	3.67	3.50	3.36	3.26
	0.995	9.6	6.66	5.52	4.89	4.49	4.20	3.99	3.83	3.69

Table T-4: Quantiles, $f_q(df_1, df_2)$, for the F distribution with df_1 degrees of freedom in the numerator and df_2 degrees of freedom in the denominator (continued)

		df_1								
df_2	q	10	12	15	20	24	30	60	120	∞
9	**0.90**	2.42	2.38	2.34	2.30	2.28	2.25	2.21	2.18	2.16
	0.95	3.14	3.07	3.01	2.94	2.90	2.86	2.79	2.75	2.71
	0.975	3.96	3.87	3.77	3.67	3.61	3.56	3.45	3.39	3.33
	0.99	5.3	5.11	4.96	4.81	4.73	4.65	4.48	4.40	4.31
	0.995	6.4	6.2	6.03	5.83	5.73	5.62	5.41	5.30	5.19
10	**0.90**	2.32	2.28	2.24	2.20	2.18	2.16	2.11	2.08	2.06
	0.95	2.98	2.91	2.85	2.77	2.74	2.70	2.62	2.58	2.54
	0.975	3.72	3.62	3.52	3.42	3.37	3.31	3.20	3.14	3.08
	0.99	4.8	4.71	4.56	4.41	4.33	4.25	4.08	4.00	3.91
	0.995	5.8	5.66	5.47	5.27	5.17	5.07	4.86	4.75	4.64
12	**0.90**	2.19	2.15	2.10	2.06	2.04	2.01	1.96	1.93	1.90
	0.95	2.75	2.69	2.62	2.54	2.51	2.47	2.38	2.34	2.30
	0.975	3.37	3.28	3.18	3.07	3.02	2.96	2.85	2.79	2.73
	0.99	4.30	4.16	4.01	3.86	3.78	3.70	3.54	3.45	3.36
	0.995	5.1	4.91	4.72	4.53	4.43	4.33	4.12	4.01	3.90
15	**0.90**	2.06	2.02	1.97	1.92	1.90	1.87	1.82	1.79	1.76
	0.95	2.54	2.48	2.40	2.33	2.29	2.25	2.16	2.11	2.07
	0.975	3.06	2.96	2.86	2.76	2.70	2.64	2.52	2.46	2.40
	0.99	3.80	3.67	3.52	3.37	3.29	3.21	3.05	2.96	2.87
	0.995	4.4	4.25	4.07	3.88	3.79	3.69	3.48	3.37	3.26
20	**0.90**	1.94	1.89	1.84	1.79	1.77	1.74	1.68	1.64	1.61
	0.95	2.35	2.28	2.20	2.12	2.08	2.04	1.95	1.90	1.84
	0.975	2.77	2.68	2.57	2.46	2.41	2.35	2.22	2.16	2.09
	0.99	3.37	3.23	3.09	2.94	2.86	2.78	2.61	2.52	2.42
	0.995	3.8	3.68	3.50	3.32	3.22	3.12	2.92	2.81	2.69
24	**0.90**	1.88	1.83	1.78	1.73	1.70	1.67	1.61	1.57	1.53
	0.95	2.25	2.18	2.11	2.03	1.98	1.94	1.84	1.79	1.73
	0.975	2.64	2.54	2.44	2.33	2.27	2.21	2.08	2.01	1.94
	0.99	3.17	3.03	2.89	2.74	2.66	2.58	2.40	2.31	2.21
	0.995	3.6	3.42	3.25	3.06	2.97	2.87	2.66	2.55	2.43
24	**0.90**	1.88	1.83	1.78	1.73	1.70	1.67	1.61	1.57	1.53
	0.95	2.25	2.18	2.11	2.03	1.98	1.94	1.84	1.79	1.73
	0.975	2.64	2.54	2.44	2.33	2.27	2.21	2.08	2.01	1.94
	0.99	3.17	3.03	2.89	2.74	2.66	2.58	2.40	2.31	2.21
	0.995	3.6	3.42	3.25	3.06	2.97	2.87	2.66	2.55	2.43

Table T-4: Quantiles, $f_q(df_1, df_2)$, for the F distribution with df_1 degrees of freedom in the numerator and df_2 degrees of freedom in the denominator (continued)

		df_1								
df_2	q	1	2	3	4	5	6	7	8	9
30	**0.90**	2.88	2.49	2.28	2.14	2.05	1.98	1.93	1.88	1.85
	0.95	4.17	3.32	2.92	2.69	2.53	2.42	2.33	2.27	2.21
	0.975	5.57	4.18	3.59	3.25	3.03	2.87	2.75	2.65	2.57
	0.99	7.6	5.39	4.51	4.02	3.70	3.47	3.30	3.17	3.07
	0.995	9.2	6.4	5.24	4.62	4.23	3.95	3.74	3.58	3.45
60	**0.90**	2.79	2.39	2.18	2.04	1.95	1.87	1.82	1.77	1.74
	0.95	4.00	3.15	2.76	2.53	2.37	2.25	2.17	2.10	2.04
	0.975	5.29	3.93	3.34	3.01	2.79	2.63	2.51	2.41	2.33
	0.99	7.1	4.98	4.13	3.65	3.34	3.12	2.95	2.82	2.72
	0.995	8.5	5.8	4.73	4.14	3.76	3.49	3.29	3.13	3.01
120	**0.90**	2.75	2.35	2.13	1.99	1.90	1.82	1.77	1.72	1.68
	0.95	3.92	3.07	2.68	2.45	2.29	2.18	2.09	2.02	1.96
	0.975	5.15	3.80	3.23	2.89	2.67	2.52	2.39	2.30	2.22
	0.99	6.9	4.79	3.95	3.48	3.17	2.96	2.79	2.66	2.56
	0.995	8.2	5.54	4.50	3.92	3.55	3.28	3.09	2.93	2.81
∞	**0.90**	2.71	2.30	2.08	1.94	1.85	1.77	1.72	1.67	1.63
	0.95	3.84	3.00	2.60	2.37	2.21	2.10	2.01	1.94	1.88
	0.975	5.02	3.69	3.12	2.79	2.57	2.41	2.29	2.19	2.11
	0.99	6.64	4.61	3.78	3.32	3.02	2.80	2.64	2.51	2.41
	0.995	7.9	5.30	4.28	3.72	3.35	3.09	2.90	2.74	2.62

Table T-4: Quantiles, $f_q(df_1, df_2)$, for the *F* distribution with df_1 degrees of freedom in the numerator and df_2 degrees of freedom in the denominator (continued)

		df_1								
df_2	q	10	12	15	20	24	30	60	120	∞
30	0.90	1.82	1.77	1.72	1.67	1.64	1.61	1.54	1.50	1.46
	0.95	2.16	2.09	2.01	1.93	1.89	1.84	1.74	1.68	1.62
	0.975	2.51	2.41	2.31	2.20	2.14	2.07	1.94	1.87	1.79
	0.99	2.98	2.84	2.70	2.55	2.47	2.39	2.21	2.11	2.01
	0.995	3.34	3.18	3.01	2.82	2.73	2.63	2.42	2.30	2.18
60	0.90	1.71	1.66	1.60	1.54	1.51	1.48	1.40	1.35	1.29
	0.95	1.99	1.92	1.84	1.75	1.70	1.65	1.53	1.47	1.39
	0.975	2.27	2.17	2.06	1.94	1.88	1.82	1.67	1.58	1.48
	0.99	2.63	2.50	2.35	2.20	2.12	2.03	1.84	1.73	1.60
	0.995	2.90	2.74	2.57	2.39	2.29	2.19	1.96	1.83	1.69
120	0.90	1.65	1.60	1.55	1.48	1.45	1.41	1.32	1.26	1.19
	0.95	1.91	1.83	1.75	1.66	1.61	1.55	1.43	1.35	1.25
	0.975	2.16	2.05	1.94	1.82	1.76	1.69	1.53	1.43	1.31
	0.99	2.47	2.34	2.19	2.03	1.95	1.86	1.66	1.53	1.38
	0.995	2.71	2.54	2.37	2.19	2.09	1.98	1.75	1.61	1.43
∞	0.90	1.60	1.55	1.49	1.42	1.38	1.34	1.24	1.17	1.00
	0.95	1.83	1.75	1.67	1.57	1.52	1.46	1.32	1.22	1.00
	0.975	2.05	1.94	1.83	1.71	1.64	1.57	1.39	1.27	1.00
	0.99	2.32	2.18	2.04	1.88	1.79	1.70	1.47	1.32	1.00
	0.995	2.52	2.36	2.19	2.00	1.90	1.79	1.53	1.36	1.00

Table T-5: Quantiles, $f_{max,\,q}(k,\, df)$, for the F_{max} distribution

The F_{max} statistic is designed to test the hypothesis that the variances of several groups are equal.

(1) Start with n observations from each of k groups.

(2) State the null hypothesis H_0: $\sigma_1^2 = \sigma_2^2 = \ldots = \sigma_k^2$ and the alternative hypothesis H_1: $\sigma_i^2 \neq \sigma_j^2$ for at least one pair of indices i and j, where $i, j = 1, 2, \ldots, k$.

(3) Set the level of significance at α and the quantile at $q = (1 - \alpha)$.

(4) Calculate the variance for each of the k groups.

(5) Identify the largest, S_{max}^2, and the smallest, S_{min}^2, of the k variances.

(6) Calculate the test statistic as the ratio of the largest variance to the smallest: $F_{max} = S_{max}^2 / S_{min}^2$.

(7) Find the q quantile, $f_{max,\,q}(k,\, df)$, of the test statistic associated with k groups and $df = n - 1$ in Table T-5.

(8) If F_{max} exceeds $f_{max,\,q}(k,\, df)$, reject the hypothesis that the variances are equal.

A Revisit to Example 10-3: Given a set of six sample variances, {0.40, 1.49, 3.20, 1.93, 2.35, 1.58}, each with 7 degrees of freedom, $S_{max}^2 = 3.20$ and $S_{min}^2 = 0.40$. Set $\alpha = 0.05$. The value of the test statistic is $f_{max} = 3.20/0.40 = 8.00$. Table T-5 provides the critical value $f_{max,\,0.95}(6,\, 7) = 10.8$. Since $f_{max} = 8.00$ is smaller than 10.8, the hypothesis of homoscedasticity is *not* rejected.

Table T-5: Quantiles, $f_{max,\,q}(k,\,df)$, for the F_{max} distribution*

$q = 1 - \alpha = 0.95$

	k = number of groups										
df	2	3	4	5	6	7	8	9	10	11	12
2	39.0	87.5	142	202	266	333	403	475	550	626	704
3	15.4	27.8	39.2	50.7	62.0	72.9	83.5	93.9	104	114	124
4	9.60	15.5	20.6	25.2	29.5	33.6	37.5	41.1	44.6	48.0	51.4
5	7.15	10.8	13.7	16.3	18.7	20.8	22.9	24.7	26.5	28.2	29.9
6	5.82	8.38	10.4	12.1	13.7	15.0	16.3	17.5	18.6	19.7	20.7
7	4.99	6.94	8.44	9.70	10.8	11.8	12.7	13.5	14.3	15.1	15.8
8	4.43	6.00	7.18	8.12	9.03	9.78	10.5	11.1	11.7	12.2	12.7
9	4.03	5.34	6.31	7.11	7.80	8.41	8.95	9.45	9.91	10.3	10.7
10	3.72	4.85	5.67	6.34	6.92	7.42	7.87	8.28	8.66	9.01	9.34
12	3.28	4.16	4.79	5.30	5.72	6.09	6.42	6.72	7.00	7.25	7.48
15	2.86	3.54	4.01	4.37	4.68	4.95	5.19	5.40	5.59	5.77	5.93
20	2.46	2.95	3.29	3.54	3.76	3.94	4.10	4.24	4.37	4.49	4.59
30	2.07	2.40	2.61	2.78	2.91	3.02	3.12	3.21	3.29	3.36	3.39
60	1.67	1.85	1.96	2.04	2.11	2.17	2.22	2.26	2.30	2.33	2.36
∞	1.00	1.00	1.00	1.00	1.00	1.00	1.00	1.00	1.00	1.00	1.00

$q = 1 - \alpha = 0.99$

	k = number of groups										
df	2	3	4	5	6	7	8	9	10	11	12
2	199	448	729	1,036	1,362	1,705	2,063	2,432	2,813	3,204	3,605
3	47.5	85	120	151	184	216	249	281	310	337	361
4	23.2	37	49	59	69	79	89	97	106	113	120
5	14.9	22	28	33	38	42	46	50	54	57	60
6	11.1	15.5	19.1	22	25	27	30	32	34	36	37
7	8.89	12.1	14.5	16.5	18.4	20	22	23	24	26	27
8	7.50	9.9	11.7	13.2	14.5	15.8	16.9	17.9	18.9	19.8	21
9	6.54	8.5	9.9	11.1	12.1	13.1	13.9	14.7	15.3	16.0	16.6
10	5.85	7.4	8.6	9.6	10.4	11.1	11.8	12.4	12.9	13.4	13.9
12	4.91	6.1	6.9	7.6	8.2	8.7	9.1	9.5	9.9	10.2	10.6
15	4.07	4.9	5.5	6.0	6.4	6.7	7.1	7.3	7.5	7.8	8.0
20	3.32	3.8	4.3	4.6	4.9	5.1	5.3	5.5	5.6	5.8	5.9
30	2.63	3.0	3.3	3.4	3.6	3.7	3.8	3.9	4.0	4.1	4.2
60	1.96	2.2	2.3	2.4	2.4	2.5	2.5	2.6	2.6	2.7	2.7
∞	1.00	1.0	1.0	1.0	1.0	1.0	1.0	1.0	1.0	1.0	1.0

* Adapted from David, H. A., 1952, Upper 5 and 1% Points of the Maximum F-Ratio, *Biometrika*, **39**, pp. 422-424, with permission of the Biometrika Trustees and Herbert A. David.

Table T-6a: Coefficients $\{a_{n-i+1}\}$ for the *W* test for normality

The *W* test for normality is based upon the ratio of two estimates of the variance of a population: one being the square of a linear combination of the ordered sample values and the other being the conventional sum-of-squared-deviations estimate. The test statistic, *W*, can range from 0 to 1. It is designed so that *small* values reflect evidence of non-normality; thus, the *W* test's critical values are the *smaller* quantiles of the statistic's distribution. The test can be applied to sample sizes as small as 3 and as large as 50 in terms of Tables T-6a and T-6b.

A Revisit to Example 7-1: The following data are the percent uranium value for 17 cans of ammonium diuranate (ADU) scrap:

35.5 79.4 35.2 40.1 25.0 78.5 78.2 37.1 48.4 28.6 75.5 34.3 29.4 29.8 28.4 23.4 77.0.

The *W* test is detailed here as an 11-step procedure.

Step 1. Arrange the *n* sample observations in *ascending* order. Using established convention, for a sample $\{y_1, y_2, \ldots, y_i, \ldots, y_n\}$ of size *n*, let $y_{(1)}$ denote the smallest observation, $y_{(2)}$ the second smallest observation, ..., $y_{(i)}$ the ith ordered observation, ..., and $y_{(n)}$ the largest observation. The ordered observations are given in the second column of the table headed **A Table for a Revisit to Example 7-1** on Page ST-19. In that table, the *rank* of an observation refers to its numeric-order assignment.

A Table for a Revisit to Example 7-1

Rank (i)	Ascending ordered data $y_{(i)}$	Descending order data $y_{(n-i+1)}$	Difference $y_{(n-i+1)} - y_{(i)}$	Table T-6a coefficients for $k = 8$ a_i	$a_i(y_{(n-i+1)} - y_{(i)})$
1	23.4	79.4	56.0	0.4968	27.8208
2	25.0	78.5	53.5	0.3273	17.5106
3	28.4	78.2	49.8	0.2540	12.6492
4	28.6	77.0	48.4	0.1988	9.6219
5	29.4	75.5	46.1	0.1524	7.0256
6	29.8	48.4	18.6	0.1109	2.0627
7	34.3	40.1	5.8	0.0725	0.4205
8	35.2	37.1	1.9	0.0359	0.0682
9	35.5				
10	37.1				
11	40.1				
12	48.4				
13	75.5				
14	77.0				
15	78.2				
16	78.5				
17	79.4				
					b = 77.1796

Step 2. If n is even, set $k = n/2$; if n is odd, set $k = (n-1)/2$. For this example, in which $n = 17$, set $k = 8$.

Step 3. Rearrange the observations in *descending* order of magnitude, and enter the first k of them as shown in the third column of **A Table for a Revisit to Example 7-1**.

Step 4. Calculate the differences between the corresponding entries of the third and the second columns of **A Table for a Revisit to Example 7-1** and enter those differences in the fourth column.

Step 5. From Table T-6a, copy the k coefficients $\{a_1, a_2, \ldots, a_i, \ldots, a_k\}$ associated with sample size n into the fifth column of **A Table for a Revisit to Example 7-1**.

Step 6. Multiply the associated elements of the fourth and the fifth columns of **A Table for a Revisit to Example 7-1**. Enter the corresponding products in the sixth column of the table.

Step 7. Sum the last column of **A Table for a Revisit to Example 7-1**. Denote the sum by B. In Table T-6a Example, $b = 77.1796$.

Step 8. Calculate S^2, the sample variance of the n observations. For the data in Table T-6a Example, $s^2 = 476.5968$.

Step 9. Calculate the test statistic, W, where

$$W = \frac{B^2}{(n - 1)S^2}.$$

In **A Table for a Revisit to Example 7-1**,

$$w = \frac{(77.1796)^2}{(16)(476.5968)} = 0.7811.$$

Turn to Page *ST*-24 for the final two steps of this example.

Table T-6a: Coefficients $\{a_{n-i+1}\}$ for the W test for normality*

$i \backslash n$	3	4	5	6	7	8	9	10
1	0.7071	0.6872	0.6646	0.6431	0.6233	0.6052	0.5888	0.5739
2		0.1677	0.2413	0.2806	0.3031	0.3164	0.3244	0.3291
3				0.0875	0.1401	0.1743	0.1976	0.2141
4						0.0561	0.0947	0.1224
5								0.0399

$i \backslash n$	11	12	13	14	15	16	17	18	19	20
1	0.5601	0.5475	0.5359	0.5251	0.5150	0.5056	0.4968	0.4886	0.4808	0.4734
2	0.3315	0.3325	0.3325	0.3318	0.3306	0.3290	0.3273	0.3253	0.3232	0.3211
3	0.2260	0.2347	0.2412	0.2460	0.2495	0.2521	0.2540	0.2553	0.2561	0.2565
4	0.1429	0.1586	0.1707	0.1802	0.1878	0.1939	0.1988	0.2027	0.2059	0.2085
5	0.0695	0.0922	0.1099	0.1240	0.1353	0.1447	0.1524	0.1587	0.1641	0.1686
6		0.0303	0.0539	0.0727	0.0880	0.1005	0.1109	0.1197	0.1271	0.1334
7				0.0240	0.0433	0.0593	0.0725	0.0837	0.0932	0.1013
8						0.0196	0.0359	0.0496	0.0612	0.0711
9								0.0163	0.0303	0.0422
10										0.0140

$i \backslash n$	21	22	23	24	25	26	27	28	29	30
1	0.4643	0.4590	0.4542	0.4493	0.4450	0.4407	0.4366	0.4328	0.4291	0.4254
2	0.3185	0.3156	0.3126	0.3098	0.3069	0.3043	0.3018	0.2992	0.2968	0.2944
3	0.2578	0.2571	0.2563	0.2554	0.2543	0.2533	0.2522	0.2510	0.2499	0.2487
4	0.2119	0.2131	0.2139	0.2145	0.2148	0.2151	0.2152	0.2151	0.2150	0.2148
5	0.1736	0.1764	0.1787	0.1807	0.1822	0.1836	0.1848	0.1857	0.1864	0.1870
6	0.1399	0.1443	0.1480	0.1512	0.1539	0.1563	0.1584	0.1601	0.1616	0.1630
7	0.1092	0.1150	0.1201	0.1245	0.1283	0.1316	0.1346	0.1372	0.1395	0.1415
8	0.0804	0.0878	0.0941	0.0997	0.1046	0.1089	0.1128	0.1162	0.1192	0.1219
9	0.0530	0.0618	0.0696	0.0764	0.0823	0.0876	0.0923	0.0965	0.1002	0.1036
10	0.0263	0.0368	0.0459	0.0539	0.0610	0.0672	0.0728	0.0778	0.0822	0.0862
11		0.0122	0.0228	0.0321	0.0403	0.0476	0.0540	0.0598	0.0650	0.0697
12				0.0107	0.0200	0.0284	0.0358	0.0424	0.0483	0.0537
13						0.0094	0.0178	0.0253	0.0320	0.0381
14								0.0084	0.0159	0.0227
15										0.0076

* Adapted from Shapiro, S. S., and M. B. Wilk, 1965, An Analysis of Variance Test for Normality (Complete Samples), *Biometrika*, **52**, pp. 591-611, with permission of the Biometrika Trustees and Samuel Shapiro.

Table T-6a: Coefficients $\{a_{n-i+1}\}$ for the W test for normality (continued)

i\n	31	32	33	34	35	36	37	38	39	40
1	0.4220	0.4188	0.4156	0.4127	0.4096	0.4068	0.4040	0.4015	0.3989	0.3964
2	0.2921	0.2898	0.2876	0.2854	0.2834	0.2813	0.2794	0.2774	0.2755	0.2737
3	0.2475	0.2463	0.2451	0.2439	0.2427	0.2415	0.2403	0.2391	0.2380	0.2368
4	0.2145	0.2141	0.2137	0.2132	0.2127	0.2121	0.2116	0.2110	0.2104	0.2098
5	0.1874	0.1878	0.1880	0.1882	0.1883	0.1883	0.1883	0.1881	0.1880	0.1878
6	0.1641	0.1651	0.1660	0.1667	0.1673	0.1678	0.1683	0.1686	0.1689	0.1691
7	0.1433	0.1449	0.1463	0.1475	0.1487	0.1496	0.1505	0.1513	0.1520	0.1526
8	0.1243	0.1265	0.1284	0.1301	0.1317	0.1331	0.1344	0.1356	0.1366	0.1376
9	0.1066	0.1093	0.1118	0.1140	0.1160	0.1179	0.1196	0.1211	0.1225	0.1237
10	0.0899	0.0931	0.0961	0.0988	0.1013	0.1036	0.1056	0.1075	0.1092	0.1108
11	0.0739	0.0777	0.0812	0.0844	0.0873	0.0900	0.0924	0.0947	0.0967	0.0986
12	0.0585	0.0629	0.0669	0.0706	0.0739	0.0770	0.0798	0.0824	0.0848	0.0870
13	0.0435	0.0485	0.0530	0.0572	0.0610	0.0645	0.0677	0.0706	0.0733	0.0759
14	0.0289	0.0344	0.0395	0.0441	0.0484	0.0523	0.0559	0.0592	0.0622	0.0651
15	0.0144	0.0206	0.0262	0.0314	0.0361	0.0404	0.0444	0.0481	0.0515	0.0546
16		0.0068	0.0131	0.0187	0.0239	0.0287	0.0331	0.0372	0.0409	0.0444
17				0.0062	0.0119	0.0172	0.0220	0.0264	0.0305	0.0343
18						0.0057	0.0110	0.0158	0.0203	0.0244
19								0.0053	0.0101	0.0146
20										0.0049

Table T-6a: Coefficients $\{a_{n-i+1}\}$ for the *W* test for normality (continued)

i\n	41	42	43	44	45	46	47	48	49	50
1	0.3940	0.3817	0.3894	0 .3872	0.3850	0.3830	0.3808	0.3789	0.3770	0.3751
2	0.2719	0.2701	0.2684	0.2667	0.2651	0.2635	0.2620	0.2604	0.2589	0.2574
3	0.2357	0.2345	0.2334	0.2323	0.2313	0.2302	0.2291	0.2281	0.2271	0.2260
4	0.2091	0.2085	0.2078	0.2072	0.2065	0.2058	0.2052	0.2045	0.2038	0.2032
5	0.1876	0.1874	0.1871	0.1868	0.1865	0.1862	0.1859	0.1855	0.1851	0.1847
6	0.1693	0.1694	0.1695	0.1695	0.1695	0.1695	0.1695	0.1693	0.1692	0.1691
7	0.1531	0.1535	0.1539	0.1542	0.1545	0.1548	0.1550	0.1551	0.1553	0.1554
8	0.1384	0.1392	0.1398	0.1405	0.1410	0.1415	0.1420	0.1423	0.1427	0.1430
9	0.1249	0.1259	0.1269	0.1278	0.1286	0.1293	0.1300	0.1306	0.1312	0.1317
10	0.1123	0.1136	0.1149	0.1160	0.1170	0.1180	0.1189	0.1197	0.1205	0.1212
11	0.1004	0.1020	0.1035	0.1049	0.1062	0.1073	0.1085	0.1095	0.1105	0.1113
12	0.0891	0.0909	0.0927	0.0943	0.0959	0.0972	0.0986	0.0998	0.1010	0.1020
13	0.0782	0.0804	0.0824	0.0842	0.0860	0.0876	0.0892	0.0906	0.0919	0.0932
14	0.0677	0.0701	0.0724	0.0745	0.0765	0.0783	0.0801	0.0817	0.0832	0.0846
15	0.0575	0.0602	0.0628	0.0651	0.0673	0.0694	0.0713	0.0731	0.0748	0.0764
16	0.0476	0.0506	0.0534	0.0560	0.0584	0.0607	0.0628	0.0648	0.0667	0.0685
17	0.0379	0.0411	0.0442	0.0471	0.0497	0.0522	0.0546	0.0568	0.0588	0.0608
18	0.0283	0.0318	0.0352	0.0383	0.0412	0.0439	0.0465	0.0489	0.0511	0.0532
19	0.0188	0.0227	0.0263	0.0296	0.0328	0.0357	0.0385	0.0411	0.0436	0.0459
20	0.0094	0.0136	0.0175	0.0211	0.0245	0.0277	0.0307	0.0335	0.0361	0.0386
21		0.0045	0.0087	0.0126	0.0163	0.0197	0.0229	0.0259	0.0288	0.0314
22				0.0042	0.0081	0.0118	0.0153	0.0185	0.0215	0.0244
23						0.0039	0.0076	0.0111	0.0143	0.0174
24								0.0037	0.0071	0.0104
25										0.0035

Table T-6b: Quantiles, $w_q(n)$, for the W test for normality

Table T-6b contains the quantiles, $w_q(n)$, which provide the critical values for the W test for normality. Because small values of the test statistic indicate non-normality, values of the test statistic that are *less than* the critical values indicate significance for the particular choice of the level of significance α.

Steps 10 and **11** complete the illustrative example started in the description of Table T-6a.

Step 10. From Table T-6b, obtain the critical point $w_\alpha(n)$ for the corresponding sample size and the appropriate level of significance. For $n = 17$ and $\alpha = 0.05$, $W_{0.05}(17) = 0.892$.

Step 11. Compare w from **Step 9** to w_α in **Step 10**. If w is *smaller* than $w_{0.05}$, the hypothesis about normality is rejected. In **A Revisit to Example 7-1**, $w = 0.7811$ is less than $w_{0.95}(17) = 0.892$; thus, these data yield sufficient evidence to reject normality for $\alpha = 0.05$.

Table T-6b: Quantiles, $w_q(n)$, for the *W* test for normality

n	$q = 0.01$	$q = 0.02$	$q = 0.05$	$q = 0.10$	$q = 0.50$
3	0.753	0.756	0.767	0.789	0.959
4	0.687	0.707	0.748	0.792	0.935
5	0.686	0.715	0.762	0.806	0.927
6	0.713	0.743	0.788	0.826	0.927
7	0.730	0.760	0.803	0.838	0.928
8	0.749	0.778	0.818	0.851	0.932
9	0.764	0.791	0.829	0.859	0.935
10	0.781	0.806	0.842	0.869	0.938
11	0.792	0.817	0.850	0.876	0.940
12	0.805	0.828	0.859	0.883	0.943
13	0.814	0.837	0.866	0.889	0.945
14	0.825	0.846	0.874	0.895	0.947
15	0.835	0.855	0.881	0.901	0.950
16	0.844	0.863	0.887	0.906	0.952
17	0.851	0.869	0.892	0.910	0.954
18	0.858	0.874	0.897	0.914	0.956
19	0.863	0.879	0.901	0.917	0.957
20	0.868	0.884	0.905	0.920	0.959
21	0.873	0.888	0.908	0.923	0.960
22	0.878	0.892	0.911	0.926	0.961
23	0.881	0.895	0.914	0.928	0.962
24	0.884	0.898	0.916	0.930	0.963
25	0.888	0.901	0.918	0.931	0.964
26	0.891	0.904	0.920	0.933	0.965
27	0.894	0.906	0.923	0.935	0.965
28	0.896	0.908	0.924	0.936	0.966
29	0.898	0.910	0.926	0.937	0.966
30	0.900	0.912	0.927	0.939	0.967
32	0.904	0.915	0.930	0.941	0.968
34	0.908	0.919	0.933	0.943	0.969
36	0.912	0.922	0.935	0.945	0.970
38	0.916	0.925	0.938	0.947	0.971
40	0.919	0.928	0.940	0.949	0.972
42	0.922	0.930	0.942	0.951	0.972
44	0.924	0.933	0.944	0.952	0.973
46	0.927	0.935	0.945	0.953	0.974
48	0.929	0.937	0.947	0.954	0.974
50	0.930	0.938	0.947	0.955	0.974

* Adapted from Shapiro, S. S., and M. B. Wilk, 1965, An Analysis of Variance Test for Normality (Complete Samples), *Biometrika*, **52**, pp. 591-611, with permission of the Biometrika Trustees and Samuel Shapiro.

Table T-7: Significant ranges, $q_{0.95}(p, N - k)$, for Duncan's multiple-range test

Duncan's multiple-range test (Duncan, 1955) is one of a class of procedures designed to determine which of a set of k means are different from the others— once an analysis of variance declares significance among those means.

A Revisit to Example 12-1c: This discussion gives both the symbolic algorithm for the procedure and an illustration using the analysis of variance results in Table 12-1b in which four production lines (called **Line 1**, **Line 2**, **Line 3**, and **Line 4**) are compared. To apply Duncan's multiple-range test, assemble and record several "building- block" values.

The experiment size: $N = 35$;

the $k = 4$ sample sizes: $n_1 = 3$, $n_2 = 11$, $n_3 = 15$, $n_4 = 6$;

the $k = 4$ means: $\bar{y}_1 = 87.37$, $\bar{y}_2 = 87.85$, $\bar{y}_3 = 88.57$, $\bar{y}_4 = 87.48$;

and the Within-groups Mean square (from Table 12-3b): $MS_W = 0.0528$.

Calculate the harmonic mean of the k sample sizes:

$$\begin{aligned}\bar{n}_h &= k/(1/n_1 + 1/n_2 + \dots 1/n_k)\\ &= 4/(1/3 + 1/11 + 1/15 + 1/6)\\ &= 6.0829.\end{aligned}$$

[Note that, if the sample sizes are all the same (i.e., $n_1 = n_2 = \dots = n_k = n$), then $\bar{n}_h = n$, say, and no special calculation is required.]

Calculate a *pseudo* standard deviation of each sample mean:

$$S^*_{\bar{y}_i} = \sqrt{\frac{MS_W}{\bar{n}_h}} = \sqrt{\frac{0.0528}{6.0829}} = 0.0932.$$

From Table T-7 with $\alpha = 0.05$ and $N - k = 35 - 4 = 31$, record the values of $q_{0.95}(p, N - k)$ for each $p = 2, 3, \dots, k$; for this example, after some interpolating, these are:

$q_{0.95}(2, 31) = 2.89$
$q_{0.95}(3, 31) = 3.04$
$q_{0.95}(4, 31) = 3.12.$

Calculate $R_p^* = q_{0.95}(p, N - k)S_{\bar{y}_i}^*$ for each $p = 2, 3, \ldots, k$. These are the critical values against which differences in means are compared; for this example, these are:

$$R_2^* = 2.89(0.0932) = 0.2693$$
$$R_3^* = 3.04(0.0932) = 0.2832$$
$$R_4^* = 3.12(0.0932) = 0.2907.$$

Arrange the sample means in ascending order; for this example:

$\bar{y}_1 = 87.37$
$\bar{y}_4 = 87.48$
$\bar{y}_2 = 87.85$
$\bar{y}_3 = 88.57.$

Denote the *rank* of the smallest mean by 1, the next smallest by 2, ⋯ , and the largest by k. Next, for any two means, let p denote the difference between the corresponding ranks plus 1. Thus, for comparing the largest and the smallest means among k means, $p = (k - 1) + 1 = k$.

Compare the *difference between the largest mean and each of the other means* to the value of the corresponding R_p^* , beginning with the *smallest*, and determine which of the differences are statistically significant; for this example:

$$\bar{y}_3 - \bar{y}_1 = 1.20 > R_4^* = 0.2907$$
$$\bar{y}_3 - \bar{y}_4 = 1.09 > R_3^* = 0.2832$$
$$\bar{y}_3 - \bar{y}_2 = 0.72 > R_2^* = 0.2693.$$

Here, all three differences are significantly different. Conclude that **Line 3**'s mean is larger than any of the means of **Line 1**, **Line 2**, and **Line 4**.

Next, compare the *second largest* mean with each of the *smaller* means, beginning with the smallest, and determine which of the differences are statistically significant; for this example:

$$\bar{y}_2 - \bar{y}_1 = 0.48 > R_3{}^* = 0.2832$$
$$\bar{y}_2 - \bar{y}_4 = 0.37 > R_2{}^* = 0.2693.$$

Here, both differences are significantly different. Conclude that **Line 2**'s mean is larger than either **Line 1**'s or **Line 4**'s.

Continue in this fashion until the two smallest means are compared; for this example, compare the *third largest* mean with the *smallest* mean and determine if the difference is statistically significant:

$$\bar{y}_1 - \bar{y}_4 = 0.11 < R_2{}^* = 0.2693.$$

Here, the difference is *not* significantly different. Conclude that **Line 1**'s mean is *not* larger than that of **Line 4**.

A graphical display, described by Duncan (1974, p. 704), among others, is useful for describing multiple-range test results. Write the means in an increasing sequence, left-to-right, and underline those collections of means that are *not declared to be significant* by the multiple-range test; for this example, the display

Group:	**Line 1**	**Line 4**	**Line 2**	**Line 3**
Mean:	**87.37**	**87.48**	**87.85**	**88.57**

indicates that only the means of **Line 4** and **Line 1** are *not statistically significantly different*; i.e., they form a group, as do each of **Line 2** and **Line 3** individually.

Table T-7: Significant ranges, $q_{0.95}(p, N - k)$, for Duncan's multiple-range test

	p											
N - k	2	3	4	5	6	7	8	9	10	20	50	100
1	18.0	18.0	18.0	18.0	18.0	18.0	18.0	18.0	18.0	18.0	18.0	18.0
2	6.09	6.09	6.09	6.09	6.09	6.09	6.09	6.09	6.09	6.09	6.09	6.09
3	4.50	4.50	4.50	4.50	4.50	4.50	4.50	4.50	4.50	4.50	4.50	4.50
4	3.93	4.01	4.02	4.02	4.02	4.02	4.02	4.02	4.02	4.02	4.02	4.02
5	3.64	3.74	3.79	3.83	3.83	3.83	3.83	3.83	3.83	3.83	3.83	3.83
6	3.46	3.58	3.64	3.68	3.68	3.68	3.68	3.68	3.68	3.68	3.68	3.68
7	3.35	3.47	3.54	3.58	3.60	3.61	3.61	3.61	3.61	3.61	3.61	3.61
8	3.26	3.39	3.47	3.52	3.55	3.56	3.56	3.56	3.56	3.56	3.56	3.56
9	3.20	3.34	3.41	3.47	3.50	3.52	3.52	3.52	3.52	3.52	3.52	3.52
10	3.15	3.30	3.37	3.43	3.46	3.47	3.47	3.47	3.47	3.48	3.48	3.48
11	3.11	3.27	3.35	3.39	3.43	3.44	3.45	3.46	3.46	3.48	3.48	3.48
12	3.08	3.23	3.33	3.36	3.40	3.42	3.44	3.44	3.46	3.48	3.48	3.48
13	3.06	3.21	3.30	3.35	3.38	3.41	3.42	3.44	3.45	3.47	3.47	3.47
14	3.03	3.18	3.27	3.33	3.37	3.39	3.41	3.42	3.44	3.47	3.47	3.47
15	3.01	3.16	3.25	3.31	3.36	3.38	3.40	3.42	3.43	3.47	3.47	3.47
16	3.00	3.15	3.23	3.30	3.34	3.37	3.39	3.41	3.43	3.47	3.47	3.47
17	2.98	3.13	3.22	3.28	3.33	3.36	3.38	3.40	3.42	3.47	3.47	3.47
18	2.97	3.12	3.21	3.27	3.32	3.35	3.37	3.39	3.41	3.47	3.47	3.47
19	2.96	3.11	3.19	3.26	3.31	3.35	3.37	3.39	3.41	3.47	3.47	3.47
20	2.95	3.10	3.18	3.25	3.30	3.34	3.36	3.38	3.40	3.47	3.47	3.47
30	2.89	3.04	3.12	3.20	3.25	3.29	3.32	3.35	3.37	3.47	3.47	3.47
40	2.86	3.01	3.10	3.17	3.22	3.27	3.30	3.33	3.35	3.47	3.47	3.47
60	2.83	2.98	3.08	3.14	3.20	3.24	3.28	3.31	3.33	3.47	3.48	3.48
100	2.80	2.95	3.05	3.12	3.18	3.22	3.26	3.29	3.32	3.47	3.53	3.53
∞	2.77	2.92	3.02	3.09	3.15	3.19	3.23	3.26	3.29	3.47	3.61	3.67

Table T-8: Selected binomial probabilities

A binomial variable, B, has a density function given by

$$Pr\{B = b; n, \pi\} = \binom{n}{b} \pi^b (1 - \pi)^{n-b} = \frac{n!}{b!(n-b)!} \pi^b (1 - \pi)^{n-b}$$

$b = 1, 2, \ldots, n; \quad n = 1, 2, 3, \ldots; 0 \leq \pi \leq 1.$

Its two parameters are n and π.

Table T-8 gives values of the binomial density function for $b = 0, 1, \ldots, 9$ when $n = 1, 2, \ldots, 30, 32, 34, 36, 38, 40$ and $\pi = 0.01, 0.05, 0.10, 0.25$, and 0.50.

Example 1: The probability that $B = 3$ when $n = 7$ and $\pi = 0.25$ is found on page ST-34 at the intersection of the row labeled **7** and the column labeled **3**. The entry there is 0.1730. Formally: $Pr\{B = 3; 7, 0.25\} = 0.1730$.

Example 2: The probability that $B \leq 2$ when $n = 14$ and $\pi = 0.25$ can be found by summing three values from page ST-34. In terms of Chapter 17's notational shorthand, $Pr\{B \leq 2\} = Pr\{B = 0\} + Pr\{B = 1\} + Pr\{B = 2\}$ $= 0.0178 + 0.0832 + 0.1802 = 0.2812$.

Example 1: The probability that $B \geq 2$ when $n = 14$ and $\pi = 0.25$ can be found by recalling the identity that $Pr\{B \geq b; n, \pi\} = 1 - Pr\{B < b; n, \pi\}$. In terms of Chapter 17's notational shorthand, $Pr\{B \geq 2\} = 1 - Pr\{B < 2\}$ $= 1 - [Pr\{B = 0\} + Pr\{B = 1\}] = 1 - [0.0178 + 0.0832]$ $= 1 - 0.1010 = 0.8990$.

Table T-8: Selected binomial probabilities*

	b = number of items of interest in a binomial sample of size n with π = 0.01									
n	0	1	2	3	4	5	6	7	8	9
1	0.9900	0.0100								
2	0.9801	0.0198	0.0001							
3	0.9703	0.0294	0.0003	0.0000						
4	0.9606	0.0388	0.0006	0.0000	0.0000					
5	0.9510	0.0480	0.0010	0.0000	0.0000	0.0000				
6	0.9415	0.0571	0.0014	0.0000	0.0000	0.0000	0.0000			
7	0.9321	0.0659	0.0020	0.0000	0.0000	0.0000	0.0000	0.0000		
8	0.9227	0.0746	0.0026	0.0001	0.0000	0.0000	0.0000	0.0000	0.0000	
9	0.9135	0.0830	0.0034	0.0001	0.0000	0.0000	0.0000	0.0000	0.0000	0.0000
10	0.9044	0.0914	0.0042	0.0001	0.0000	0.0000	0.0000	0.0000	0.0000	0.0000
11	0.8953	0.0995	0.0050	0.0002	0.0000	0.0000	0.0000	0.0000	0.0000	0.0000
12	0.8864	0.1074	0.0060	0.0002	0.0000	0.0000	0.0000	0.0000	0.0000	0.0000
13	0.8775	0.1152	0.0070	0.0003	0.0000	0.0000	0.0000	0.0000	0.0000	0.0000
14	0.8687	0.1229	0.0081	0.0003	0.0000	0.0000	0.0000	0.0000	0.0000	0.0000
15	0.8601	0.1303	0.0092	0.0004	0.0000	0.0000	0.0000	0.0000	0.0000	0.0000
16	0.8515	0.1376	0.0104	0.0005	0.0000	0.0000	0.0000	0.0000	0.0000	0.0000
17	0.8429	0.1447	0.0117	0.0006	0.0000	0.0000	0.0000	0.0000	0.0000	0.0000
18	0.8345	0.1517	0.0130	0.0007	0.0000	0.0000	0.0000	0.0000	0.0000	0.0000
19	0.8262	0.1586	0.0144	0.0008	0.0000	0.0000	0.0000	0.0000	0.0000	0.0000
20	0.8179	0.1652	0.0159	0.0010	0.0000	0.0000	0.0000	0.0000	0.0000	0.0000
21	0.8097	0.1718	0.0173	0.0011	0.0001	0.0000	0.0000	0.0000	0.0000	0.0000
22	0.8016	0.1781	0.0189	0.0013	0.0001	0.0000	0.0000	0.0000	0.0000	0.0000
23	0.7936	0.1844	0.0205	0.0014	0.0001	0.0000	0.0000	0.0000	0.0000	0.0000
24	0.7857	0.1905	0.0221	0.0016	0.0001	0.0000	0.0000	0.0000	0.0000	0.0000
25	0.7778	0.1964	0.0238	0.0018	0.0001	0.0000	0.0000	0.0000	0.0000	0.0000
26	0.7700	0.2022	0.0255	0.0021	0.0001	0.0000	0.0000	0.0000	0.0000	0.0000
27	0.7623	0.2079	0.0273	0.0023	0.0001	0.0000	0.0000	0.0000	0.0000	0.0000
28	0.7547	0.2135	0.0291	0.0025	0.0002	0.0000	0.0000	0.0000	0.0000	0.0000
29	0.7472	0.2189	0.0310	0.0028	0.0002	0.0000	0.0000	0.0000	0.0000	0.0000
30	0.7397	0.2242	0.0328	0.0031	0.0002	0.0000	0.0000	0.0000	0.0000	0.0000
32	0.7250	0.2343	0.0367	0.0037	0.0003	0.0000	0.0000	0.0000	0.0000	0.0000
34	0.7106	0.2440	0.0407	0.0044	0.0003	0.0000	0.0000	0.0000	0.0000	0.0000
36	0.6964	0.2532	0.0448	0.0051	0.0004	0.0000	0.0000	0.0000	0.0000	0.0000
38	0.6826	0.2620	0.0490	0.0059	0.0005	0.0000	0.0000	0.0000	0.0000	0.0000
40	0.6690	0.2703	0.0532	0.0068	0.0006	0.0000	0.0000	0.0000	0.0000	0.0000

* Prepared by the authors.

Table T-8: Selected binomial probabilities (continued)

b = number of items of interest in a binomial sample of size n with $\pi = 0.05$

n	0	1	2	3	4	5	6	7	8	9
1	0.9500	0.0500								
2	0.9025	0.0950	0.0025							
3	0.8574	0.1354	0.0071	0.0001						
4	0.8145	0.1715	0.0135	0.0005	0.0000					
5	0.7738	0.2036	0.0214	0.0011	0.0000	0.0000				
6	0.7351	0.2321	0.0305	0.0021	0.0001	0.0000	0.0000			
7	0.6983	0.2573	0.0406	0.0036	0.0002	0.0000	0.0000	0.0000		
8	0.6634	0.2793	0.0515	0.0054	0.0004	0.0000	0.0000	0.0000	0.0000	
9	0.6302	0.2985	0.0629	0.0077	0.0006	0.0000	0.0000	0.0000	0.0000	0.0000
10	0.5987	0.3151	0.0746	0.0105	0.0010	0.0001	0.0000	0.0000	0.0000	0.0000
11	0.5688	0.3293	0.0867	0.0137	0.0014	0.0001	0.0000	0.0000	0.0000	0.0000
12	0.5404	0.3413	0.0988	0.0173	0.0021	0.0002	0.0000	0.0000	0.0000	0.0000
13	0.5133	0.3512	0.1109	0.0214	0.0028	0.0003	0.0000	0.0000	0.0000	0.0000
14	0.4877	0.3593	0.1229	0.0259	0.0037	0.0004	0.0000	0.0000	0.0000	0.0000
15	0.4633	0.3658	0.1348	0.0307	0.0049	0.0006	0.0000	0.0000	0.0000	0.0000
16	0.4401	0.3706	0.1463	0.0359	0.0061	0.0008	0.0001	0.0000	0.0000	0.0000
17	0.4181	0.3741	0.1575	0.0415	0.0076	0.0010	0.0001	0.0000	0.0000	0.0000
18	0.3972	0.3763	0.1683	0.0473	0.0093	0.0014	0.0002	0.0000	0.0000	0.0000
19	0.3774	0.3774	0.1787	0.0533	0.0112	0.0018	0.0002	0.0000	0.0000	0.0000
20	0.3585	0.3774	0.1887	0.0596	0.0133	0.0022	0.0003	0.0000	0.0000	0.0000
21	0.3406	0.3764	0.1981	0.0660	0.0156	0.0028	0.0004	0.0000	0.0000	0.0000
22	0.3235	0.3746	0.2070	0.0726	0.0182	0.0034	0.0005	0.0001	0.0000	0.0000
23	0.3074	0.3721	0.2154	0.0794	0.0209	0.0042	0.0007	0.0001	0.0000	0.0000
24	0.2920	0.3688	0.2232	0.0862	0.0238	0.0050	0.0008	0.0001	0.0000	0.0000
25	0.2774	0.3650	0.2305	0.0930	0.0269	0.0060	0.0010	0.0001	0.0000	0.0000
26	0.2635	0.3606	0.2372	0.0999	0.0302	0.0070	0.0013	0.0002	0.0000	0.0000
27	0.2503	0.3558	0.2434	0.1068	0.0337	0.0082	0.0016	0.0002	0.0000	0.0000
28	0.2378	0.3505	0.2490	0.1136	0.0374	0.0094	0.0019	0.0003	0.0000	0.0000
29	0.2259	0.3448	0.2541	0.1204	0.0412	0.0108	0.0023	0.0004	0.0001	0.0000
30	0.2146	0.3389	0.2586	0.1270	0.0451	0.0124	0.0027	0.0005	0.0001	0.0000
32	0.1937	0.3263	0.2662	0.1401	0.0535	0.0158	0.0037	0.0007	0.0001	0.0000
34	0.1748	0.3128	0.2717	0.1525	0.0622	0.0196	0.0050	0.0011	0.0002	0.0000
36	0.1578	0.2990	0.2753	0.1642	0.0713	0.0240	0.0065	0.0015	0.0003	0.0000
38	0.1424	0.2848	0.2773	0.1751	0.0807	0.0289	0.0084	0.0020	0.0004	0.0001
40	0.1285	0.2706	0.2777	0.1851	0.0901	0.0342	0.0105	0.0027	0.0006	0.0001

Table T-8: Selected binomial probabilities (continued)

	b = number of items of interest in a binomial sample of size n with π = 0.10									
n	**0**	**1**	**2**	**3**	**4**	**5**	**6**	**7**	**8**	**9**
1	0.9000	0.1000								
2	0.8100	0.1800	0.0100							
3	0.7290	0.2430	0.0270	0.0010						
4	0.6561	0.2916	0.0486	0.0036	0.0001					
5	0.5905	0.3281	0.0729	0.0081	0.0005	0.0000				
6	0.5314	0.3543	0.0984	0.0146	0.0012	0.0001	0.0000			
7	0.4783	0.3720	0.1240	0.0230	0.0026	0.0002	0.0000	0.0000		
8	0.4305	0.3826	0.1488	0.0331	0.0046	0.0004	0.0000	0.0000	0.0000	
9	0.3874	0.3874	0.1722	0.0446	0.0074	0.0008	0.0001	0.0000	0.0000	0.0000
10	0.3487	0.3874	0.1937	0.0574	0.0112	0.0015	0.0001	0.0000	0.0000	0.0000
11	0.3138	0.3835	0.2131	0.0710	0.0158	0.0025	0.0003	0.0000	0.0000	0.0000
12	0.2824	0.3766	0.2301	0.0852	0.0213	0.0038	0.0005	0.0000	0.0000	0.0000
13	0.2542	0.3672	0.2448	0.0997	0.0277	0.0055	0.0008	0.0001	0.0000	0.0000
14	0.2288	0.3559	0.2570	0.1142	0.0349	0.0078	0.0013	0.0002	0.0000	0.0000
15	0.2059	0.3432	0.2669	0.1285	0.0428	0.0105	0.0019	0.0003	0.0000	0.0000
16	0.1853	0.3294	0.2745	0.1423	0.0514	0.0137	0.0028	0.0004	0.0001	0.0000
17	0.1668	0.3150	0.2800	0.1556	0.0605	0.0175	0.0039	0.0007	0.0001	0.0000
18	0.1501	0.3002	0.2835	0.1680	0.0700	0.0218	0.0052	0.0010	0.0002	0.0000
19	0.1351	0.2852	0.2852	0.1796	0.0798	0.0266	0.0069	0.0014	0.0002	0.0000
20	0.1216	0.2702	0.2852	0.1901	0.0898	0.0319	0.0089	0.0020	0.0004	0.0001
21	0.1094	0.2553	0.2837	0.1996	0.0998	0.0377	0.0112	0.0027	0.0005	0.0001
22	0.0985	0.2407	0.2808	0.2080	0.1098	0.0439	0.0138	0.0035	0.0007	0.0001
23	0.0886	0.2265	0.2768	0.2153	0.1196	0.0505	0.0168	0.0045	0.0010	0.0002
24	0.0798	0.2127	0.2718	0.2215	0.1292	0.0574	0.0202	0.0058	0.0014	0.0003
25	0.0718	0.1994	0.2659	0.2265	0.1384	0.0646	0.0239	0.0072	0.0018	0.0004
26	0.0646	0.1867	0.2592	0.2304	0.1472	0.0720	0.0280	0.0089	0.0023	0.0005
27	0.0581	0.1744	0.2520	0.2333	0.1555	0.0795	0.0324	0.0108	0.0030	0.0007
28	0.0523	0.1628	0.2442	0.2352	0.1633	0.0871	0.0371	0.0130	0.0038	0.0009
29	0.0471	0.1518	0.2361	0.2361	0.1705	0.0947	0.0421	0.0154	0.0047	0.0012
30	0.0424	0.1413	0.2277	0.2361	0.1771	0.1023	0.0474	0.0180	0.0058	0.0016
32	0.0343	0.1221	0.2103	0.2336	0.1882	0.1171	0.0585	0.0242	0.0084	0.0025
34	0.0278	0.1051	0.1926	0.2283	0.1966	0.1311	0.0704	0.0313	0.0117	0.0038
36	0.0225	0.0901	0.1752	0.2206	0.2023	0.1438	0.0826	0.0393	0.0158	0.0055
38	0.0182	0.0770	0.1584	0.2112	0.2053	0.1551	0.0948	0.0481	0.0207	0.0077
40	0.0148	0.0657	0.1423	0.2003	0.2059	0.1647	0.1068	0.0576	0.0264	0.0104

Table T-8: Selected binomial probabilities (continued)

	b = number of items of interest in a binomial sample of size n with $\pi = 0.25$									
n	**0**	**1**	**2**	**3**	**4**	**5**	**6**	**7**	**8**	**9**
1	0.7500	0.2500								
2	0.5625	0.3750	0.0625							
3	0.4219	0.4219	0.1406	0.0156						
4	0.3164	0.4219	0.2109	0.0469	0.0039					
5	0.2373	0.3955	0.2637	0.0879	0.0146	0.0010				
6	0.1780	0.3560	0.2966	0.1318	0.0330	0.0044	0.0002			
7	0.1335	0.3115	0.3115	0.1730	0.0577	0.0115	0.0013	0.0001		
8	0.1001	0.2670	0.3115	0.2076	0.0865	0.0231	0.0038	0.0004	0.0000	
9	0.0751	0.2253	0.3003	0.2336	0.1168	0.0389	0.0087	0.0012	0.0001	0.0000
10	0.0563	0.1877	0.2816	0.2503	0.1460	0.0584	0.0162	0.0031	0.0004	0.0000
11	0.0422	0.1549	0.2581	0.2581	0.1721	0.0803	0.0268	0.0064	0.0011	0.0001
12	0.0317	0.1267	0.2323	0.2581	0.1936	0.1032	0.0401	0.0115	0.0024	0.0004
13	0.0238	0.1029	0.2059	0.2517	0.2097	0.1258	0.0559	0.0186	0.0047	0.0009
14	0.0178	0.0832	0.1802	0.2402	0.2202	0.1468	0.0734	0.0280	0.0082	0.0018
15	0.0134	0.0668	0.1559	0.2252	0.2252	0.1651	0.0917	0.0393	0.0131	0.0034
16	0.0100	0.0535	0.1336	0.2079	0.2252	0.1802	0.1101	0.0524	0.0197	0.0058
17	0.0075	0.0426	0.1136	0.1893	0.2209	0.1914	0.1276	0.0668	0.0279	0.0093
18	0.0056	0.0338	0.0958	0.1704	0.2130	0.1988	0.1436	0.0820	0.0376	0.0139
19	0.0042	0.0268	0.0803	0.1517	0.2023	0.2023	0.1574	0.0974	0.0487	0.0198
20	0.0032	0.0211	0.0669	0.1339	0.1897	0.2023	0.1686	0.1124	0.0609	0.0271
21	0.0024	0.0166	0.0555	0.1172	0.1757	0.1992	0.1770	0.1265	0.0738	0.0355
22	0.0018	0.0131	0.0458	0.1017	0.1611	0.1933	0.1826	0.1391	0.0869	0.0451
23	0.0013	0.0103	0.0376	0.0878	0.1463	0.1853	0.1853	0.1500	0.1000	0.0555
24	0.0010	0.0080	0.0308	0.0752	0.1316	0.1755	0.1853	0.1588	0.1125	0.0667
25	0.0008	0.0063	0.0251	0.0641	0.1175	0.1645	0.1828	0.1654	0.1241	0.0781
26	0.0006	0.0049	0.0204	0.0544	0.1042	0.1528	0.1782	0.1698	0.1344	0.0896
27	0.0004	0.0038	0.0165	0.0459	0.0917	0.1406	0.1719	0.1719	0.1432	0.1008
28	0.0003	0.0030	0.0133	0.0385	0.0803	0.1284	0.1641	0.1719	0.1504	0.1114
29	0.0002	0.0023	0.0107	0.0322	0.0698	0.1164	0.1552	0.1699	0.1558	0.1212
30	0.0002	0.0018	0.0086	0.0269	0.0604	0.1047	0.1455	0.1662	0.1593	0.1298
32	0.0001	0.0011	0.0055	0.0185	0.0446	0.0832	0.1249	0.1546	0.1610	0.1431
34	0.0001	0.0006	0.0035	0.0125	0.0324	0.0647	0.1042	0.1390	0.1564	0.1506
36	0.0000	0.0004	0.0022	0.0084	0.0231	0.0493	0.0849	0.1213	0.1466	0.1520
38	0.0000	0.0002	0.0014	0.0056	0.0163	0.0369	0.0677	0.1032	0.1333	0.1481
40	0.0000	0.0001	0.0009	0.0037	0.0113	0.0272	0.0530	0.0857	0.1179	0.1397

Table T-8: Selected binomial probabilities (continued)

	b = number of items of interest in a binomial sample of size n with $\pi = 0.50$									
n	**0**	**1**	**2**	**3**	**4**	**5**	**6**	**7**	**8**	**9**
1	0.5000	0.5000								
2	0.2500	0.5000	0.2500							
3	0.1250	0.3750	0.3750	0.1250						
4	0.0625	0.2500	0.3750	0.2500	0.0625					
5	0.0313	0.1563	0.3125	0.3125	0.1563	0.0313				
6	0.0156	0.0938	0.2344	0.3125	0.2344	0.0938	0.0156			
7	0.0078	0.0547	0.1641	0.2734	0.2734	0.1641	0.0547	0.0078		
8	0.0039	0.0313	0.1094	0.2188	0.2734	0.2188	0.1094	0.0313	0.0039	
9	0.0020	0.0176	0.0703	0.1641	0.2461	0.2461	0.1641	0.0703	0.0176	0.0020
10	0.0010	0.0098	0.0439	0.1172	0.2051	0.2461	0.2051	0.1172	0.0439	0.0098
11	0.0005	0.0054	0.0269	0.0806	0.1611	0.2256	0.2256	0.1611	0.0806	0.0269
12	0.0002	0.0029	0.0161	0.0537	0.1208	0.1934	0.2256	0.1934	0.1208	0.0537
13	0.0001	0.0016	0.0095	0.0349	0.0873	0.1571	0.2095	0.2095	0.1571	0.0873
14	0.0001	0.0009	0.0056	0.0222	0.0611	0.1222	0.1833	0.2095	0.1833	0.1222
15	0.0000	0.0005	0.0032	0.0139	0.0417	0.0916	0.1527	0.1964	0.1964	0.1527
16	0.0000	0.0002	0.0018	0.0085	0.0278	0.0667	0.1222	0.1746	0.1964	0.1746
17	0.0000	0.0001	0.0010	0.0052	0.0182	0.0472	0.0944	0.1484	0.1855	0.1855
18	0.0000	0.0001	0.0006	0.0031	0.0117	0.0327	0.0708	0.1214	0.1669	0.1855
19	0.0000	0.0000	0.0003	0.0018	0.0074	0.0222	0.0518	0.0961	0.1442	0.1762
20	0.0000	0.0000	0.0002	0.0011	0.0046	0.0148	0.0370	0.0739	0.1201	0.1602
21	0.0000	0.0000	0.0001	0.0006	0.0029	0.0097	0.0259	0.0554	0.0970	0.1402
22	0.0000	0.0000	0.0001	0.0004	0.0017	0.0063	0.0178	0.0407	0.0762	0.1186
23	0.0000	0.0000	0.0000	0.0002	0.0011	0.0040	0.0120	0.0292	0.0584	0.0974
24	0.0000	0.0000	0.0000	0.0001	0.0006	0.0025	0.0080	0.0206	0.0438	0.0779
25	0.0000	0.0000	0.0000	0.0001	0.0004	0.0016	0.0053	0.0143	0.0322	0.0609
26	0.0000	0.0000	0.0000	0.0000	0.0002	0.0010	0.0034	0.0098	0.0233	0.0466
27	0.0000	0.0000	0.0000	0.0000	0.0001	0.0006	0.0022	0.0066	0.0165	0.0349
28	0.0000	0.0000	0.0000	0.0000	0.0001	0.0004	0.0014	0.0044	0.0116	0.0257
29	0.0000	0.0000	0.0000	0.0000	0.0000	0.0002	0.0009	0.0029	0.0080	0.0187
30	0.0000	0.0000	0.0000	0.0000	0.0000	0.0001	0.0006	0.0019	0.0055	0.0133
32	0.0000	0.0000	0.0000	0.0000	0.0000	0.0000	0.0002	0.0008	0.0024	0.0065
34	0.0000	0.0000	0.0000	0.0000	0.0000	0.0000	0.0001	0.0003	0.0011	0.0031
36	0.0000	0.0000	0.0000	0.0000	0.0000	0.0000	0.0000	0.0001	0.0004	0.0014
38	0.0000	0.0000	0.0000	0.0000	0.0000	0.0000	0.0000	0.0000	0.0002	0.0006
40	0.0000	0.0000	0.0000	0.0000	0.0000	0.0000	0.0000	0.0000	0.0001	0.0002

Table T-9: Selected Poisson probabilities

A Poisson variable, P, has a density function given by

$$Pr\{P = p\} = \frac{e^{-\lambda}\lambda^{p}}{p!} \; ; \quad \lambda > 0 \; ; \; p = 0, 1, 2, \ldots .$$

Its one parameter is λ.

Table T-9 gives values of the Poisson density function for $p = 0, 1, 2, \ldots$ and $\lambda = 0.1$ to 10 in steps of 0.1 and $\lambda = 11$ to 20 in steps of 1—for values which are greater than zero when rounded to four decimal places.

Example 1: The probability that $P = 2$ when $\lambda = 2.3$ is found on page ST-37 at the intersection of the row labeled **2** and the column labeled $\boldsymbol{\lambda = 2.3}$. The entry there is 0.2652. Formally, this is $Pr\{P = 2; 2.3\} = 0.2652$.

Example 2: The probability that $P \leq 1$ when $\lambda = 2.3$ can be found by summing two values from page ST-37. In terms of Chapter 18's notational shorthand, $Pr\{P \leq 1\} = Pr\{P = 0\} + Pr\{P = 1\} = 0.1003 + 0.2306$ $= 0.3309$.

Example 3: The probability that $P \geq 2$ when $\lambda = 2.3$ can be found by recalling the identity that $Pr\{P \geq p; \lambda\} = 1 - Pr\{P < p; \lambda\}$. In terms of Chapter 17's notational shorthand, $Pr\{P \geq 2\} = 1 - Pr\{P < 2\}$
$= 1 - [Pr\{P = 0\} + Pr\{P = 1\}] = 1 - [0.1003 + 0.2306]$
$= 1 - 0.3309 = 0.6691$.

Table T-9: Selected Poisson probabilities*

	λ									
p	**0.1**	**0.2**	**0.3**	**0.4**	**0.5**	**0.6**	**0.7**	**0.8**	**0.9**	**1.0**
0	0.9048	0.8187	0.7408	0.6703	0.6065	0.5488	0.4966	0.4493	0.4066	0.3679
1	0.0905	0.1637	0.2222	0.2681	0.3033	0.3293	0.3476	0.3595	0.3659	0.3679
2	0.0045	0.0164	0.0333	0.0536	0.0758	0.0988	0.1217	0.1438	0.1647	0.1839
3	0.0002	0.0011	0.0033	0.0072	0.0126	0.0198	0.0284	0.0383	0.0494	0.0613
4	0.0000	0.0001	0.0003	0.0007	0.0016	0.0030	0.0050	0.0077	0.0111	0.0153
5	0.0000	0.0000	0.0000	0.0001	0.0002	0.0004	0.0007	0.0012	0.0020	0.0031
6	0.0000	0.0000	0.0000	0.0000	0.0000	0.0000	0.0001	0.0002	0.0003	0.0005
7	0.0000	0.0000	0.0000	0.0000	0.0000	0.0000	0.0000	0.0000	0.0000	0.0001

	λ									
p	**1.1**	**1.2**	**1.3**	**1.4**	**1.5**	**1.6**	**1.7**	**1.8**	**1.9**	**2.0**
0	0.3329	0.3012	0.2725	0.2466	0.2231	0.2019	0.1827	0.1653	0.1496	0.1353
1	0.3662	0.3614	0.3543	0.3452	0.3347	0.3230	0.3106	0.2975	0.2842	0.2707
2	0.2014	0.2169	0.2303	0.2417	0.2510	0.2584	0.2640	0.2678	0.2700	0.2707
3	0.0738	0.0867	0.0998	0.1128	0.1255	0.1378	0.1496	0.1607	0.1710	0.1804
4	0.0203	0.0260	0.0324	0.0395	0.0471	0.0551	0.0636	0.0723	0.0812	0.0902
5	0.0045	0.0062	0.0084	0.0111	0.0141	0.0176	0.0216	0.0260	0.0309	0.0361
6	0.0008	0.0012	0.0018	0.0026	0.0035	0.0047	0.0061	0.0078	0.0098	0.0120
7	0.0001	0.0002	0.0003	0.0005	0.0008	0.0011	0.0015	0.0020	0.0027	0.0034
8	0.0000	0.0000	0.0001	0.0001	0.0001	0.0002	0.0003	0.0005	0.0006	0.0009
9	0.0000	0.0000	0.0000	0.0000	0.0000	0.0000	0.0001	0.0001	0.0001	0.0002

	λ									
p	**2.1**	**2.2**	**2.3**	**2.4**	**2.5**	**2.6**	**2.7**	**2.8**	**2.9**	**3.0**
0	0.1225	0.1108	0.1003	0.0907	0.0821	0.0743	0.0672	0.0608	0.0550	0.0498
1	0.2572	0.2438	0.2306	0.2177	0.2052	0.1931	0.1815	0.1703	0.1596	0.1494
2	0.2700	0.2681	0.2652	0.2613	0.2565	0.2510	0.2450	0.2384	0.2314	0.2240
3	0.1890	0.1966	0.2033	0.2090	0.2138	0.2176	0.2205	0.2225	0.2237	0.2240
4	0.0992	0.1082	0.1169	0.1254	0.1336	0.1414	0.1488	0.1557	0.1622	0.1680
5	0.0417	0.0476	0.0538	0.0602	0.0668	0.0735	0.0804	0.0872	0.0940	0.1008
6	0.0146	0.0174	0.0206	0.0241	0.0278	0.0319	0.0362	0.0407	0.0455	0.0504
7	0.0044	0.0055	0.0068	0.0083	0.0099	0.0118	0.0139	0.0163	0.0188	0.0216
8	0.0011	0.0015	0.0019	0.0025	0.0031	0.0038	0.0047	0.0057	0.0068	0.0081
9	0.0003	0.0004	0.0005	0.0007	0.0009	0.0011	0.0014	0.0018	0.0022	0.0027
10	0.0001	0.0001	0.0001	0.0002	0.0002	0.0003	0.0004	0.0005	0.0006	0.0008
11	0.0000	0.0000	0.0000	0.0000	0.0000	0.0001	0.0001	0.0001	0.0002	0.0002
12	0.0000	0.0000	0.0000	0.0000	0.0000	0.0000	0.0000	0.0000	0.0000	0.0001

* Prepared by the authors.

Table T-9: Selected Poisson probabilities (continued)

p	λ 3.1	3.2	3.3	3.4	3.5	3.6	3.7	3.8	3.9	4.0
0	0.0450	0.0408	0.0369	0.0334	0.0302	0.0273	0.0247	0.0224	0.0202	0.0183
1	0.1397	0.1304	0.1217	0.1135	0.1057	0.0984	0.0915	0.0850	0.0789	0.0733
2	0.2165	0.2087	0.2008	0.1929	0.1850	0.1771	0.1692	0.1615	0.1539	0.1465
3	0.2237	0.2226	0.2209	0.2186	0.2158	0.2125	0.2087	0.2046	0.2001	0.1954
4	0.1733	0.1781	0.1823	0.1858	0.1888	0.1912	0.1931	0.1944	0.1951	0.1954
5	0.1075	0.1140	0.1203	0.1264	0.1322	0.1377	0.1429	0.1477	0.1522	0.1563
6	0.0555	0.0608	0.0662	0.0716	0.0771	0.0826	0.0881	0.0936	0.0989	0.1042
7	0.0246	0.0278	0.0312	0.0348	0.0385	0.0425	0.0466	0.0508	0.0551	0.0595
8	0.0095	0.0111	0.0129	0.0148	0.0169	0.0191	0.0215	0.0241	0.0269	0.0298
9	0.0033	0.0040	0.0047	0.0056	0.0066	0.0076	0.0089	0.0102	0.0116	0.0132
10	0.0010	0.0013	0.0016	0.0019	0.0023	0.0028	0.0033	0.0039	0.0045	0.0053
11	0.0003	0.0004	0.0005	0.0006	0.0007	0.0009	0.0011	0.0013	0.0016	0.0019
12	0.0001	0.0001	0.0001	0.0002	0.0002	0.0003	0.0003	0.0004	0.0005	0.0006
13	0.0000	0.0000	0.0000	0.0000	0.0001	0.0001	0.0001	0.0001	0.0002	0.0002
14	0.0000	0.0000	0.0000	0.0000	0.0000	0.0000	0.0000	0.0000	0.0000	0.0001

p	λ 4.1	4.2	4.3	4.4	4.5	4.6	4.7	4.8	4.9	5.0
0	0.0166	0.0150	0.0136	0.0123	0.0111	0.0101	0.0091	0.0082	0.0074	0.0067
1	0.0679	0.0630	0.0583	0.0540	0.0500	0.0462	0.0427	0.0395	0.0365	0.0337
2	0.1393	0.1323	0.1254	0.1188	0.1125	0.1063	0.1005	0.0948	0.0894	0.0842
3	0.1904	0.1852	0.1798	0.1743	0.1687	0.1631	0.1574	0.1517	0.1460	0.1404
4	0.1951	0.1944	0.1933	0.1917	0.1898	0.1875	0.1849	0.1820	0.1789	0.1755
5	0.1600	0.1633	0.1662	0.1687	0.1708	0.1725	0.1738	0.1747	0.1753	0.1755
6	0.1093	0.1143	0.1191	0.1237	0.1281	0.1323	0.1362	0.1398	0.1432	0.1462
7	0.0640	0.0686	0.0732	0.0778	0.0824	0.0869	0.0914	0.0959	0.1002	0.1044
8	0.0328	0.0360	0.393	0.0428	0.0463	0.0500	0.0537	0.0575	0.0614	0.0653
9	0.0150	0.0168	0.0188	0.0209	0.0232	0.0255	0.0281	0.0307	0.0334	0.0363
10	0.0061	0.0071	0.0081	0.0092	0.0104	0.0118	0.0132	0.0147	0.0164	0.0181
11	0.0023	0.0027	0.0032	0.0037	0.0043	0.0049	0.0056	0.0064	0.0073	0.0082
12	0.0008	0.0009	0.0011	0.0013	0.0016	0.0019	0.0022	0.0026	0.0030	0.0034
13	0.0002	0.0003	0.0004	0.0005	0.0006	0.0007	0.0008	0.0009	0.0011	0.0013
14	0.0001	0.0001	0.0001	0.0001	0.0002	0.0002	0.0003	0.0003	0.0004	0.0005
15	0.0000	0.0000	0.0000	0.0000	0.0001	0.0001	0.0001	0.0001	0.0001	0.0002

Table T-9: Selected Poisson probabilities (continued)

p	λ 5.1	5.2	5.3	5.4	5.5	5.6	5.7	5.8	5.9	6.0
0	0.0061	0.0055	0.0050	0.0045	0.0041	0.0037	0.0033	0.0030	0.0027	0.0025
1	0.0311	0.0287	0.0265	0.0244	0.0225	0.0207	0.0191	0.0176	0.0162	0.0149
2	0.0793	0.0746	0.0701	0.0659	0.0618	0.0580	0.0544	0.0509	0.0477	0.0446
3	0.1348	0.1293	0.1239	0.1185	0.1133	0.1082	0.1033	0.0985	0.0938	0.0892
4	0.1719	0.1681	0.1641	0.1600	0.1558	0.1515	0.1472	0.1428	0.1383	0.1339
5	0.1753	0.1748	0.1740	0.1728	0.1714	0.1697	0.1678	0.1656	0.1632	0.1606
6	0.1490	0.1515	0.1537	0.1555	0.1571	0.1584	0.1594	0.1601	0.1605	0.1606
7	0.1086	0.1125	0.1163	0.1200	0.1234	0.1267	0.1298	0.1326	0.1353	0.1377
8	0.0692	0.0731	0.0771	0.0810	0.0849	0.0887	0.0925	0.0962	0.0998	0.1033
9	0.0392	0.0423	0.0454	0.0486	0.0519	0.0552	0.0586	0.0620	0.0654	0.0688
10	0.0200	0.0220	0.0241	0.0262	0.0285	0.0309	0.0334	0.0359	0.0386	0.0413
11	0.0093	0.0104	0.0116	0.0129	0.0143	0.0157	0.0173	0.0190	0.0207	0.0225
12	0.0039	0.0045	0.0051	0.0058	0.0065	0.0073	0.0082	0.0092	0.0102	0.0113
13	0.0015	0.0018	0.0021	0.0024	0.0028	0.0032	0.0036	0.0041	0.0046	0.0052
14	0.0006	0.0007	0.0008	0.0009	0.0011	0.0013	0.0015	0.0017	0.0019	0.0022
15	0.0002	0.0002	0.0003	0.0003	0.0004	0.0005	0.0006	0.0007	0.0008	0.0009
16	0.0001	0.0001	0.0001	0.0001	0.0001	0.0002	0.0002	0.0002	0.0003	0.0003
17	0.0000	0.0000	0.0000	0.0000	0.0000	0.0001	0.0001	0.0001	0.0001	0.0001

p	λ 6.1	6.2	6.3	6.4	6.5	6.6	6.7	6.8	6.9	7.0
0	0.0022	0.0020	0.0018	0.0017	0.0015	0.0014	0.0012	0.0011	0.0010	0.0009
1	0.0137	0.0126	0.0116	0.0106	0.0098	0.0090	0.0082	0.0076	0.0070	0.0064
2	0.0417	0.0390	0.0364	0.0340	0.0318	0.0296	0.0276	0.0258	0.0240	0.0223
3	0.0848	0.0806	0.0765	0.0726	0.0688	0.0652	0.0617	0.0584	0.0552	0.0521
4	0.1294	0.1249	0.1205	0.1162	0.1118	0.1076	0.1034	0.0992	0.0952	0.0912
5	0.1579	0.1549	0.1519	0.1487	0.1454	0.1420	0.1385	0.1349	0.1314	0.1277
6	0.1605	0.1601	0.1595	0.1586	0.1575	0.1562	0.1546	0.1529	0.1511	0.1490
7	0.1399	0.1418	0.1435	0.1450	0.1462	0.1472	0.1480	0.1486	0.1489	0.1490
8	0.1066	0.1099	0.1130	0.1160	0.1188	0.1215	0.1240	0.1263	0.1284	0.1304
9	0.0723	0.0757	0.0791	0.0825	0.0858	0.0891	0.0923	0.0954	0.0985	0.1014
10	0.0441	0.0469	0.0498	0.0528	0.0558	0.0588	0.0618	0.0649	0.0679	0.0710
11	0.0244	0.0265	0.0285	0.0307	0.0330	0.0353	0.0377	0.0401	0.0426	0.0452
12	0.0124	0.0137	0.0150	0.0164	0.0179	0.0194	0.0210	0.0227	0.0245	0.0263
13	0.0058	0.0065	0.0073	0.0081	0.0089	0.0099	0.0108	0.0119	0.0130	0.0142
14	0.0025	0.0029	0.0033	0.0037	0.0041	0.0046	0.0052	0.0058	0.0064	0.0071
15	0.0010	0.0012	0.0014	0.0016	0.0018	0.0020	0.0023	0.0026	0.0029	0.0033
16	0.0004	0.0005	0.0005	0.0006	0.0007	0.0008	0.0010	0.0011	0.0013	0.0014
17	0.0001	0.0002	0.0002	0.0002	0.0003	0.0003	0.0004	0.0004	0.0005	0.0006
18	0.0000	0.0001	0.0001	0.0001	0.0001	0.0001	0.0001	0.0002	0.0002	0.0002
19	0.0000	0.0000	0.0000	0.0000	0.0000	0.0000	0.0001	0.0001	0.0001	0.0001

Table T-9: Selected Poisson probabilities (continued)

	λ									
p	7.1	7.2	7.3	7.4	7.5	7.6	7.7	7.8	7.9	8.0
0	0.0008	0.0007	0.0007	0.0006	0.0006	0.0005	0.0005	0.0004	0.0004	0.0003
1	0.0059	0.0054	0.0049	0.0045	0.0041	0.0038	0.0035	0.0032	0.0029	0.0027
2	0.0208	0.0194	0.0180	0.0167	0.0156	0.0145	0.0134	0.0125	0.0116	0.0107
3	0.0492	0.0464	0.0438	0.0413	0.0389	0.0366	0.0345	0.0324	0.0305	0.0286
4	0.0874	0.0836	0.0799	0.0764	0.0729	0.0696	0.0663	0.0632	0.0602	0.0573
5	0.1241	0.1204	0.1167	0.1130	0.1094	0.1057	0.1021	0.0986	0.0951	0.0916
6	0.1468	0.1445	0.1420	0.1394	0.1367	0.1339	0.1311	0.1282	0.1252	0.1221
7	0.1489	0.1486	0.1481	0.1474	0.1465	0.1454	0.1442	0.1428	0.1413	0.1396
8	0.1321	0.1337	0.1351	0.1363	0.1373	0.1381	0.1388	0.1392	0.1395	0.1396
9	0.1042	0.1070	0.1096	0.1121	0.1144	0.1167	0.1187	0.1207	0.1224	0.1241
10	0.0740	0.0770	0.0800	0.0829	0.0858	0.0887	0.0914	0.0941	0.0967	0.0993
11	0.0478	0.0504	0.0531	0.0558	0.0585	0.0613	0.0640	0.0667	0.0695	0.0722
12	0.0283	0.0303	0.0323	0.0344	0.0366	0.0388	0.0411	0.0434	0.0457	0.0481
13	0.0154	0.0168	0.0181	0.0196	0.0211	0.0227	0.0243	0.0260	0.0278	0.0296
14	0.0078	0.0086	0.0095	0.0104	0.0113	0.0123	0.0134	0.0145	0.0157	0.0169
15	0.0037	0.0041	0.0046	0.0051	0.0057	0.0062	0.0069	0.0075	0.0083	0.0090
16	0.0016	0.0019	0.0021	0.0024	0.0026	0.0030	0.0033	0.0037	0.0041	0.0045
17	0.0007	0.0008	0.0009	0.0010	0.0012	0.0013	0.0015	0.0017	0.0019	0.0021
18	0.0003	0.0003	0.0004	0.0004	0.0005	0.0006	0.0006	0.0007	0.0008	0.0009
19	0.0001	0.0001	0.0001	0.0002	0.0002	0.0002	0.0003	0.0003	0.0003	0.0004
20	0.0000	0.0000	0.0001	0.0001	0.0001	0.0001	0.0001	0.0001	0.0001	0.0002
21	0.0000	0.0000	0.0000	0.0000	0.0000	0.0000	0.0000	0.0000	0.0001	0.0001

Table T-9: Selected Poisson probabilities (continued)

	λ									
p	8.1	8.2	8.3	8.4	8.5	8.6	8.7	8.8	8.9	9.0
0	0.0003	0.0003	0.0002	0.0002	0.0002	0.0002	0.0002	0.0002	0.0001	0.0001
1	0.0025	0.0023	0.0021	0.0019	0.0017	0.0016	0.0014	0.0013	0.0012	0.0011
2	0.0100	0.0092	0.0086	0.0079	0.0074	0.0068	0.0063	0.0058	0.0054	0.0050
3	0.0269	0.0252	0.0237	0.0222	0.0208	0.0195	0.0183	0.0171	0.0160	0.0150
4	0.0544	0.0517	0.0491	0.0466	0.0443	0.0420	0.0398	0.0377	0.0357	0.0337
5	0.0882	0.0849	0.0816	0.0784	0.0752	0.0722	0.0692	0.0663	0.0635	0.0607
6	0.1191	0.1160	0.1128	0.1097	0.1066	0.1034	0.1003	0.0972	0.0941	0.0911
7	0.1378	0.1358	0.1338	0.1317	0.1294	0.1271	0.1247	0.1222	0.1197	0.1171
8	0.1395	0.1392	0.1388	0.1382	0.1375	0.1366	0.1356	0.1344	0.1332	0.1318
9	0.1256	0.1269	0.1280	0.1290	0.1299	0.1306	0.1311	0.1315	0.1317	0.1318
10	0.1017	0.1040	0.1063	0.1084	0.1104	0.1123	0.1140	0.1157	0.1172	0.1186
11	0.0749	0.0776	0.0802	0.0828	0.0853	0.0878	0.0902	0.0925	0.0948	0.0970
12	0.0505	0.0530	0.0555	0.0579	0.0604	0.0629	0.0654	0.0679	0.0703	0.0728
13	0.0315	0.0334	0.0354	0.0374	0.0395	0.0416	0.0438	0.0459	0.0481	0.0504
14	0.0182	0.0196	0.0210	0.0225	0.0240	0.0256	0.0272	0.0289	0.0306	0.0324
15	0.0098	0.0107	0.0116	0.0126	0.0136	0.0147	0.0158	0.0169	0.0182	0.0194
16	0.0050	0.0055	0.0060	0.0066	0.0072	0.0079	0.0086	0.0093	0.0101	0.0109
17	0.0024	0.0026	0.0029	0.0033	0.0036	0.0040	0.0044	0.0048	0.0053	0.0058
18	0.0011	0.0012	0.0014	0.0015	0.0017	0.0019	0.0021	0.0024	0.0026	0.0029
19	0.0005	0.0005	0.0006	0.0007	0.0008	0.0009	0.0010	0.0011	0.0012	0.0014
20	0.0002	0.0002	0.0002	0.0003	0.0003	0.0004	0.0004	0.0005	0.0005	0.0006
21	0.0001	0.0001	0.0001	0.0001	0.0001	0.0002	0.0002	0.0002	0.0002	0.0003
22	0.0000	0.0000	0.0000	0.0000	0.0001	0.0001	0.0001	0.0001	0.0001	0.0001

Table T-9 Selected Poisson probabilities (continued)

	λ									
p	9.1	9.2	9.3	9.4	9.5	9.6	9.7	9.8	9.9	10.0
0	0.0001	0.0001	0.0001	0.0001	0.0001	0.0001	0.0001	0.0001	0.0001	0.0000
1	0.0010	0.0009	0.0009	0.0008	0.0007	0.0007	0.0006	0.0005	0.0005	0.0005
2	0.0046	0.0043	0.0040	0.0037	0.0034	0.0031	0.0029	0.0027	0.0025	0.0023
3	0.0140	0.0131	0.0123	0.0115	0.0107	0.0100	0.0093	0.0087	0.0081	0.0076
4	0.0319	0.0302	0.0285	0.0269	0.0254	0.0240	0.0226	0.0213	0.0201	0.0189
5	0.0581	0.0555	0.0530	0.0506	0.0483	0.0460	0.0439	0.0418	0.0398	0.0378
6	0.0881	0.0851	0.0822	0.0793	0.0764	0.0736	0.0709	0.0682	0.0656	0.0631
7	0.1145	0.1118	0.1091	0.1064	0.1037	0.1010	0.0982	0.0955	0.0928	0.0901
8	0.1302	0.1286	0.1269	0.1251	0.1232	0.1212	0.1191	0.1170	0.1148	0.1126
9	0.1317	0.1315	0.1311	0.1306	0.1300	0.1293	0.1284	0.1274	0.1263	0.1251
10	0.1198	0.1210	0.1219	0.1228	0.1235	0.1241	0.1245	0.1249	0.1250	0.1251
11	0.0991	0.1012	0.1031	0.1049	0.1067	0.1083	0.1098	0.1112	0.1125	0.1137
12	0.0752	0.0776	0.0799	0.0822	0.0844	0.0866	0.0888	0.0908	0.0928	0.0948
13	0.0526	0.0549	0.0572	0.0594	0.0617	0.0640	0.0662	0.0685	0.0707	0.0729
14	0.0342	0.0361	0.0380	0.0399	0.0419	0.0439	0.0459	0.0479	0.0500	0.0521
15	0.0208	0.0221	0.0235	0.0250	0.0265	0.0281	0.0297	0.0313	0.0330	0.0347
16	0.0118	0.0127	0.0137	0.0147	0.0157	0.0168	0.0180	0.0192	0.0204	0.0217
17	0.0063	0.0069	0.0075	0.0081	0.0088	0.0095	0.0103	0.0111	0.0119	0.0128
18	0.0032	0.0035	0.0039	0.0042	0.0046	0.0051	0.0055	0.0060	0.0065	0.0071
19	0.0015	0.0017	0.0019	0.0021	0.0023	0.0026	0.0028	0.0031	0.0034	0.0037
20	0.0007	0.0008	0.0009	0.0010	0.0011	0.0012	0.0014	0.0015	0.0017	0.0019
21	0.0003	0.0003	0.0004	0.0004	0.0005	0.0006	0.0006	0.0007	0.0008	0.0009
22	0.0001	0.0001	0.0002	0.0002	0.0002	0.0002	0.0003	0.0003	0.0004	0.0004
23	0.0000	0.0001	0.0001	0.0001	0.0001	0.0001	0.0001	0.0001	0.0002	0.0002
24	0.0000	0.0000	0.0000	0.0000	0.0000	0.0000	0.0000	0.0001	0.0001	0.0001

Table T-9 Selected Poisson probabilities (continued)

	λ									
p	11.0	12.0	13.0	14.0	15.0	16.0	17.0	18.0	19.0	20.0
0	0.0000	0.0000	0.0000	0.0000	0.0000	0.0000	0.0000	0.0000	0.0000	0.0000
1	0.0002	0.0001	0.0000	0.0000	0.0000	0.0000	0.0000	0.0000	0.0000	0.0000
2	0.0010	0.0004	0.0002	0.0001	0.0000	0.0000	0.0000	0.0000	0.0000	0.0000
3	0.0037	0.0018	0.0008	0.0004	0.0002	0.0001	0.0000	0.0000	0.0000	0.0000
4	0.0102	0.0053	0.0027	0.0013	0.0006	0.0003	0.0001	0.0001	0.0000	0.0000
5	0.0224	0.0127	0.0070	0.0037	0.0019	0.0010	0.0005	0.0002	0.0001	0.0001
6	0.0411	0.0255	0.0152	0.0087	0.0048	0.0026	0.0014	0.0007	0.0004	0.0002
7	0.0646	0.0437	0.0281	0.0174	0.0104	0.0060	0.0034	0.0019	0.0010	0.0005
8	0.0888	0.0655	0.0457	0.0304	0.0194	0.0120	0.0072	0.0042	0.0024	0.0013
9	0.1085	0.0874	0.0661	0.0473	0.0324	0.0213	0.0135	0.0083	0.0050	0.0029
10	0.1194	0.1048	0.0859	0.0663	0.0486	0.0341	0.0230	0.0150	0.0095	0.0058
11	0.1194	0.1144	0.1015	0.0844	0.0663	0.0496	0.0355	0.0245	0.0164	0.0106
12	0.1094	0.1144	0.1099	0.0984	0.0829	0.0661	0.0504	0.0368	0.0259	0.0176
13	0.0926	0.1056	0.1099	0.1060	0.0956	0.0814	0.0658	0.0509	0.0378	0.0271
14	0.0728	0.0905	0.1021	0.1060	0.1024	0.0930	0.0800	0.0655	0.0514	0.0387
15	0.0534	0.0724	0.0885	0.0989	0.1024	0.0992	0.0906	0.0786	0.0650	0.0516
16	0.0367	0.0543	0.0719	0.0866	0.0960	0.0992	0.0963	0.0884	0.0772	0.0646
17	0.0237	0.0383	0.0550	0.0713	0.0847	0.0934	0.0963	0.0936	0.0863	0.0760
18	0.0145	0.0255	0.0397	0.0554	0.0706	0.0830	0.0909	0.0936	0.0911	0.0844
19	0.0084	0.0161	0.0272	0.0409	0.0557	0.0699	0.0814	0.0887	0.0911	0.0888
20	0.0046	0.0097	0.0177	0.0286	0.0418	0.0559	0.0692	0.0798	0.0866	0.0888
21	0.0024	0.0055	0.0109	0.0191	0.0299	0.0426	0.0560	0.0684	0.0783	0.0846
22	0.0012	0.0030	0.0065	0.0121	0.0204	0.0310	0.0433	0.0560	0.0676	0.0769
23	0.0006	0.0016	0.0037	0.0074	0.0133	0.0216	0.0320	0.0438	0.0559	0.0669
24	0.0003	0.0008	0.0020	0.0043	0.0083	0.0144	0.0226	0.0328	0.0442	0.0557
25	0.0001	0.0004	0.0010	0.0024	0.0050	0.0092	0.0154	0.0237	0.0336	0.0446
26	0.0000	0.0002	0.0005	0.0013	0.0029	0.0057	0.0101	0.0164	0.0246	0.0343
27	0.0000	0.0001	0.0002	0.0007	0.0016	0.0034	0.0063	0.0109	0.0173	0.0254
28	0.0000	0.0000	0.0001	0.0003	0.0009	0.0019	0.0038	0.0070	0.0117	0.0181
29	0.0000	0.0000	0.0001	0.0002	0.0004	0.0011	0.0023	0.0044	0.0077	0.0125
30	0.0000	0.0000	0.0000	0.0001	0.0002	0.0006	0.0013	0.0026	0.0049	0.0083
31	0.0000	0.0000	0.0000	0.0000	0.0001	0.0003	0.0007	0.0015	0.0030	0.0054
32	0.0000	0.0000	0.0000	0.0000	0.0001	0.0001	0.0004	0.0009	0.0018	0.0034
33	0.0000	0.0000	0.0000	0.0000	0.0000	0.0001	0.0002	0.0005	0.0010	0.0020
34	0.0000	0.0000	0.0000	0.0000	0.0000	0.0000	0.0001	0.0002	0.0006	0.0012
35	0.0000	0.0000	0.0000	0.0000	0.0000	0.0000	0.0000	0.0001	0.0003	0.0007
36	0.0000	0.0000	0.0000	0.0000	0.0000	0.0000	0.0000	0.0001	0.0002	0.0004
37	0.0000	0.0000	0.0000	0.0000	0.0000	0.0000	0.0000	0.0000	0.0001	0.0002
38	0.0000	0.0000	0.0000	0.0000	0.0000	0.0000	0.0000	0.0000	0.0000	0.0001
39	0.0000	0.0000	0.0000	0.0000	0.0000	0.0000	0.0000	0.0000	0.0000	0.0001

Table T-10: Confidence limits for the Poisson parameter λ

Confidence limits for the Poisson parameter $\lambda$are derived using the expressions given in Chapter 18. The details here show the use of Table T-10.

(1) Record P events in a total time interval of length t.

(2) Select a confidence level.

(3) For the observed number of events in the first column of Table T-10, find the lower and upper values corresponding to the level of significance.

(4) Given that $\lambda = t\theta$, the results of step (3) lead to a 100(1 - α)% confidence interval on θ. Writing $\theta = \lambda/t$, $\theta_L = \lambda_L/t$ and $\theta_U = \lambda_U/t$.

Example 1: A two-sided interval:

(1) $p = 7$ events recorded in $t = 12$ minutes.
(2) 95% confidence level.
(3) From Table T-10, $\lambda_L = 2.81$ and $\lambda_U = 14.42$.
(4) For $t = 12$, $\theta_L = \lambda_L/t = 2.81/12 = 0.234$ and $\theta_U = \lambda_U/t = 14.42/12 = 1.202$.

Example 2: An upper one-sided interval:

(1) $p = 7$ events recorded in $t = 12$ minutes.
(2) 95% confidence level.
(3) From Table T-10, $\lambda_U = 13.15$.
(4) For $t = 12$, $\theta_U = \lambda_U/t = 13.15/12 = 1.096$.

Example 3: A lower one-sided interval:

(1) $p = 7$ events recorded in $t = 12$ minutes.
(2) 99.9% confidence level.
(3) From Table T-10, $\lambda_L = 1.52$.
(4) For $t = 12$, $\theta_L = \lambda_L/t = 1.52/12 = 0.127$.

Table T-10: Confidence limits for the Poisson parameter λ

$1-\alpha$	0.998		0.99		0.95		0.90	
$\alpha/2$	0.001		0.005		0.025		0.05	
p	λ_L	λ_U	λ_L	λ_U	λ_L	λ_U	λ_L	λ_U
0	0	6.9075	0	5.2983	0	3.6889	0	2.9957
1	0.0011	9.2331	0.0051	7.4301	0.0251	5.5721	0.0511	4.7441
2	0.0454	11.2287	0.1035	9.2738	0.2422	7.2247	0.3554	6.2958
3	0.191	13.062	0.338	10.997	0.619	8.767	0.818	7.754
4	0.429	14.794	0.672	12.594	1.090	10.242	1.366	9.154
5	0.739	16.45	1.08	14.15	1.62	11.67	1.97	10.51
6	1.11	18.06	1.54	15.66	2.20	13.06	2.61	11.84
7	1.52	19.63	2.04	17.13	2.81	14.42	3.29	13.15
8	1.97	21.16	2.57	18.58	3.45	15.76	3.98	14.43
9	2.45	22.66	3.13	20.00	4.12	17.08	4.70	15.71
10	2.96	24.13	3.72	21.40	4.80	18.39	5.43	16.96
11	3.49	25.59	4.32	22.78	5.49	19.68	6.17	18.21
12	4.04	27.03	4.94	24.14	6.20	20.96	6.92	19.44
13	4.61	28.45	5.58	25.50	6.92	22.23	7.69	20.67
14	5.20	29.85	6.23	26.84	7.65	23.49	8.46	21.89
15	5.79	31.24	6.89	28.16	8.40	24.74	9.25	23.10
16	6.41	32.62	7.57	29.48	9.15	25.98	10.04	24.30
17	7.03	33.99	8.25	30.79	9.90	27.22	10.83	25.50
18	7.66	35.35	8.94	32.09	10.67	28.45	11.63	26.69
19	8.31	36.70	9.64	33.38	11.44	29.67	12.44	27.88
20	8.96	38.04	10.35	34.67	12.22	30.89	13.25	29.06
21	9.62	39.37	11.07	35.95	13.00	32.10	14.07	30.24
22	10.29	40.70	11.79	37.22	13.79	33.31	14.89	31.41
23	10.96	42.02	12.52	38.48	14.58	34.51	15.72	32.59
24	11.65	43.33	13.25	39.74	15.38	35.71	16.55	33.75
25	12.34	44.64	14.00	41.00	16.18	36.90	17.38	34.92
26	13.03	45.94	14.74	42.25	16.98	38.10	18.22	36.08
27	13.73	47.23	15.49	43.50	17.79	39.28	19.06	37.23
28	14.44	48.52	16.25	44.74	18.61	40.47	19.90	38.39
29	15.15	49.80	17.00	45.98	19.42	41.65	20.75	39.54
30	15.87	51.08	17.77	47.21	20.24	42.83	21.59	40.69
35	19.52	57.42	21.64	53.32	24.38	48.68	25.87	46.40
40	23.26	63.66	25.59	59.36	28.58	54.47	30.20	52.07
45	27.08	69.83	29.60	65.34	32.82	60.21	34.56	57.69
50	30.96	75.94	33.66	71.27	37.11	65.92	38.96	63.29

* Prepared by the authors.

Table T-11a: Two-sided tolerance limit factors for a normal distribution

Statistical tolerance limits are values derived from sample data in such a manner as to encompass a specified fraction of a population's values with a prescribed level of confidence. Table 11-a provides the factors needed to produce two-sided intervals from samples of size n of normally distributed variables. Factors are given for three fractions—π = 0.90, 0.95, and 0.99—and for three levels of confidence—γ = 0.90, 0.95, and 0.99. For other combinations of n, π, and γ, refer to Odeh and Owen (1980, pp. 85-113).

Because statistical tolerance intervals are functions of the sample's mean and standard deviation, the intervals themselves are random variables: they change their calculated endpoints and their resulting lengths with each new sample. But, by the nature of their construction, $100\gamma\%$ of them will contain $100\pi\%$ of the population from which their samples are drawn.

Example: Consider a sample of size $n = 7$ that yields $\bar{x} = 37.28$ and $s = 3.45$. To create an interval that contains 99% of the population with 90% confidence, turn to Table T-11a. From the three columns headed by γ = **0.90**, find the column headed by π = **0.99**. At the intersection of that column with the row labelled **7**, note the value $k = 4.508$. The interval's endpoints are found by calculating $\bar{x} \pm ks = 37.28 \pm 4.508(3.45) = 37.28 \pm 15.55$. Thus, with 90% confidence, the interval (21.73, 52.83) contains 99% of the population.

Table T-11a: Two-sided tolerance limit factors for a normal distribution*

	$\gamma = 0.90$			$\gamma = 0.95$			$\gamma = 0.99$		
n	$\pi = 0.90$	$\pi = 0.95$	$\pi = 0.99$	$\pi = 0.90$	$\pi = 0.95$	$\pi = 0.99$	$\pi = 0.90$	$\pi = 0.95$	$\pi = 0.99$
2	15.512	18.221	23.423	31.092	36.519	46.944	155.569	182.720	234.877
3	5.788	6.823	8.819	8.306	9.789	12.647	18.782	22.131	28.586
4	4.157	4.913	6.372	5.368	6.341	8.221	9.416	11.118	14.405
5	3.499	4.142	5.387	4.291	5.077	6.598	6.655	7.870	10.220
6	3.141	3.723	4.850	3.733	4.422	5.758	5.383	6.376	8.292
7	2.913	3.456	4.508	3.390	4.020	5.241	4.658	5.520	7.191
8	2.754	3.270	4.271	3.156	3.746	4.889	4.189	4.968	6.479
9	2.637	3.132	4.094	2.986	3.546	4.633	3.860	4.581	5.980
10	2.546	3.026	3.958	2.856	3.393	4.437	3.617	4.294	5.610
12	2.414	2.871	3.759	2.670	3.175	4.156	3.279	3.896	5.096
14	2.322	2.762	3.620	2.542	3.024	3.962	3.054	3.631	4.753
16	2.254	2.682	3.517	2.449	2.913	3.819	2.89	3.441	4.507
18	2.201	2.620	3.436	2.376	2.828	3.709	2.771	3.297	4.321
20	2.158	2.570	3.372	2.319	2.760	3.621	2.675	3.184	4.175
22	2.123	2.528	3.318	2.272	2.705	3.549	2.598	3.092	4.056
24	2.094	2.494	3.274	2.232	2.658	3.489	2.534	3.017	3.958
26	2.069	2.464	3.235	2.199	2.619	3.437	2.480	2.953	3.875
28	2.048	2.439	3.202	2.170	2.585	3.393	2.434	2.898	3.804
30	2.029	2.417	3.173	2.145	2.555	3.355	2.394	2.851	3.742
35	1.991	2.371	3.114	2.094	2.495	3.276	2.314	2.756	3.618
40	1.961	2.336	3.069	2.055	2.448	3.216	2.253	2.684	3.524
45	1.938	2.308	3.032	2.024	2.412	3.168	2.205	2.627	3.450
50	1.918	2.285	3.003	1.999	2.382	3.129	2.166	2.580	3.390
60	1.888	2.250	2.956	1.960	2.335	3.068	2.106	2.509	3.297
70	1.866	2.224	2.922	1.931	2.300	3.023	2.062	2.457	3.228
80	1.849	2.203	2.895	1.908	2.274	2.988	2.028	2.416	3.175
90	1.835	2.186	2.873	1.890	2.252	2.959	2.001	2.384	3.133
100	1.823	2.172	2.855	1.875	2.234	2.936	1.978	2.357	3.098
150	1.786	2.128	2.796	1.826	2.176	2.859	1.906	2.271	2.985
200	1.764	2.102	2.763	1.798	2.143	2.816	1.866	2.223	2.921
250	1.750	2.085	2.741	1.780	2.121	2.788	1.839	2.191	2.880
300	1.740	2.073	2.725	1.767	2.106	2.767	1.820	2.169	2.850
350	1.732	2.064	2.713	1.757	2.094	2.752	1.806	2.152	2.828
400	1.726	2.057	2.703	1.749	2.084	2.739	1.794	2.138	2.810
500	1.717	2.046	2.689	1.737	2.070	2.721	1.777	2.117	2.783
1,000	1.695	2.019	2.654	1.709	2.036	2.676	1.736	2.068	2.718
∞	1.645	1.960	2.576	1.645	1.960	2.576	1.645	1.960	2.576

* Adapted from Odeh, R. E., and D. B. Owen, 1980, *Tables for Normal Tolerance Limits, Sampling Plans, and Screening*, Marcel Dekker, Inc., New York, NY, Table 3, pp. 85-113, courtesy of Marcel Dekker, Inc.

Table T-11b: One-sided tolerance limit factors for a normal distribution

Statistical tolerance limits are values derived from sample data in such a manner as to encompass a specified fraction of a population's values and to do so with a prescribed level of confidence. Table 11-b provides the factors needed to produce one-sided lower *or* upper bounds from samples of size n of normally distributed variables. Factors are given for three fractions—$\pi = 0.90$, 0.95, and 0.99—and for three levels of confidence—$\gamma = 0.90$, 0.95, and 0.99. For other combinations of n, π, and γ, refer to Odeh and Owen (1980, pp. 17-69).

Because statistical tolerance intervals are functions of the sample's mean and standard deviation, the intervals themselves are random variables: they change their calculated endpoints with each new sample. But, because of the nature of their construction, $100\gamma\%$ of them will contain $100\pi\%$ of the population from which their samples are drawn.

Example 1: Consider a sample of size $n = 10$ that yields $\bar{x} = 143.2$ and $s = 13.9$. To create a one-sided *upper* interval that contains 90% of the population with 95% confidence, turn to Table T-11b. From the three columns headed by γ = **0.95**, find the column headed by π = **0.90**. At the intersection of that column with the row labelled **10**, note the value $k = 2.355$. The interval's upper endpoint is found by calculating $\bar{x} + ks = 143.2 + 2.355(13.9) = 143.2 + 32.7 = 175.9$. Thus, with 95% confidence, the interval $(-\infty, 175.9)$ contains 90% of the population.

Example 2: Consider a sample of size $n = 10$ that yields $\bar{x} = 143.2$ and $s = 13.9$. To create a one-sided *lower* interval that contains 90% of the population with 95% confidence, turn to Table T-11b. From the three columns headed by γ = **0.95**, find the column headed by π = **0.90**. At the intersection of that column with the row labelled **10**, note the value $k = 2.355$. The interval's upper endpoint is found by calculating $\bar{x} - ks = 143.2 - 2.355(13.9) = 143.2 - 32.7 = 110.5$. Thus, with 95% confidence, the interval $(110.5, -\infty)$ contains 90% of the population.

Table T-11b: One-sided tolerance limit factors for a normal distribution*

	$\gamma = 0.90$			$\gamma = 0.95$			$\gamma = 0.99$		
n	$\pi = 0.90$	$\pi = 0.95$	$\pi = 0.99$	$\pi = 0.90$	$\pi = 0.95$	$\pi = 0.99$	$\pi = 0.90$	$\pi = 0.95$	$\pi = 0.99$
2	10.253	13.090	18.500	20.581	26.260	37.094	103.029	131.426	185.617
3	4.258	5.311	7.340	6.155	7.656	10.553	13.995	17.370	23.896
4	3.188	3.957	5.438	4.162	5.144	7.042	7.380	9.083	12.387
5	2.742	3.400	4.666	3.407	4.203	5.741	5.362	6.578	8.939
6	2.494	3.092	4.243	3.006	3.708	5.062	4.411	5.406	7.335
7	2.333	2.894	3.972	2.755	3.399	4.642	3.859	4.728	6.412
8	2.219	2.754	3.783	2.582	3.187	4.354	3.497	4.285	5.812
9	2.133	2.650	3.641	2.454	3.031	4.143	3.240	3.972	5.389
10	2.066	2.568	3.532	2.355	2.911	3.981	3.048	3.738	5.074
12	1.966	2.448	3.371	2.210	2.736	3.747	2.777	3.410	4.633
14	1.895	2.363	3.257	2.109	2.614	3.585	2.596	3.189	4.337
16	1.842	2.299	3.172	2.033	2.524	3.464	2.459	3.028	4.123
18	1.800	2.249	3.105	1.974	2.453	3.370	2.357	2.905	3.960
20	1.765	2.208	3.052	1.926	2.396	3.295	2.276	2.808	3.832
22	1.737	2.174	3.007	1.886	2.349	3.233	2.209	2.729	3.727
24	1.712	2.145	2.969	1.853	2.309	3.181	2.154	2.662	3.640
26	1.691	2.120	2.937	1.824	2.275	3.136	2.106	2.606	3.566
28	1.673	2.099	2.909	1.799	2.246	3.098	2.065	2.558	3.502
30	1.657	2.080	2.884	1.777	2.220	3.064	2.030	2.515	3.447
35	1.624	2.041	2.833	1.732	2.167	2.995	1.957	2.430	3.334
40	1.598	2.010	2.793	1.697	2.125	2.941	1.902	2.364	3.249
45	1.577	1.986	2.761	1.669	2.092	2.898	1.857	2.312	3.180
50	1.559	1.965	2.735	1.646	2.065	2.862	1.821	2.269	3.125
60	1.532	1.933	2.694	1.609	2.022	2.807	1.764	2.202	3.038
70	1.511	1.909	2.662	1.581	1.990	2.765	1.722	2.153	2.974
80	1.495	1.890	2.638	1.559	1.964	2.733	1.688	2.114	2.924
90	1.481	1.874	2.618	1.542	1.944	2.706	1.661	2.082	2.883
100	1.470	1.861	2.601	1.527	1.927	2.684	1.639	2.056	2.850
150	1.433	1.818	2.546	1.478	1.870	2.611	1.566	1.971	2.740
200	1.411	1.793	2.514	1.450	1.837	2.570	1.524	1.923	2.679
250	1.397	1.777	2.493	1.431	1.815	2.542	1.496	1.891	2.638
300	1.386	1.765	2.477	1.417	1.800	2.522	1.475	1.868	2.608
350	1.378	1.755	2.466	1.406	1.787	2.506	1.461	1.850	2.585
400	1.372	1.748	2.456	1.398	1.778	2.494	1.448	1.836	2.567
500	1.362	1.736	2.442	1.385	1.763	2.475	1.430	1.814	2.540
1,000	1.338	1.709	2.407	1.354	1.727	2.430	1.385	1.762	2.475
∞	1.282	1.645	2.326	1.282	1.645	2.326	1.282	1.645	2.326

* Adapted from Odeh, R. E., and D. B. Owen, 1980, *Tables for Normal Tolerance Limits, Sampling Plans, and Screening*, Marcel Dekker, Inc., New York, NY, Table 1, pp. 17-69, courtesy of Marcel Dekker, Inc.

Table T-12: Two thousand random digits

The 2,000 digits arranged in 40 rows and 50 columns in Table T-12 are called *random* because they were generated by a computer algorithm that has itself satisfied a sufficient variety of criteria that it is called a *random number generator*. In particular, Table T-12 was constructed by using the RND function in Microsoft's *Quick Basic*, version 4.5, with a random starting point.

The basic idea behind a table of random digits is that knowledge of the value and the row-and-column position of any one of the digits provides *no* information about any other single digit in the table (or its position). Thus, the single digits can be combined to form 2-, 3-, ..., and *k*-digit random numbers.

To use Table T-12 requires a process for locating a starting position in terms of the row-column coordinates and then for moving around the table until the desired set of random numbers is acquired. Numerous processes are possible, including using the table as a target and throwing a dart at it or blindly stabbing at the page with one's forefinger. The accompanying example illustrates these requirements with a U.S. currency note.

Example: Use Table T-12 to select a random sample of size $n = 5$ from a set of $N = 52$ distinguishable items.

(1) Assign a number to each the **52** distinguishable items, so that each is uniquely tagged with **01**, **02** , ..., or **52**.
(2) Find the serial number on a U.S. currency note. Say it is **L92332902P**.
(3) Record the serial number's *digits* in reverse order: **20923329**.
(4) Use the first 2 digits to determine the starting row. Here, it is **20**.
(5) Use the next two digits to determine the starting column. Here, it is **92**. But Table T-12 contains only **50** columns. Divide **92** by **50** and use the remainder, **42**, to designate the starting column. The digit in the **20**th row and **42**nd column is **9**.
(6) Assign the digits **1**, **2**, **3**, and **4** to Left, Right, Up, and Down, respectively. Divide the **5**th digit in the reversed set by **4**, and use the remainder to determine movement left, right, up, or down. That digit is **3**, so the remainder is **3**. Movement is Up.[1]
(7) Beginning with the **9** found in (5), record the digits in pairs: **99**, **96**, **35**, **46**, **20**. If any of these numbers is larger than **52**, divide it by **52** and replace it by the remainder; this yields: **47**, **43**, **35**, **46**, **20**.
(8) Thus, a random sample of size n = **5** of the **52** items is provided by taking the **20**th, **35**th, **43**rd, **46**th, and **49**th of them.

[1] Certain details in Steps (2)-(6) can be replaced by any alternative that produces a random row *and* a random column as a starting place *and* a random direction. Indeed, the algorithm here ignores diagonal movements which could have been included in the set of directions.

Table T-12: Two thousand random digits*

	1-5	6-10	11-15	16-20	21-25	26-30	31-35	36-40	41-45	46-50
1	56580	93353	14830	88611	77491	94159	55682	61651	76278	90039
2	82694	44512	56868	96549	97676	81145	51299	78324	87458	19158
3	24300	89591	51838	78203	68168	64742	93755	56080	61768	16029
4	69791	64849	47370	41245	59336	91731	78722	07645	56793	21081
5	93678	10246	68835	27682	60318	07379	19157	12037	67776	07248
6	07473	88205	27403	13619	74578	16637	14566	40858	58759	24621
7	93255	94101	44641	53302	91743	92258	18179	07676	39974	51887
8	28910	26516	25706	48056	50645	72581	84289	31997	18444	10330
9	70419	62779	69059	74384	06975	92192	23660	33342	31465	83723
10	95145	92735	91272	57287	53395	79920	93979	99456	85530	79258
11	27054	06908	12502	66882	86941	37710	17092	53887	80494	09245
12	60999	30405	16956	04938	51713	78369	91595	37502	22165	91337
13	50513	14966	28936	51542	27276	40725	81142	10178	36861	32862
14	14918	14251	63794	60699	44952	74709	24733	50845	64032	26115
15	83143	72762	92850	34146	69102	21201	18647	69705	05843	24621
16	44762	01384	61844	28710	93492	04594	99063	72617	63939	59104
17	90525	13775	15676	30909	49632	93762	20406	59516	06116	45980
18	65288	57500	42177	82255	52023	99471	15693	20134	89639	96221
19	62309	36394	77163	92427	65271	89899	93288	15417	89998	13986
20	90434	39040	29549	48332	12172	32711	98444	67332	49853	76770
21	71192	17298	21629	28567	45628	99871	76063	03132	69163	92841
22	39191	99240	40029	32771	39050	95144	07049	36518	08289	92136
23	94055	84500	26272	27985	91858	60511	91802	13735	95525	04157
24	10718	72291	26193	79285	29209	87962	13485	53738	08642	22828
25	83134	62665	17823	13358	55677	84591	81232	50910	83995	93294
26	93392	93965	88188	82628	68956	39247	89597	40521	17850	20716
27	33193	11074	13299	70680	95082	67903	50078	55113	00057	29045
28	65846	08032	65737	45903	25773	17815	58805	48617	93974	06308
29	41773	34637	59350	31782	57933	15605	13648	93406	89275	97494
30	43908	04150	37788	53587	05483	53569	96579	52470	42818	62589
31	72367	74503	61417	67453	83903	43326	52742	77183	66826	28232
32	93703	43522	26709	43846	47587	45688	67433	15227	37428	37437
33	23622	70833	97297	07793	97656	24860	74883	77142	31451	12219
34	06461	72197	26511	74964	54490	67611	54022	16892	80799	65864
35	50175	29889	18630	20552	60114	20652	22881	79254	88917	85399
36	10942	67178	93214	09890	51054	29666	10428	96478	88082	41255
37	91159	14499	26915	90942	69215	23544	57355	11040	83574	10788
38	88368	26257	43969	58835	92663	54980	50501	32510	94197	51546
39	09587	55021	86738	51441	25297	44086	80012	07801	11944	44102
40	46036	67486	53662	41233	76645	30871	17030	86671	49669	01711

* Prepared by the authors.

Index[1]

[1] For the most part, page references to indexed entries point to the first occurences in any chapters in which there is a substantive discussion of the entries.

Federal Recycling Program

☆GPO : 1994 O - 365-366 QL 3

Distribution	**Variable symbol**	**Quantile** $0 \leq q \leq 1$	**Observed or calculated value**
F **Parameters:** df_1, $df_1 = 1, 2, \ldots$ df_2, $df_2 = 1, 2, \ldots$ **Mean:** $df_2/(df_2 - 2)$, $df_2 > 2$ **Variance:** $\frac{2(df_2)^2(df_1 + df_2 - 2)}{df_1(df_2 - 2)^2(df_2 - 4)}$, $df_2 > 4$ *Cf.* Chapters 10, 12, 14; Table T-4	F **Domain:** $0 \leq F \leq +\infty$	$f_q(df_1, df_2)$	f
Hypergeometric **Parameters:** N, $N = 1, 2, \ldots$ (the number of elements in the population) M, $M = 0, 1, 2, \ldots, \leq N$ (the number of elements of interest in the population) n, $n = 1, 2, \ldots, \leq N$ (the sample size) **Mean:** nM/N **Variance:** $(\frac{nM}{N})(\frac{N - M}{N})(\frac{N - n}{N - 1})$ *Cf.* Chapter 16	H **Domain:** $H = 0, 1, 2, \ldots, M$	not used	h